E. G. Love

Pavements and Roads

Their Construction and Maintenance, Reprinted from The Engineering and Building

Record

E. G. Love

Pavements and Roads
Their Construction and Maintenance, Reprinted from The Engineering and Building Record

ISBN/EAN: 9783337250928

Printed in Europe, USA, Canada, Australia, Japan

Cover: Foto ©berggeist007 / pixelio.de

More available books at **www.hansebooks.com**

PAVEMENTS AND ROADS.

Pavements and Roads

THEIR

CONSTRUCTION and MAINTENANCE.

REPRINTED FROM

The Engineering and Building Record.

(Prior to 1887, *The Sanitary Engineer.*)

COMPILED BY E. G. LOVE, PH.D.

1890.

THE ENGINEERING AND BUILDING RECORD.

NEW YORK.

THE ENGINEERING & BUILDING RECORD PRESS,
277 Pearl Street, N. Y.

PREFACE.

THIS book is a compilation of articles which have appeared in recent volumes of *The Engineering and Building Record*, edited with a view of eliminating whatever was of timely or local interest, and arranged by divisions for convenience of use.

The science of paving and the need of proper maintenance of pavements is yet comparatively little understood in this country, and the same is true in even greater degree with regard to roads. The editor of the journal named was led to give the matter special attention by seeing what was done in Europe, during his visits there, and finally began an investigation of work on streets and roads in England, France and other countries, the result of which was the gathering of a large and very valuable mass of information in regard to the subject. Of this everything likely to be of practical use in America was printed in *The Engineering and Building Record*, and is given here in more convenient shape. With it appears a large quantity of matter from American sources, including the prize essays on Road Construction and Maintenance submitted in the competition instituted by the journal named in December, 1889.

It will be seen that the great bulk of the book is made up of records of experience and statements of cost in different places. The comments are based on this experience.

TABLE OF CONTENTS.

PART I.

PART II.

PART III.

INDEX.

NOTE.

The foot notes refer to the volume and page of
THE ENGINEERING AND BUILDING RECORD.

CHAPTER I.

STONE PAVEMENTS.

STONE PAVEMENTS IN LIVERPOOL.*

THE introduction of impervious pavement in Liverpool began in 1872, the Health Committee in that year approving of the recommendation of their engineer (G. F. Deacon, M. Inst. C. E.) relative to the laying in the city generally of this class of pavement. From that date the work has been prosecuted without intermission. Up to the end of 1879 there were laid 371,500 yards, and since then up to 1886, 1,000,000 yards have been laid. We have in another article described the method of laying tramways through paved streets. From the nearly fifty miles of these now laid, the city derives a rental of $150,000 per annum, and the annual cost of maintenance is found to be but about $7,500, leaving a handsome balance toward a sinking fund for repayment of cost.

Through the courtesy of Mr. Clement Dunscombe, M. Inst. C. E. present City Engineer, we have obtained the following information respecting the pavements now being laid down :

The streets of Liverpool are divided into first, second, and third class.

Paving "First-Class" Streets. Foundation.—First-class streets, or main lines of communication, are paved as follows : First, a Portland cement concrete foundation is laid six inches deep. The concrete consists of one part by measure of cement, five to six parts gravel, and seven to eight of broken stone ; the gravel and cement being thoroughly mixed dry, and only enough water then allowed to flow on it to make the material damp enough after it is incorporated to retain its form when a portion is taken in the hand and pressed.

The ground having been excavated, thoroughly consolidated, and properly graded to the requisite depth and shape, a layer of broken stone (or other material) is spread evenly over the surface and thoroughly wet from the rose of a watering-can. A stratum of mortar, mixed as described above, is spread over this and a second layer of stone added. The stone is then "beaten in with a heavy flat beater." Other layers of mortar and stone are added and thoroughly beaten in, until the required thickness is obtained, the final layer of cement mortar being smoothed off to an even and uniform surface.

---

*xiv. 589.

Material.—It is required that the broken stone shall all be capable of passing in any direction through a 2½-inch ring and be clean and free from foreign matter. The so-called "gravel" must be "free from all stones that will not pass through an inch sieve," and clean and free from foreign matter. No inferior limit is given. The term gravel, as used in the specifications, is *not*, as we would understand it, gravel containing no sand, but gravel and sand as they would come from the pit. Its fineness would depend upon the size of mesh through which it had been riddled, so that "fine gravel" would mean that which had passed through a half-inch mesh or less.

"In Liverpool we are very particular as regards the quality of the materials forming concrete—viz.: broken stone, gravel and sand, all of which are perfectly clean, free from all foreign matter, and the best of their respective kinds that can be obtained. We have exceptional facilities for getting a supply of these materials from the River Dee, the Wyre, and the Isle of Man mines, which supply a gravel, granitic or otherwise, and sand of a superior quality, and in addition much good material is brought as ship's ballast."

The cement must not leave more than ten per cent. residuum on a No. 50 wire sieve, and pure cement setting in water must after seven days give a tensile strength of 400 pounds per square inch.

The concrete must *set ten days* before paving is begun upon it.

The Blocks.—On the foundation thus prepared is spread a layer of "fine gravel" not exceeding a half an inch in thickness, on which the granite or syenite sets are laid in "regular, straight, and properly bonded courses, with close joints. The sets are 3¼″x3¼″ blocks 6¼″ deep on a 6-inch cement concrete foundation. In streets with heavy traffic the depth of the blocks is 7¼ inches. The joints of the sets are then filled with hard, clean, dry shingle; the sets then thoroughly rammed and additional shingle added until the joints are full. The joints are then carefully filled with a hot mixture composed of coal-pitch and creosote oil, and the whole pavement covered with "half an inch of sharp gravel."

The channel-stones (or gutter-stones, as we call them) are of granite or syenite, and are 3 inches thick, 16 inches wide, and not less than 3 feet long, with parallel beds and faces, and square sides and ends. They are laid in cement concrete, and have joints filled as before described.

The curb-stones are 6 inches thick at top, 7 inches thick at 5 inches below the top, and not less than that below, nor less than 3 feet long or 12 inches deep, whether straight or curved. To be smoothly dressed on top, 8 inches down the face and 3 inches down the back; the remainder of the stone to be hammer dressed, and the joints square for the entire depth.

The footways are laid with 3-inch flags of best quality ; no flags to be less than 2 feet wide or 3 feet long. Joints to be squared for the whole thickness. Stones laid on a bed of "fine gravel," and joints flushed with cement mortar.

In all streets the height of crown above the channel is proportioned to the width of the street, the rate of inclination in the cross-section being 1 to 36. The channel-stone is laid level (in a direction crossing the street), and between it and the centre the section curves, so that, calling the centre height 1, the heights at one-fourth, one-half, and three-fourths the distance to it will be respectively 0.35, 0.65, and 0.87 (see cross-sections).

Crossings consist of three rows of 16x8-inch granite crossing-stones, with sides and joints dressed to full depth. All bedded in concrete, and joints filled same as the paving joints. No piece less than 3 feet long. A groove 1 inch long and ¼-inch deep is cut along this middle line of upper surface of each stone.

Second-Class Streets.—For *second*-class streets the foundation is either as has just been described, or it may consist of "6 inches of bituminous concrete, made of clean, angular, broken stone, grouted with a hot mixture of coal-pitch and creosote oil, covered with chippings and thoroughly consolidated by rolling with a roller of sufficient weight."

The paving-blocks are granite or syenite, and the blocks are either "4-inch cubes, or 3″x5″ to 7″ and 6¼″ deep, or 3″x5″ to 7″ and 5″ deep." Where wood is used the blocks are 4″x5″ to 7″ and 6″ deep. Where cubes are used starting blocks are set at the beginning of each course across the street, so as to insure a perfect breaking of the joints in passing from row to row.

Third-Class Streets.—For *third*-class streets a foundation of hand-placed rock 10 inches deep and set on edge is laid. Over this enough gravel is spread to fill the spaces and form a smooth surface, and this is then compacted by a steam-roller. The blocks consist of 4-inch cubes of granite or syenite, set and finished in the same manner as first-class on ½-inch of "fine gravel."

Specifications for Blocks.—The specifications for sets require that they shall be equal in quality and toughness to the standard samples, and shall be dressed and gauged with equal accuracy. The maximum deviations allowed are *one-quarter* inch in depth and breadth respectively. This is readily determined by placing several of them in close contact side by side on a board.

The sizes of the sets are often varied, inasmuch as a large number of set-dressers are engaged to re-dress the sets taken up from a street requiring repaving for use in other streets, and what the old sets will re-dress too often regulates the size of sets used in certain streets.

The accurate gauging of sets is a matter of which the advantage is perhaps not fully realized outside of Liverpool, but it should be noted that if any good work is to be executed the sets when laid should be in parallel and even courses, and if the sets be not accurately gauged to one uniform size the result is either a badly paved street, with the courses running unevenly and bad joints, or the sets have to be picked on the ground, put into courses of various widths, and to be laid in such courses. The cost of this is far in excess of any additional price which would have to be paid at the quarry for gauging the sets to one uniform width in the first instance, and my experience has always been that although gauged sets may cost 1s. 6d. per ton more than ungauged sets, there is eventually a considerable saving by specifying that they be accurately gauged prior to leaving the quarry. As regards the width of the joints of the sets, these cannot, in my opinion, be too small. It is a mistaken notion to suppose that the width of the joints makes the pavement in any way safer, as the inequalities of the sets give sufficient foot-hold. Wide joints are open to very great objection on many grounds, and where these joints are grouted in lime or cement instead of pitch and creosote oil they often become the receptacles for all filth. Payment is made by *weight*, and not by number, as with us.

The specifications for paving contain an important provision often overlooked—viz.: "suitable screens to be provided wherever stones are being chipped or dressed, for the purpose of protecting pedestrians."

Labor and Material.—The whole of the work in Liverpool of paving the streets and the construction of the sewers is executed by corporation workmen skilled in their respective callings, at the lowest possible prices consistent with good work, and the material is of the best quality and selected with the greatest care. The most durable syenite only is used, and the work is such that the maintenance upon it for a series of years, and perhaps long after the original debt has been recouped, will be trifling.

Traffic.—The traffic in certain of the streets in Liverpool is very exceptional, and ranges as follows:

Name of street.	Vehicular traffic in tons per yard in width of carriage-way per annum.	Remarks.
Great Howard Street (over wooden pavement)..............	301,849 Tons.	....
Bath Street and New Quay)...............	360,000 "	....
North John Street......	216,570 "	....
Lord Street.............	137,484 "	Exclusive of an aggregate tramway traffic of 348,816 tons per annum.
Church Street..........	148,995 "	

The above figures were taken in 1879. If the same were taken at the present time they would show a considerable increase.

Durability.—As an instance of the quality of the material used it may be stated that sets have been taken up in North John Street which were laid in 1872, with a traffic of 216,570 tons per yard in width of carriageway per annum, and the wear was not measurable. Although the work executed by the Health Committee of the Corporation of Liverpool may often be described as expensive and extravagant, it can be conclusively shown that it is for a public body which never dies the truest economy, and far cheaper in the long run than the system which too often obtains under the plea of economy of perpetually expending small sums in maintenance. The result of this temporizing policy is that the work is never complete or as satisfactory as by the process adopted and already described of executing public works in the best possible manner in the first instance, and thus reducing the annual maintenance charges upon them. By this means the sinking fund necessary to pay off the original debt is provided without in any way increasing the burden of the ratepayers in the shape of increased taxation.

No Boulder Pavements.—Up to the year 1872 many of the streets of the city were private and mostly paved with boulders, but since that time *all* such paving has been prohibited, and no private street has been adopted by the city unless paved, flagged, curbed, channeled, and otherwise completed in accordance with the specifications applicable to the street due to its circumstances of traffic.

Within the last few years over a million and a quarter superficial yards of set paving on a concrete foundation have been laid in streets either macadamized or badly paved, the boulders, inferior paving materials, and other stones being utilized for concrete. The present cost of the syenite set pavement is as follows: .

Cost of Construction.—First-class specification: 3x6-inch sets on a 6-inch cement concrete foundation. The cost per superficial yard, 13s. 6d.

Second-class specification: Four-inch cubes or 3x5-inch sets on a 4 to 5-inch concrete foundation. Price per superficial yard, 9s. 9d.

Third-class specification: Four-inch cubes on a local hand-pitched foundation 10 to 12 inches in depth. Price per superficial yard, 8s. 4d.

Cost of Maintenance.—The length of adopted streets in Liverpool under maintenance by the Corporation is about 249 miles, and the cost of their maintenance per annum in the best possible manner is £7,400, or $37,000; showing that the superior mode of executing the paving-works in Liverpool has a resulting maintenance charge of the smallest amount. A considerable portion of this

small annual charge is due to the cost of maintaining certain macadamized roads, which, on account of convenience, it is considered
advisable to maintain as such.

In 1871, when the mileage of streets under maintenance was
considerably less, these charges were estimated at £22,000 or
$110,000.

Disposal of Refuse.—As the wear of the pavement is practically
nothing, the scavenging of the streets is reduced to a minimum, the
material removed from the streets consisting principally of manure,
which has a ready sale. In a city such as Liverpool this is a matter
of the greatest importance, as the collection and removal of road
refuse is a most expensive process, inasmuch as the material has to
be collected at a considerable cost and barged away in specially
constructed steam hopper-barges to sea and there deposited. The
reduction of the quantity of unsalable road refuse effects in itself a
considerable annual saving, and is an important element in showing
that the best-class pavements are the cheapest under the conditions
stated.

The drawing published herewith shows the various cross-sections
of the streets as paved in Liverpool, with the respective materials
used. The footways are also sometimes completed with concrete
slabs especially laid.

The area of the city is 5,210 acres, and the population is approximately 586, 320, or at the rate of 112.5 per acre.

Sanitary Conditions.—Liverpool being so densely built upon
and being the largest seaport, it is especially liable to the importation of disease or the occurrence of epidemics. The perfect sewerage and the internal drainage of the city, together with the impervious nature of the streets and surroundings and the rapid collection
of all refuse, has had an important bearing on the health of the
community, and it is this, together with the precautions taken by the
Medical Officer of Health's Department and other agencies, that
has tended to create the satisfactory health condition of the city, the
death-rate being at present 23 $\frac{7}{10}$ per 1,000.

REINSTATING STONE PAVEMENT.[*]

We are glad to see the *Times* call attention to the fact that the
Fifth Avenue pavement is taken up in spots, the foundation destroyed, and when the blocks are replaced they are simply put back
in the sand. We noticed an instance of this kind in front of the
residence of Mr. Cornelius Vanderbilt last week. On inquiry, the
man in charge stated that the blocks were put back temporarily.

---

[*] Ed., xvii. 67.

CITY of LIVERPOOL

CARRIAGEWAY PAVEMENTS.

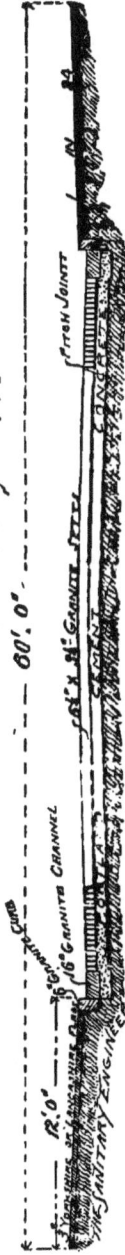

CROSS SECTION OF STREET SHOWING PAVING WHEN FINISHED IN ACCORDANCE WITH FIRST CLASS SPECIFICATION.

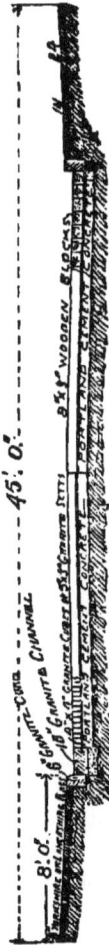

60'. 0"

8'. 0"

R. 0"

6" Granite Curb

16" Granite Channel

THE SANITARY ENGINEER

6"x4" N° Granite Setts

Pitch Joints

HALF CROSS SECTION OF STREET SHOWING PAVING WHEN FINISHED IN ACCORDANCE WITH SECOND CLASS SPECIFICATION.

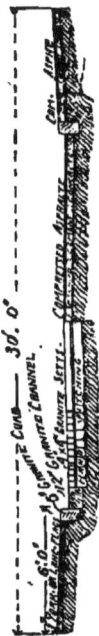

45'. 0"

8'. 0"

6" Granite Curb

16" Granite Channel

8"x4" WOODEN BLOCKS

Portland Cement

HALF CROSS SECTION OF STREET WHEN PAVED WITH WOODEN-BLOCKS

HALF CROSS SECTION OF STREET SHOWING PAVING WHEN FINISHED IN ACCORDANCE WITH THIRD CLASS SPECIFICATION.

30'. 0"

6'. 0"

6" Granite Curb

18" Granite Channel

6"x2" Granite Setts

COMPRESSED ASPHALTE

HALF CROSS SECTION OF STREET WHEN FINISHED WITH NATURAL COMPRESSED ASPHALTE IN PORTLAND CEMENT.

Scale - of - Feet

0   5   10   15   20   25   30

Clement Dunscombe
City Engineer
Liverpool

Note.—

The height of the Crown of the Street above the Channel is proportional to the width of Carriageway & the rate of inclination being 1 in 35 * The Channel is laid perfectly level; and the surface between it and the centre of Street to a curve having ordinates of .35 .65 .87 at equal distances from the Channel 1 being taken as the total rise *

We shall watch with interest to see when the concrete foundation is properly restored.   Our opinion is that every opening of this kind would better be inclosed and travel over it prohibited until the department can restore the foundation and pavement properly. These temporary repairs simply result in making a series of depressions in the whole street, and breaking down the edges of the sound pavements.

If such makeshifts were not tolerated at all, but the disturbed places kept inclosed until perfectly restored, it would hasten such restoration and in the end save much inconvenience and expense.

INADEQUATE METHOD OF PAVING INSPECTION IN NEW YORK.*

Part of the three million dollars recently appropriated for improving the pavements of New York City is being expended in replacing the old Belgium paving in Chambers Street, between Park Row and Greenwich Street, by a pavement of granite blocks, in accordance with the specifications of the Department of Public Works.

These specifications require, among other things, that the blocks shall measure on their upper surface not less than 8 nor more than 10 inches in length and not less than $3\frac{1}{2}$ nor more than $4\frac{1}{2}$ inches in width.   A representative of *The Engineering and Building Record* recently observed in some of the pavement just laid several blocks 13 to $15\frac{1}{2}$ inches in length, and 5 to $5\frac{1}{2}$ inches in width.

This clearly shows lack of a proper method of inspection which, however, it is difficult to secure if improper material is allowed to be brought on the ground ; only the exterior blocks can be inspected in the piles, and when several pavers are at work it is impossible for one inspector to watch them all, and after the blocks have been laid only their faces can be seen and that with difficulty, as the blocks are required to be covered with sand as laid, and the replacement of a block, too large for instance, might require the disturbance of several adjacent blocks, so that a block that would have been promptly rejected if seen in time is apt, after it has been laid, to be allowed to remain because the public is so much incommoded by the accumulation of rejected blocks and delays incurred, all of which is taken advantage of by some contractors.

The simple and obvious remedy, as we have pointed out on previous occasions, is to have the blocks carefully inspected and culled at the city dock.   This will insure the use of proper material and leave the inspector on the work free to give his entire attention

---

* Ed., xx 313.

to the way in which that work is performed, with the result, as we believe, of securing a much needed improvement both in material and workmanship. Inspection before shipment is common in most engineering work, and is usually preferred by the contractor, as it saves him the expense of forwarding and returning unsatisfactory material.

COMMISSIONER GILROY ON PAVEMENT SPECIFICATIONS.*

Specifications and Inspection of Stone Paving Blocks in New York.—In our issue of November 2, 1889, we called attention to the necessity for having the inspection of stone paving blocks for New York City at the docks instead of in the street where the paving is done, giving as an illustration in the case the fact that blocks of dimensions other than those fixed by the specifications were used in recent work on Chambers Street. We attributed this lax enforcement of specifications to methods hitherto in vogue in this city, partly due to the trouble occasioned store-keepers and the public by stoppage of traffic and delays resulting from rejections of defective material on the ground, which circumstance is taken advantage of by contractors, whether they supply the stone or lay them. A *Commercial Advertiser* reporter called the attention of Public Works Commissioner Gilroy to our article, and the Commissioner's explanation, as reported in that journal, is as follows:

The variations from the standard fixed, which are pointed out, if they exist, are very exceptional, I believe. Two inspectors are employed whose sole business is to see that the blocks used are of the right kind. You see that quite a margin of variation is allowed in the specifications, and it really does not matter if the blocks are somewhat longer or wider than the specifications, provided that only those of the same length and width are used in the same course. In some European cities blocks 18 inches long are regularly used, and they might be 25 inches without harm, provided the size of the blocks in the same course was uniform. Often, too, longer blocks are needed in filling out a course owing to irregularities in the street ; so, a comparatively few blocks deviating from the fixed standard are really needed.

Mr. McManus, the Chief Clerk of the Water Purveyor's Office, under whose supervision the repaving is being done, said, according to the same journal:

Some irregularity of size is expected, as can be seen from the specifications. The standard of length quoted by the *Record* is incorrect and does not give the right idea of the sanction permitted. Instead of being fixed at 8 to 10 inches, it is from 8 to 12, giving a greater margin for the contractor. The reason for allowing such a margin is that a difference of a few inches does not make much difference, if blocks of different lengths are not mixed up in the same course. In that case the joints are irregular.

---

* Ed., xx. 347.

A course is technically one line of blocks laid across the street. A considerable allowance for variation is really necessitated by the irregularities in the size and direction of the street, a lengthening of the course, in one case, calling for longer blocks, and the shortening, in another, of shorter blocks ; while the curve of a street may call for a corresponding variation in width.

These remarkable explanations amount to saying that the specifications are as inadequate as the inspection was. Or, in other words, the inspection has to be lax because the specifications are too rigid, and the fact that they allow a wide margin of variation is gravely put forth as a reason for permitting a great deal more. This is obvious nonsense. A slip of the pen made us give 10 inches as the specified length of the block, when it should have been twelve, but this is of little consequence and does not affect the main question, since the important practical point lies in the width of the block rather than the length. This fact the Commissioner and his clerk do not seem to understand. In our first reference to this matter we made no criticism of the specifications, merely calling attention to the fact that compliance with them was not enforced. We were not, however, prepared to hear the Commissioner of Public Works publicly discredit his own new specifications, and openly advocate a departure from them on such absurdly inadequate and imaginary grounds. Since, however, he has not only done so, but sought to justify his course by reference to the practice of some European cities, not named, it may be worth while to tell him what the best European practice really is—especially as to width of blocks and joints, which is the most important matter—since it is that which so directly affects the ability of a horse to get a foot-hold to haul a load and avoid slipping. For this purpose we quote from the series of articles on "Pavements and Street Railroads," appearing in *The Engineering and Building Record.*

Stone Pavements in Liverpool.—In our account of the stone block pavements in Liverpool, recognized as models the world over, we published a description of the present practice, prepared for us by Clement Dunscombe, M. Inst., C. E., the City Engineer, in which, after mentioning the preparation of the foundation of Portland cement concrete, something not done in the New York, except in an inadequate manner on Fifth Avenue, it is stated that the granite or syenite blocks, or "sets," as they are there called, are laid thereon with *close* joints. They are $3\frac{1}{4}$ inches by $3\frac{1}{4}$ inches, and $6\frac{1}{4}$ inches deep, or in streets with heavy traffic $7\frac{1}{4}$ inches. The maximum deviation allowed in depth and breadth is *one-quarter of an inch.* The description continues :

The accurate guaging of sets is a matter of which the advantage is perhaps not fully realized outside of Liverpool, but it should be noted that if any good work is to be executed, the sets when laid should be parallel and in even courses ; and if the sets be not accurately guaged to one uniform size the result is either a badly paved street, with the courses running unevenly and bad joints, or the sets have to be picked on the ground, put into courses of various widths, and to be laid in such courses. The cost of this is far in excess of any additional price which would have to be paid at the quarry for gauging the sets to one uniform width in the first instance, and my experience has always been, *that although gauged sets may cost* 1s. 6d. *per ton more than ungauged sets, there is eventually a considerable saving in specifying that they be accurately gauged prior to leaving the quarry.* As regards the width of the joints of the sets, these *cannot, in my opinion, be too small.* It is a mistaken notion to suppose that the width of the joints makes the pavement in any way safer, as the inequalities of the sets give sufficient foot-hold. Wide joints are open to very great objection on many grounds, and when these joints are grouted in lime or cement instead of pitch and creosote oil they often become the receptacles for all filth. [The italics are ours.]

New York Specifications.—The New York specifications permit 1-inch joints, which are not grouted at all, while the blocks, as already stated, are specified to be from 8 to 12 inches long, 3¼ to 4¼ inches wide, and 7 to 8 inches deep.

And these specifications, it is officially announced, are not expected to be adhered to. Probably no city in the world has heavier traffic or better stone pavements than Liverpool, and it would be well to profit by the experience there gained. There is every facility for doing equally good work in New York, and the generous appropriation recently made for improving the pavements shows that the city is willing to pay for the best and expects to get it. And it is to be regretted that so competent an executive officer as Commissioner Gilroy should so far forget himself as to talk on a subject on which he can not be expected to be an expert, and evidently has yet a great deal to learn, and announce the dangerous doctrine that the department's specifications are not expected to be enforced.

VIOLATION OF SPECIFICATIONS ON THE CEDAR STREET PAVEMENT WORK.*

As we go to press a member of our staff has discovered that in the new paving work which is to be done under the recent three million dollars appropriation, in Cedar Street, New York, a large number of the blocks on the ground, *i. e.*, sidewalk, exceed the size called for in the specification, which permits a variation of 8 to 12 inches in length, 3¼ to 4¼ in width, and 7 to 8 in depth. A number of these blocks measure from 12 to 19 inches in length and from 4¼ to 5¼ inches in width, and this extra width is a most serious defect. The specifications, based upon the experience of engineers in all the

---

* xx, 346.

cities that have decent pavements, explicitly provide (the clause being in italics), that none of the old pavement can be disturbed while any unfit or rejected material remains on the ground. In defiance of this clear and most important provision the street is torn up from Nassau to Pearl Street, traffic is suspended, the street is a mud-hole, and store-keepers are subjected to loss. We suggest that our contemporaries of the daily press give this matter their attention. Merchants on the street should be able to recover damages from parties responsible for causing this needless interruption to their business.

GRANITE BLOCKS VS. ASPHALT IN CINCINNATI.*

About $2,000,000 will be spent on paving the streets of Cincinnati during 1887, with either granite blocks or asphalt. About one mile and a half of one street has been laid in asphalt, while several miles have been laid in granite blocks. The latter is deemed best for heavy hauling, but the former shows no sign of wear, although put down last summer, and since then subjected to the severest test, as, owing to its smoothness, all sorts of vehicles have crowded upon it. Citizens living on streets yet to be paved are petitioning for asphalt instead of granite blocks. If the same care is taken with laying the streets next year as was this, Cincinnati will have the smoothest roadways of any city in the United States.

GRANITE, SANDSTONE AND LIMESTONE FOR PAVING.†

A report on pavements, made to the Common Council of Topeka, Kan., refers to stone pavements as follows :

For this use, Medina and Colorado sandstone, they say, take first rank as to durability. Omaha and Kansas City are using Sioux Falls and Missouri granite. While the granite may be somewhat more lasting, the sandstone is very durable, is less noisy, and has the advantage of not becoming polished or glossy in use. Hard Argentine limestone was tried in Kansas City on a concrete foundation, but, being set on edge, it wore unevenly, and in a year or two was shivered and split by the frost, and has had to be replaced by granite and limestone. This is the universal experience of all cities using limestone blocks.

Captain Greene, of Washington, states that granite pavements there have given satisfaction, except on the score of noise. This has led to its restriction to streets having exceptionally heavy traffic, and a desire for the substitution of asphalt even on these.

---

* xv, 20.   † xv, 375.

CHAPTER II.

WOOD PAVEMENTS.

WOOD PAVEMENT IN THE METROPOLIS.*

SINCE wood and asphalt pavements have been extensively adopted in the best paved thoroughfares of London and Paris, and since the American experience with wood pavements has been unsatisfactory because of the improper methods adopted in laying them, and lack of care after being laid, we have thought it well to reprint in this series the paper by George Henry Stayton, Assoc. M. Inst. C.E., with an abstract of the discussion thereon, which was had before the Institution in 1884.

Though the paper deals with wood pavements, for which the author apparently has a preference, yet with the discussion and criticism it elicited it is a valuable contribution on this subject, which our American readers will, we believe, appreciate.

The necessity for stimulating the efforts of those persons who are actively engaged in the construction and maintenance of street carriageway pavements in the metropolis and large cities was strenuously urged during the discussions at the Institution in 1879, when the subject was brought forward by Mr. Deacon and by Mr. Howard. The object of this paper is to call attention to the various wood-pavement works recently executed in the metropolis, and to a comparison of the results obtained thereby. Although the paper may not contain much that is new, the author ventures to think that the general interest evinced in works which tend to the efficient and economical maintenance of the carriageways of important thoroughfares, and the direct bearing which such works have upon the comfort and convenience of a community, may be sufficient to justify a review of the progress in this system of pavement,

It may not be uninteresting to consider for one moment the extent of the streets of the metropolis, and the nature of the materials of which the carriageways thereof are formed; and the author desires to tender his warmest acknowledgments to the president, who in his official capacity as the Chief Engineer to the Metropolitan Board of Works furnished him with valuable data, and to Mr. W. Haywood, M. Inst. C. E., the city engineer, and to forty-three chief surveyors of parishes and districts of the metropolis, for their courtesy in replying to his communications thereon. The information thus obtained has enabled him to present it in the tabulated form which will be found in Table I. in the appendix, according to which it appears that at the commencement of the present year the aggregate length

---

*xvi, 322. A paper by George Henry Stayton, Assoc. M. Inst. C. E., and printed in the Minutes of the Proceedings.

of the streets of London amounted to 1,966 miles. Of that length, however, 248 miles are at present "new" street, inasmuch as they have not
been adopted by a local authority; consequently there are 1,718 miles of
public streets under the maintenance of the various authorities, the carriageways of which consist of the following materials—viz. :

	Miles.
Macadam	573
Granite	280
Wood	53
Asphalt	13½
Flints or gravel	798½
Total	1,718

The extent of the vehicular traffic is equally remarkable, the result of
inquiries instituted by the author showing that in the metropolis alone, at
the present time, there are approximately 100,000 horses and 40,000 vehicles,
the licensed cabs numbering 10,381, and omnibuses, etc., 2,223, and the estimated value of the horses, harness, and vehicles amounts to no less than
£5,000,000 sterling. Obviously the ordinary wear and tear of these
vehicles must in a great measure depend upon the condition of the street
carriageways.

No doubt many members of the Institution have a vivid recollection of
the extremely unsatisfactory state of those macadamized carriageways in
leading West-end thoroughfares, which have given place to wood pavement. It was rarely the case that such roadways were in a good state of
repair; on a hot summer day they invariably emitted disagreeable smells
and frequently gave off a great amount of dust; and it is scarcely possible
to conceive anything more deplorable than the state of such streets whenever the surface became greasy or sloppy after rain. All things considered,
there was not only undoubted cause for dissatisfaction, but ample justification in the outcry against the former state of things ; as, what with damage
to horses, harness, vehicles, and pedestrians' clothing, together with the
sheer waste of money in laying down broken granite to be ground into
mud, an alteration was most necessary.

The efficient condition of street carriageways is essentially a ratepayers' question, and the unprecedented adoption of wood as a paving material in substitutions of macadam, proves that several of the metropolitan
vestries and other authorities have taken a new departure, and apparently a
step in the right direction. In expressing this opinion it will be readily understood that the author in no way desires to pass over the respective merits of granite, asphalt, or bituminous concrete pavements, and of roadways
formed with broken granite, flints, or gravel, as they are undoubtedly suitable
for certain localities, and in many cases are economical as compared with
wood. While therefore it is asserted that a properly constructed wood pavement possesses the advantages of noiselessness, surface elasticity, safety,
and cleanliness, and is pre-eminently suitable and economical for business
and residential thoroughfares having a high traffic standard, it should not
be forgotten that in the case of narrow business streets leading out of main
thoroughfares, wood pavement might be unsuitable, as by reason of the un-
importance of the vehicular traffic, the blocks would probably decay inter-

nally long before they were worn out by the traffic. For such reasons, and from a sanitary point of view, it would appear that asphalt would not only be preferable, but also eminently suitable.

Large sums of money were expended during the early years of wood pavement revival, in the acquisition of patent rights of doubtful value, in experimenting thereon, and in foolish contracts. These stages have been surmounted, and the result is that street paving in the West-end has been almost revolutionized within a few years. Wood pavment has been termed a " West-end luxury," and in one sense the assertion is perhaps justifiable, inasmuch as not more than 4.38 per cent. of the wood pavement in London is east of the city, or south of the Thames.

Extent and Construction.—The superficial area of wood pavement laid in London during the last ten years has been 980,533 square yards, its length 53¼ miles, and the cost of its construction, together with subsidiary works, has probably involved an outlay of £600,000. It is now possible to drive through wood-paved main thoroughfares for a distance of several miles ; (*e.g.*) wood pavement is practically continuous from London Bridge, the exception being the eastern part of the Strand, to any of the stations between Chelsea and Uxbridge Road on the West London Railway.

Obviously, much valuable experience and data have thus been gained as to the best mode of construction and maintenance, and as the author ventures to think that a paper on wood pavement would be incomplete unless it embraced every detail, however simple, he will endeavor to discuss the various points which arise, with a view to ascertaining whether wood has practically realized the expectation that it would prove to be a safe, convenient, and economical material for street carriageway pavements.

The suitability of " wood as a paving material under heavy traffic " was so fully treated by Mr. Howarth in 1879, that the author thinks it unnecessary to refer in detail to questions which have reference to the proper growth of wood, the cause and effect of wear and tear, and the method adopted for recording the traffic ; the object being to describe at length the various points of construction which have hitherto been only partially considered, together with particulars of recent modifications.

Excavation.—The suggestion has been made that macadamized carriageways might be broken up expeditiously with the aid of explosive mixtures, but the author would hesitate to try it ; as, however slight the concussion and vibration might be, there is little doubt that gas and water mains and services, especially those which have been in existence for a considerable period, would be injured. Practically there is little variation in the method adopted, the main object being to get the work done as quickly as possible. In preparing for the construction of the wood-pavement works in Chelsea, the method of operation consisted in breaking up the macadamized carriageways by driving steel wedges into and through the consolidated layer of broken granite, commonly called mac, at intervals of a few feet. The surface or crust being thus "started," its removal was effected by prizing it upwards in lumps of a square yard or more with stout hand-levers, 6 to 8 feet long, after which it was easily disintegrated. As a rule, a depth of 9 inches of macadam was thus removed prior to the excavation of some five or six inches of foundation.

In those instances where the carriageway had previously been paved with granite upon a concrete foundation, as in Oxford Street, it was found unnecessary to disturb the concrete, since by adding about 2 inches thereto, and floating the surface, a satisfactory foundation was obtained, and considerable expenditure avoided.

Levels and Contour.—When the width of a street is irregular, and the levels of the footways on either side vary considerably, it is sometimes not a little complex to satisfactorily determine the question of level and contour; but when the longitudinal inclination is naturally slight and uniform, the width parallel, and the footways nearly correspond in level, as in Sloane Street, a very simple rule may be observed.

The practice of the author has been to first determine the level of the crown or vertex of the carriageway to be paved, and next to set out the levels of the channels, by allowing a rise to the crown equivalent to 1 inch in 3 feet ($\frac{1}{36}$) above the mean channel level. By slightly flattening the crown, Fig. 1, it will be observed that a contour is obtained which not only renders traffic easy, but is of pleasing appearance, and satisfactory in other points.

Whenever practicable, the longitudinal inclinations of the channels should not exceed 1 in 150, and it is desirable that the minimum depth of curb exposed at the summit of a channel should be 2½ inches, with a maximum depth of 6½ inches at a gulley. By the observance of this rule it necessarily follows, that in a tolerably level street the provision of street-gulleys becomes an important item, but the extra cost, about 4 per cent., which their construction entails, is amply repaid, not only by the convenient and uniform appearance of the carriageway, but in the prompt and effectual removal of rain-water from the surface of the pavement. The author submits that this is an element of success in the construction of wood pavement which unfortunately, is too often overlooked.

The longitudinal crown-level should be uniformly sustained from street to street whenever practicable, so as to prevent undulations; and it is likewise important that the crown should be extended transversely at all intersections, partly for the sake of appearance, but mainly to obviate the unpleasant effect which is caused by driving over a channel. The neglect of this rule is very apparent upon observing the effect of vehicular traffic over the crossings to Rutland Gate and Princess Gate on the south side of the Kensington High Road.

Foundation.—It is satisfactory to note that foundations consisting of single or double planks placed upon a bed of sand have been completely discarded, and that the fallacies of the so-called elastic foundation have given place to a more permanent system. It was only necessary, shortly after a fall of rain, to witness the effect of a vehicle being rapidly driven over a pavement which had been laid on the former system for a period of two or three years, to have ocular demonstration of its utter unworthiness. The series of little fountains of dirty water which have spurted up in the wheel tracks from the open joints would soon have dispelled the hopes of the most ardent believer in the theory; and the result not only made it inconvenient for pedestrians, but very soon caused the pavement to go to pieces.

Although lias-lime concrete has been used as a foundation, it may safely be asserted that 90 per cent. of the existing wood pavement is laid upon Portland cement concrete. The latter, properly prepared, is absolutely impervious, and as a solid and sound foundation is essential for a first-class pavement, the author fails to see that a more suitable material can be substituted. From a variety of causes the strength of the concrete actually used appears to have varied from 5 to 7 parts of Thames ballast to 1 part of Portland cement. In the commencement of the Chelsea works the proportion was 6 to 1 ; but it was soon found desirable to provide a quicker-setting concrete, to enable the street to be reopened at the earliest possible moment. The exact proportions, as measured out in boxes, subsequently

FORM OF STUD

FIG·4

FIG·3· MODE OF JOINTING

FIG·1

FIG·5

FIG·6

FOOTWAY

MODE OF BLOCKING AT INTERSECTION OF STEETS

FIG·2

WOOD PAVEMENT ·IN THE· METROPOLIS

LINE OF

consisted of 33 cubic feet of ballast to 6.4 cubic feet sand, 5 bushels of cement, or about 5¼ to 1.

It is almost superfluous to state that too much care cannot be taken to insure the use of none but good cement. With this object in view, numerous samples of the cement supplied for the Chelsea works were tested, the average result of 197 tests by Mr. H. Faija, M. Inst. C. E., showing the strength and quality to be :

Weight per imperial striked bushel ....115.45 lbs.
Specific gravity............................ .... 2.98
Fineness (25 gauge-sieve).... . ... 4.18 per cent.
   "   (50   "   )................... .. .... 26.91
Tensile strength per square inch at 7 days........ ... 571 lbs.
   "   "   "   28 " .... 677 "

The full particulars of the tests are given in the Appendix to the Proceedings (Table II.).

The ground having been carefully bottomed up, regulated, punned, and when necessary watered, small ridges of concrete, technically called " screeds," were formed to the required levels and contour. The ballast and cement for the concrete were measured out and mixed upon a platform, and twice turned over dry, water being subsequently added by means of a small hose attached to a stand-pipe. Just sufficient water was used to obtain a fair consistency, and as soon as the materials were well incorporated, the concrete was either wheeled or thrown into place, to a minimum depth of 6 inches. To obtain a perfectly smooth and uniform surface, the concrete was floated and " ruled " transversely to the required contour by means of a curved rule, and in four or five days it became sufficiently hard and ready to receive the wood blocks.

That an excellent foundation was obtained has been frequently ascertained by means of the numerous gas and water trenches which have been made. To break through the concrete necessitates considerable force, and the use of sledge-hammers and steel wedges, and specimens prove that it is perfectly sound and good. The entire cost *in situ* averaged 2s. 3½d. per square yard, an amount which compares favorably with other concretes ; and if time were not so important an element in street closing, it might have been executed at a cheaper rate by decreasing the proportion of cement. The cost per square yard is made up of the following items—viz.:

	s.	d.
0.166 cubic yard Thames ballast, at 3s. 4d....	0	6½
0.74 bushel Portland cement, including extra for facing, at 1s. 11d..............................	1	5
Labor in measuring, mixing, wheeling, laying and ruling...	0	4
Total per square yard...................	2	3½

In the concrete foundation for Fulham Road a considerable quantity of the old broken granite was screened and mixed with Thames ballast, with a view to practically testing the assertion that an equally serviceable but cheaper concrete could thus be obtained. A similar system has been somewhat largely adopted in other metropolitan districts, and notwithstanding that a saving of about 3d. per square yard was effected in the Fulham Road pavement, the author is not encouraged to view the practice favorably, and does not propose to revert to it in future works. He ventures to assert that not only is the old granite of greater value for street repairs in less important thoroughfares, but that consequent upon its dirty condition, the efficacy of the cement is somewhat impaired ; that the concrete so formed is not so homogeneous as pure ballast concrete, and that it will eventually prove to be less durable than the latter. In support of this view it was recently ascertained by a sample taken up from Fulham Road that such concrete is undoubtedly of inferior quality, and the labor exerted in breaking through it was remarkably small as compared with the former material.

Blocks.—The best form and material for wood-pavement foundation having been ascertained, a question of the utmost importance arises— namely, which of the various kinds of wood available is the most durable and economical? Until this question has been satisfactorily answered it cannot be maintained that the difficulty has been surmounted ; in fact, to the absence of reliable information thereon, it is undoubtedly possible to trace the cause of many wood pavements becoming prematurely "worn out" after four or five years' wear. Many theories have been asserted as to the merits of various woods, and a large number of the latter have received a practical test, among them being Baltic and Dantzic fir, pitch pine, spruce, beech, larch, oak, elm, and ash. The shape of the blocks has received no little attention, inasmuch as under various systems they have been cut into rectangular, oblique, hexagonal, octagonal, square, and other forms, and of varying dimensions.

Considerable diversity of practice has also existed with regard to the condition in which wood has been laid, some blocks having been laid in the natural state of the wood, whilst others have either been "dipped" in creosote oil, or "dressed" with pitch, and in a few instances they have been properly creosoted. Within the last six or seven years the greater portion of the wood pavement laid in the metropolis has consisted of rectangular blocks of Swedish yellow deal, or red deal, or pitch pine, but it cannot be denied that thousands of blocks of an inferior nature have also been laid, after being subjected to the process of "dipping" or "pickling."

Prior to adopting wood pavement in Chelsea, the author inspected nearly the whole of the various systems then laid in London (February, 1879), and gave their respective merits every consideration, the outcome thereof being that a plain and substantial system was considered the most desirable. The blocks found most suitable, and of which there is an abundant supply, were those cut from Swedish yellow deals (Gothenburg thirds), and if blocks cut from close and evenly grained, well-seasoned, and thoroughly bright and sound deals of that description were always used, the author thinks that they would not fail to give satisfaction. In the construction of wood pavement, it is of the greatest importance that constant supervision should be exercised with a view to insuring the rejection of improper blocks, or blocks which possess even a suspicion of being sappy, knotty, badly cut, or otherwise unsound, or which have been cut of dead wood, or wood which has been stacked in dock for too long a period. If this system is faithfully carried out, it is doubtful whether there is at present any better wood in the market, of which sufficient quantities can be promptly obtained, and which so satisfactorily stands the wear and tear of traffic and the changes of atmosphere and climate, than Swedish yellow deal, when laid in its natural state.

The result of heavy traffic upon various kinds of wood unquestionably demonstrates that of hard woods pitch-pine takes a high place. The price, however, is considerably in advance of fir, and the irregular sizes of the deals creates much difficulty ; but if a satisfactory supply from South America in prime condition could always be relied upon, it would no doubt be

largely adopted.   A section comprising 756 square yards was laid with
pitch-pine blocks in King's Road, four and a half years since, as an experiment, and the ascertained depth of the annual vertical wear of the wood
during that period, as will be seen by Table III. in the Appendix, is 0.055
inch only.   Experience proves that its abrasive wear is better than yellow
deal, and, as a rule, it drys quicker and is cleaner ; and although pitch-pine
takes the best position as regards vertical wear, its extreme hardness may
be considered a drawback.   This is especially noticeable in the case of rapid
traffic, when the wheels of vehicles have lost the rolling motion and strike
the surface with great force.   It is then that the smallness of the elasticity
attached to pitch-pine becomes apparent, by a jarring, bumping motion,
which is far from agreeable.

Dantzic fir is durable, but would probably be attended with inconvenience and waste in converting the balks into ordinary-sized paving blocks.
Neither elm nor oak blocks would withstand the atmospheric changes to
which street surfaces are exposed.   Larch would probably take a high position as a street-paving material on account of its hardness and durability,
but the available supply is apparently too limited to permit its extensive
adoption.   There is little doubt that American spruce has been laid in London streets under certain conditions, but such wood cannot be regarded as
suitable or durable.

Preparatory Treatment of the Blocks.—A brief reference has already
been made to the preparatory treatment which blocks have received.   In
many instances they have been subjected to the process of being "dipped " in
a mixture of creosote oil and other ingredients, while in a few instances they
have been creosoted or mineralized, but at least one-third have been laid in a
plain or natural state.   The author has little faith in the ordinary " dipping "
process as a means of preserving the wood ; and upon severing " dipped "
blocks which have been laid for two or three years, it will be found that a
mere external discoloration only exists.   Indeed it is difficult to ascertain
the nature of the merit which the process is supposed to possess, and there
may even be truth in the assertion that it is worse than useless, as unfortunately, unscrupulous persons have taken advantage of it as a means of
covering up the shortcomings of defective or inferior blocks.

A great deal has been written as to the value of creosoted blocks, and,
by way of experiment, a short section upon this system was also laid in
King's Road, the blocks creosoted being selected yellow deal.   They were
subjected to Bethell's patent process, 7 pounds of hot creosote oil per cubic
foot, equal to 10.66 blocks, being injected ; and the ascertained depth of
annual wear averages 0.139 inch against 0.055 inch for pitch-pine, and 0.144
inch for plain yellow deal.   The creosoted blocks when taken up for examination were moist internally, and came up easily, whereas the plain blocks
and the pitch-pine blocks required considerable force to remove them.   The
author has invariably noticed that the creosoted section is less clean than
any other, and he certainly doubts the wisdom of a system which is not only
20 per cent. more costly, but to a certain extent closes the fibres of the
wood, and tends to produce premature internal decay.   When the latter
sets up, it will very soon be found that the traffic causes the fibres to
spread, which is quickly followed by the complete destruction of the
wood.

There can be little doubt that the reputation of wood pavement has
seriously suffered on account of the use of unsound and inferior wood.   This

may be traced to several causes, either the acceptance of extremely low tenders from inexperienced persons, want of sound judgment in selecting the deals, carelessness, insufficient supervision, especially when the work is hurriedly done, or downright neglect. The experience of the author is, that even with fairly good supervision it is possible for defective blocks to be laid, and it cannot therefore be surprising that unsound wood is used when carelessness, indifference, or neglect is exhibited. As an instance of the success of close supervision, the short section of wood pavement at the extreme north end of Sloane Street may be mentioned. The ascertained traffic, which consists in a great measure of carriages and cabs, is equal to 371 tons per yard width per day, or upwards of 100,000 tons per annum; and although the wood has been laid for four and a half years, the blocks have only worn $\frac{5}{8}$-inch, or 0.083 inch per annum, and the surface is in excellent condition. It cannot, therefore, be too strongly urged that street authorities should not only invariably pay a fair price for work, but should provide ample and competent supervision, if they desire to prevent wood pavement from being brought into disrepute.

Size of the Blocks.—There is little variation in the dimensions of the blocks, the general size being 9 inches long, 6 inches deep, and 3 inches wide. The lengths have frequently varied from 8 inches to 11 inches, in several instances the depth has varied from 5 inches to 7 inches, but the width, 3 inches, has almost been universally adhered to, it being apparently the best size for foot-hold, and most convenient to procure. Upon measuring blocks supplied by timber merchants, it will frequently be found that the depth is a trifle short of the specified dimensions; for instance, a 6-inch block seldom exceeds $5\frac{7}{8}$ inch, it being understood to be the custom of the trade to make an allowance for the thickness of the saw-cut. The question has often been raised whether 7-inch blocks on the one hand or 5-inch on the other, would not be more economical than 6-inch; and although there is very little experience thereon, the subject will be alluded to under the heading of durability.

Laying, Jointing, and Completion.—Great diversity of practice has existed in the mode of laying and jointing, and a great deal of money has been lost in the acquisition of patented inventions. Under some of the latter systems the blocks were laid on boards upon a foundation of sand, in others the blocks were laid upon a layer of asphalt, or upon tarred felt placed upon a concrete foundation. Blocks have been laid in transverse courses, and under Lloyd's "keyed" system, diagonally. Various modes of jointing have been adopted, technically known as mastic, semi-mastic, cement, lime, or felt, the latter being a close joint, whilst the former varied from $\frac{1}{4}$ inch to 1 inch in width.

In the Chelsea pavements the dimensions of the blocks were 3 inches by 9 inches by 6 inches, 40.5 being required for each square yard, and they were laid upon the concrete in their natural state, with the fibres vertical, and with intervening spaces $\frac{3}{8}$-inch wide, which were filled with cement grout. The surface of the concrete having become hard and smooth, the method of operating consisted in laying two longitudinal courses of blocks next to the curb to form the channels, and filling in with straight courses transversely, the ends of the blocks being in contact. At the intersections of streets the courses were laid V-shape, so as to insure the traffic passing over the blocks at an angle of 90°, or as nearly as possible thereto (Fig. 2). In laying the blocks the joint was kept parallel by means of three cast-iron

studs fixed in each block (Figs. 3 and 4), which materially assisted in keeping the pavement firm and steady until the grouting had thoroughly set.
By the use of studs instead of wooden strips or laths, it is certain that the
blocks are much less liable to become displaced while the work is setting,
and in practice a uniform width of joint is more easily secured.

The grout was composed of three parts of Thames sand to one part of
Portland cement, the materials being measured in boxes and mixed to a
proper consistency, in a large movable grout-truck.   The grout having been
swept out of the truck on to the surface of the wood, was swept backwards
and forwards until every joint was properly filled.   The prime cost of this
work, labor and materials, amounted to 5d. per square yard.

Top-dressing.—Before permitting vehicular traffic to pass over new
work, it is desirable to "top-dress" the surface of the wood with a fine
gritty material.   This is usually done by spreading a thin coating of very
small stones, as screened from Thames ballast, sand, or shingle, over the
entire surface, and leaving it to be crushed and ground into the fibres of the
wood by the traffic, until it has ceased to be gritty.   The cost of this work
for the Chelsea pavements amounted to 2d. per square yard for materials
and labor.   Whenever practicable, all traffic should be excluded from a
newly-laid pavement for at least a week after completion, but there is an
obvious difficulty in the enforcement of this rule, whenever a street is too
narrow to allow the works to be carried out in half-widths at a time.

The expansion of wood is a question demanding consideration in carrying out such works.   In unprepared blocks this is specially necessary, and
in proportion to the width of the street so is it desirable to leave a space
next the footways, varying from one to two inches in width.   The actual
condition of the wood and the state of the atmosphere at the time of laying
exercise a material influence on the width of the margins thus left.
Another mode is by omitting to grout the channels for a week or longer
period after the other portions have been completed ; and another by paving nearly up to the curb, and afterwards taking out the nearest course so
soon as expansion sets up.   In any case, it is necessary to temporarily fill
up the margin with sand or other suitable material, which can be raked out
from time to time until expansion ceases, which generally occurs in the
course of twelve to eighteen months from completion.   In the Chelsea
works it was then found desirable to rake out the temporary filling, and
where any space was left, to fill in with cement grout, the surface being
carefully pointed in cement.   This work effectually renders the channels
impervious to water, and tends to prevent premature decay, and was
executed at a cost equal to 0.86d. per square yard of pavement.

It frequently happens that, notwithstanding those provisions, and
whether the blocks are plain or dipped, the outer portions of the footways
become displaced during the period of expansion, but subsequent to the
above-named period wood pavement generally maintains a fairly constant
state.   The author has never observed any contraction in wood pavement
except during the severe frosts in January, 1881, at which time, when 22°
of frost were registered, several rather alarming cracks in a longitudinal
direction appeared in the Sloane Street pavement, at places where the wood
was especially exposed to the severity of the weather.   Immediately the frost
terminated the cracks appeared to close up naturally, and the pavement
assumed its former sound condition.

Advantages of Plain Blocks.—This plain system of pavement was laid in Sloane Street and King's Road in 1879, and upon closely observing the effect of nearly five years' traffic upon it, the result is eminently satisfactory. The plain type of pavement had previously been adopted by Messrs. Mowlem & Co. in several parts of the metropolis, with the exception that their joint was wider and was formed of lime grout, and that their form and manner of studding were different, and the author ventures to assert that it comprises all the essentials of a sound pavement; a quiet and smooth surface for vehicles being provided, and the pavement being sufficiently close-jointed to insure a safe foot-hold for horses. Experience proves that the cement joint is reliable, that it adheres to the blocks, effectually resists wet, and does not unduly wear down below the surface of the wood, and thereby allow dirt to accumulate in the joints, neither does it permit displacement of the blocks; the author therefore feels that he may justly pronounce it successful, and claim for it another advantage as compared with the mastic-joint—viz., the absence of the annoyances which are created by asphalt kettles in streets during the execution of the works. Upon inspecting the numerous openings which have been made for gas and water services or drainage-works, the actual condition of the wood and foundation has invariably proved satisfactory; the surface of the concrete is hard and dry, the joints are sound, and the blocks as a rule present a fairly good surface, and are not found to contain an excessive amount of moisture.

Cost and Durability.—The prime cost of the pavement in King's Road and Sloane Street amounted to 10s. 6d. per square yard, and the only repairs which have been found necessary consisted in the removal of some defective blocks during the past twelve months. The blocks were originally 5.87 inches deep, whereas their present average depth measures 5.22 inches in King's Road and 5.60 inches in Sloane Street. The probable life of the block may be taken at seven years for King's Road and eight years for Sloane Street. The annual vertical wear of wood is equal to 0.144 inch in King's Road and 0.065 inch in Sloane Street, and if reduced to a traffic standard of 750 tons per yard width daily, the following results are obtained —viz.:

King's Road........ .... 0.196 inch
Sloane Street.. .... 0.175 "

Several other systems of pavement were laid in King's Road at the same time for the purpose of practically testing their respective merits. In one instance a layer of metallic lava asphalt one-half inch in thickness was laid upon the concrete foundation, the blocks being laid and jointed as already described. Opposite the Royal Military Asylum wall creosoted blocks were adopted, one length being formed with a semi-mastic joint— namely, the lower half bituminous mastic and the upper lias lime grout— the other being entirely jointed with grout composed of lias lime and sand. The section of pitch-pine blocks previously described was also laid at this part, the joint consisting of Portland cement grout. The author has carefully examined the various experimental sections of pavement, the cost of which, together with the ascertained wear and other particulars, are as follows—viz.:

(a) *Plain Deal Blocks on Layer of Asphalt.*—The cost amounted to 12s. 4d. per square yard; the present depth of the blocks is 5¼ inches; the

foundation is in a perfect condition and the blocks are fairly good.   The estimated life of the pavement is seven years, the avarage annual wear being
0.139 inch.

(*b*) *Creosoted Blocks, with Mastic Joint.*—The cost was 14*s.* 2*d.* per
square yard, the present depth of the blocks is 5¼ inches, and the foundation is perfectly sound.   The lime grout part of the joint has perished and
almost crumbled away, consequently a good deal of dirt has been pressed
into the part above the bituminous portion, and the blocks came up very
easily.   The pavement may last for eight years, the average annual wear
being 0.139 inch.

(*c*) *Creosoted Blocks, Lime Joint.*—The cost was 11*s.* 10*d.* per square
yard ; the blocks are 5⅜ inches deep, and in a fairly good condition.   The
failure of the lime joint may possibly arise from a chemical action of the
creosote ; but it certainly has perished, and causes dirt to accumulate.   The
blocks will probably last eight years, the average annual wear being 0.111
inch.

(*d*) *Pitch-pine Blocks.*—This section costs 11*s.* 8*d.* per yard, and the
present depth of the blocks is 5⅜ inches.   The foundation is dry and sound,
and the cement-joint perfectly satisfactory.   The repairs to the blocks have
been very slight, and they will probably last eight or nine years, the average
annual wear being 0.055 inch only.

If reduced to a traffic standard of 750 tons per yard width per day, the
figures are : Pavement (*a*) 0.209 inch ; (*b*) 0.240 inch ; (*c*) 0.204 inch ; (*d*)
0.088 inch.

Practically, the plain pitch-pine pavement is cheaper than yellow deal,
and the ascertained vertical wear promises admirable results ; but it does
not appear that either of the other modes is attended with any particular
advantage.

Other Pavements.—In 1876, the Improved Wood Pavement Company
laid a short section in King's Road, near the Vestry Hall, in which planks
were laid upon the concrete foundation, and "dipped" pitch-pine blocks
were used ; but in consequence of numerous complaints of the disagreeable
rumbling and jarring noise which the rapid passage of vehicles over it
caused, it was taken up and relaid without the boards after a period of
seventeen months, an additional length being at the same time laid with
"dipped" deal blocks.   These pavements were subsequently particularly
unfortunate, inasmuch as the "mastic" joint caused the surface to become
uneven and bumpy, and rendered it necessary to execute frequent repairs,
a considerable portion being relaid in 1882.   Upon examination it transpired
that the substance which formed the "mastic" joint had partially melted,
presumably from the heat of the sun, and that the material had got underneath the blocks and forced them upwards.   In November, 1883, the pitch-
pine section was finally taken up, and the other portions were again repaired
by the company for the same cause.   Like results may be seen in other
districts where a similar joint has been adopted.

The ascertained average depth of the pitch-pine blocks at the time
of removal was 5⅜ inches, the average annual wear during a period
of seven years being 0.089 inch, or if reduced to the traffic standard,
0.119 inch.   The deal blocks were measured and found to be 5 inches
deep, the ascertained annual wear being 0.157 inch, and according
to the traffic standard 0.195 inch.   With additional repairs the latter may
last another year, making their life seven years.   The author further objects

to this form of joint, because the melted asphalt adheres so tightly to the surface of the concrete that it is necessary to chip it off before new blocks can be laid, which process destroys the smooth surface of the concrete so considerably that it necessitates a bedding of sand before new blocks can be properly laid. Further, the bitumen adheres so tightly to the blocks that to well cleanse them for reuse not only involves much labor but is injurious to them, especially if they have borne two or three years' traffic and the fibres have begun to spread. In the latter case they are not only unsound, but are also unfit to make neat or satisfactory work.

The improved Wood Pavement Company, whose experience is probably greater than that of any other company, have laid their system in many of the leading streets, among which may be mentioned King William Street, Bishopsgate Street, Queen Victoria Street, Whitehall, Bond Street, Park Lane, and Old Brompton Road. In all instances where the patented system had originally been adopted, advantage was taken to modify and simplify it whenever it became necessary to renew the blocks. The company have entirely discontinued to lay a plank in the foundation, and now limit their system to the use of "dipped" blocks, laid with a semi-mastic joint upon a concrete foundation. For the purpose of comparison, the weight of the traffic being known, the following streets are referred to :

Ludgate Hill.—Originally laid on boards in November, 1873, at a cost of 18*s.* per square yard. The terms of maintenance are, one year free, and fifteen years 1*s.* 6*d.* per square yard per annum. The blocks and plank foundation were taken up and new blocks laid on a concrete foundation and single board in 1877. The pavement was again taken up and laid with new blocks on the new principle—namely, without the intervening boards—in February, 1884, when the blocks were found to average 3 inches deep, their actual life being seven years. The author saw this pavement taken up, and was surprised to find how easily it was removed; the mastic joint had evidently lost its strength, and the result possibly points to the fact that its efficient durability does not extend beyond a limited period. The annual average wear of the last-mentioned wood was 0.428 inch, and if reduced to the traffic standard it was equal to 0.259 inch.

Aldersgate Street.—Cost 15*s.* per square yard in September, 1874 ; the maintenance terms being two years free, and fifteen years at 1*s.* per square yard per annum. It was taken up and re-laid on a concrete foundation in 1877, the blocks being merely reversed, and laid on lias lime mortar. In, March, 1884, the blocks were renewed, the average depth then being 3½ inches ; but in some parts the wood was only 2 inches thick. The annual wear was equal to 0.264 inch, and the actual life of many of the blocks was no less than nine and a half years. The blocks recently laid are 5 inches deep.

Northumberland Avenue.—Paved in February, 1876, at a cost of 15*s.* per square yard, exclusive of the concrete foundation, which had been laid by the Metropolitan Board of Works. It is maintained by the company at a rate of 13*s.* per square yard for fifteen years. Baltic pine blocks, 6 inches deep, were laid on planks, and the approximate wear is 0.125 inch per annum. The surface is uneven, and shows considerable wear at the upper end, but its life will probably be nine years.

Leadenhall Street.—In July, 1876, the western end was paved at a cost of 12*s.* 3*d.* per square yard, with two years' free maintenance, and fifteen years at 1*s.* per square yard per annum. The pavement was taken up, and 5-inch blocks laid in 1879, without the planks. The wood is considerably worn, and the surface is bumpy, probably from the mastic joint. The approximate wear of the wood is 0.200 inch per annum, which, reduced to the traffic standard, is equal to 0.186 inch per annum.

Piccadilly.—An area of over 16,000 square yards was laid in April, 1876. In 1880 nearly two-thirds were renewed, and the boards taken out, the remainder done in February, 1882. Further repairs have been executed, and the surface shows considerable signs of wear. The author has been unable to obtain information as to the depth of blocks,

Knightsbridge.—The portion east of Albert Gate was paved in April, 1878, the blocks being laid upon a single plank, and concrete foundation. The part west of Wilton Crescent was taken up in September, 1883, when the plank was removed, consequently the actual life of the wood was nearly five and a half years. The daily traffic is about 780 tons per yard width, or 250,000 tons per annum.

Parliament Street was paved in December, 1880, at a cost of 13s. 8d. per square yard, and the company maintained it for three years at an additional charge of 4d. per square yard per annum. The foundation is Portland cement concrete, 12 inches thick. The annual wear of the wood is 0.154 inch, and if reduced to the standard is equal to 0.104 inch.

Oxford Street, opposite Hereford Gardens, was originally laid on boards in November, 1874, at a cost of 16s. per square yard, with maintenance two years free and thirteen years 1s. per square yard per annum. In May, 1877, the boards were removed and new blocks laid, but they are now in a very bad condition, and are about to be taken up. Their life will have been seven years.

Hensen's patent pavement has been practically tested, having been laid in Fleet Street, Oxford Street, Brompton Road, Euston Road, Uxbridge Road, and elsewhere. The theoretical principles of this system, which consists of a cement-concrete foundation, plain deal blocks on a felt bed, a close felt joint, and a dressing or grouting of boiling or prepared tar, and the merits of the felt bed and joint were fully described in Mr. Howarth's paper, and the following particulars of ascertained wear may be useful in considering its merits.

Oxford Street, Princess Street to Marylebone Lane, was laid in November, 1875, at a cost of 16s. 6d. per square yard on existing concrete. The company offered to maintain it for fifteen years for 10s. per square yard, but the Marylebone Vestry arranged to do this work by their own staff. The pavement was repaired in 1879 and following years, the blocks being entirely renewed in September, 1883. Yellow deal blocks, 5 inches deep, were originally laid, and when taken up averaged 5½ inches thick, their actual life being 7.84 years. The annual average wear of wood was therefore 0.191 inch, which, if reduced to the traffic standard already described, gives 0.120 inch.

Oxford Street, Duke Street to Portland Street, was paved in October, 1876, at a cost of 14s. per square yard on existing concrete, the maintenance terms offered being 10s. per square yard for a period of twelve years. Six-inch blocks were laid, and they were repaired in 1880, 1881, 1882, and 1883, their present average depth being 3.60 inches. The surface of the wood is considerably worn, but upon recently inspecting an opening in the pavement the foundation and joints were in excellent condition. The probable life of the blocks is eight years. The average annual wear of the wood is equal to 0.323 inch, and as reduced to the traffic standard 0.255 inch.

Oxford Street, Hereford Gardens to Edgeware Road, was paved in December, 1875, at a cost of 16*s*. 6*d*. per square yard on existing concrete. The blocks were repaired each year from 1879 to July, 1883, when they were renewed. They were originally 6 inches deep, were worn to an average depth of 3½ inches, and had a life of 7.58 years. The probable life of the existing blocks is eight years. The annual average wear of the former wood was 0.329 inch, and as reduced to the traffic standard was equal to 0.250 inch.

Brompton Road.—The eastern half was paved in December, 1878, at a cost of 12*s*. 9*d*. per square yard, with a free guarantee of three years. Since then frequent repairs have been executed by the Kensington Vestry. Six-inch blocks were laid, the present average depth of which is 4.94 inches. The surface is worn in places, and is bumpy and somewhat unpleasant to drive over. It is very dirty at the joints. The probable life of the wood is six and a half years, the average annual wear being 0.184 inch, or according to the traffic standard 0.236 inch.

Fleet Street.—The western half was laid in September, 1877, at a cost of 16*s*. per square yard, the maintenance charge being 17*s*. per square yard for nineteen years. It has been repaired from time to time since 1881, and is now considerably worn and uneven. The probable life of wood is seven years. The average annual wear of the wood is 0.269 inch, which, according to the traffic standard, is equal to 0.173 inch.

Leadenhall Street.—In August, 1876, the eastern half was paved at 18*s*. 6*d*. per square yard, the maintenance being two years free and seventeen years at 1*s*. 6*d*. per square yard per annum. The paving has been repaired on several occasions since 1880, and parts are now in a very bad condition, the wood having quite worn through in places. The paving is to be renewed this spring, after a duration of seven and a half years ; and, taking the average depth of the block at 4 inches, the annual wear is equal to 0.264 inch, which, reduced to the traffic standard is 0.198 inch.

Euston Road, Cleveland Street to Gower Street, was paved in November, 1880, at a cost of 11*s*. 6*d*. per square yard, with three years' free and twelve years' annual maintenance at 8*d*. per square yard. The pavement is in fairly good condition, only trifling repairs having been carried out, notwithstanding that the weight of the traffic per yard width is equal to 700 tons per day.

It must be apparent to any one who has carefully noticed Henson's pavement that there is a minimum of jarring, and consequently a very steady motion in driving over it when it is in good condition ; yet experience seems to prove that after a few years' wear it is not in reality cleaner or less dusty than a plain close-jointed pavement, and a reference to Table V, in the Appendix clearly shows that in durability it does not take the highest place.

The system adopted by the Asphaltic Wood Pavement Company consists in laying a ¼-inch layer of asphalt upon the concrete foundation, upon which "dipped" blocks were placed, the lower part of the joint being of asphalt and the upper of Portland cement grout. This pavement has been laid in Fleet Street and other parts of the city, the Strand, Oxford Street, High Holborn, Hatton Garden, Brompton Road and elsewhere.

Oxford Street, Marylebone Lane to Duke Street, was laid in September, 1876, upon an existing concrete foundation, at a cost of 13*s*. 6*d*. per square yard, and offered to be maintained for fifteen years at 12*s*. per square yard. It was repaired in 1878, and each successive year to November, 1882, when the blocks were removed, and the pavement relaid upon the plain system with a cement joint. The blocks originally were 6 inches deep, and when removed had worn down to about 3 inches, their actual life being about 6.20 years. The average annual wear of wood was 0.484 inch, which, reduced to the traffic standard, is equal to 0.319 inch. The pavement suddenly broke up and went to pieces at the end of 1882, owing, it is stated to water getting underneath, and the asphalt breaking up, which was probably caused by the "open joint" of asphalt and lime grout.

The Strand was paved in February, 1877, the cost being 13s. 6d. per square yard, and maintenance undertaken at 1s. per square yard for twelve years. Six-inch Baltic red blocks were used, and the pavement lasted until 1882, when it was renewed. The actual life of the wood was a little over five years, but it should be observed that the daily traffic is upwards of 1,100 tons per yard width.

Fleet Street.—The eastern portion was paved in July, 1877, the cost being 15s. per square yard. Considerable repairs have been carried out, but it is much worn, more so than Henson's pavement, and is somewhat bumpy to drive over ; this may, perhaps, be partly accounted for by the gradient being sharp. The average annual wear of the wood is 0.456 inch, which, reduced to the traffic standard, is 0.251 inch.

Regent Street.—The portion between Piccadilly Circus and Vigo Street was paved in September, 1880, at a cost of 12s. 6d. per square yard, on a cement concrete foundation, 12 inches thick, formed of 2 parts of old macadam to 1 part of ballast. Six-inch deal blocks were laid, but they already show considerable signs of wear ; in fact, they needed repair in 1883. The ascertained annual wear is 0.286 inch, which, reduced to the traffic standard, is equal to 0.384 inch.

Brompton Road.—The eastern part was paved in December, 1878, at a cost of 13s. 9d. per square yard. The blocks have been repaired on several occasions ; the surface is in a more uneven condition than Henson's pavement, and the joints are much worn. The life of the blocks may be put at six and a half years. The ascertained annual wear is 0.373 inch, which, reduced to the traffic standard, is equal to 0.431 inch.

The author fails to observe that the asphalt bed and joint passes the merit claimed for them, and certainly the pavement in Brompton Road, the surface of which became irregular at least two years ago, is in an extremely unsatisfactory state, owing, no doubt, to the failure of the joint and quality of the wood.

Lloyd's Patent "Keyed" Pavement.—This was laid in Pall Mall in February, 1879, and, from the deplorable condition into which it had fallen in December, 1883, there can be little doubt that it has proved an utter failure. Unquestionably the work was carelessly executed, as the author found blocks varying in length from 8 inches to 15 inches, and joints from 1 inch to 2 inches in width ; and when it is considered that the blocks were laid diagonally, although the wood used was pitch-pine, it is not unreasonable to attribute the failure to the mode of jointing and diagonal form of blocking. The specimen blocks submitted were taken up in December, 1883, and they reveal the fact that the so-called "key" was no key whatever, it having entirely failed to hold the blocks together ; and considering the irregularity of the width of the joint and the poorness of the material with which it was formed, together with the uneven bedding of the blocks, it will be readily understood that the joints soon wore down, and that the blocks became rounded, thereby rendering the pavement uneven and more bumby to drive over than any other modern system. The pavement originally cost 8s. 5d. per square yard only, being laid on an existing concrete foundation. When the blocks were removed they were about 4⅛ inches deep, but being so much rounded and unevenly worn, it was found that they could not be turned and relaid ; consequently deal blocks upon the plain system were substituted. The life of a large area of the pitch-pine "keyed" blocks was, therefore, less than five years—an extremely unsatisfactory result, especially when it is remembered that the carriage-way is extremely wide.

The same system of pavement was laid in the upper part of Regent Street in September, 1880, at a cost of 13s. 9d. per square yard. It was repaired in 1883, and again this year ; it already shows considerable wear and is uneven in places. The blocks are of pitch-pine, 6 inches deep, and the ascertained annual wear is 0.214 inch, which, reduced to the traffic standard, is equal to 0.288 inch.

Carey's System.—The pavement known as Carey's system was laid in Canon Street in September, 1874, at a cost of 13s. 6d. per square yard, the maintenance terms being two years free and fifteen years at 1s. 6d. per square yard per annum. The blocks originally laid were 4 inches wide by 9 inches by 6 inches, and were shaped with alternate convex and concave ends, and laid on a thin bed of ballast, the joints being formed of lime grout. Considerable repairs and renewals have been carried out from time to time, and in March, 1884, the author saw several trenches opened near Queen Street, at which places the original blocks had worn to a depth of 3 inches, the surface being extremely uneven, not to say bad. The western section, which was renewed about a year ago, is already uneven, and begins to show signs of wear, and notwithstanding that the wood has considerable durability, yet, owing to the inferior surface of the pavement, the author would hesitate to class it among successful pavements.

The Ligno-mineral Pavement was laid throughout Coleman Street, City, in June, 1865, at a cost of 13s. 6d. per square yard. The paving consisted of mineralized pitch-pine blocks, 4 inches deep, with a "mastic" joint, the top portion being lime grout. The pavement having become worn out, owing to open joints and decayed wood, it was taken up in April, 1882, when asphalt was substituted. It is extremely probable that in consequence of the carriageway being so much in the shade, the wood was specially liable to retain moisture. Fore Street was similarly paved in December, 1874, but the wood was replaced with asphalt in July, 1883.

Messrs. Mowlem & Co.'s Pavement.—This plain system of paving has been laid in various parts of the metropolis, particularly in the city, St. Giles, St. Marylebone, St. Pancras, and Kensington. In Princess Street, Cavendish Square, blocks which were laid in September, 1874, are still in existence.

Kensington High Street.—In May, 1877, a section was laid near the Vestry Hall, at a cost of 14s. per square yard, with a three years' free maintenance. Six-inch blocks were laid in their natural state with a ¼-inch joint of lime grout. The present depth of the blocks is stated to be 3 inches, the annual average wear therefore being 0.440 inch. The surface shows considerable wear, and after rain water is retained at those points where it has worn below the channel-level. The latter evil is possibly to be attributed to the very slight rounding, one inch in four feet, to which it was laid. The blocks are to be renewed in October next, when the life of the wood will have been seven and a half years.

Fulham Road, Sydney Street to Arthur Street.—This section was paved in July, 1878, at a cost of 14*s*. per square yard, and in other respects was similar to the last-mentioned pavement. The surface is considerably worn, and the form of joint not only retains dirt but tends to round the blocks, the average depth of which is 4⅝ inches, or equal to an annual average wear of wood of 0.242 inch. The pavement will probably be relaid in 1885, when its life will have been seven years.

A large area of "plain" wood pavement has been laid in Kensington by Messrs. Nowell & Robson, who paved Kensington Road, Fulham Road, Uxbridge Road, and High Street, Notting Hill. The last-mentioned street was carried out in December, 1878, at a cost of 12*s*. 6*d*. per square yard, with a three years free maintenance. The lime-joint gave much trouble shortly after the work was completed, and in places it may be observed that it has allowed dirt to accumulate. The average wear is equal to 0.218 inch per annum.

In other metropolitan districts besides Chelsea the local authorities have successfully laid a plain system of wood pavement by means of their own staff of workmen, particularly in St. Marylebone and Paddington. The credit of introducing this method is due to the St. Marylebone Vestry, whose first work consisted in paving the portion of Oxford Street east of Regent Circus, in October, 1878. The blocks were laid upon an existing concrete foundation, and the work cost 8*s*. 4¼*d*. per square yard, exclusive of the removal of the old stones. The blocks were repaired in 1882, 1883, and in 1884, at an approximate cost of 6*d*. per square yard. Plain yellow deal blocks, 6 inches deep, were adopted, with a cement joint ¼ inch wide, and they have worn to an average depth of 3.30 inches, but in some parts of the street the thickness is 1¼ inch only. The surface shows considerable wear, and is uneven in places, the wood being so remarkably thin near the rests that it is a mere crust. The probable life of the wood is six and a half years. The average annual wear is 0.475 inch, and if reduced to the traffic standard it is equal to 0.306 inch. The heavy rate of wear is probably owing to the width of the joint, the author having taken up a piece of grout ⅜ inch thick. This irregularity would have been avoided by the use of iron studs instead of temporary strips or laths. A large area was similarly laid in Edgeware Road in October, 1880, and at the same cost. The surface is good ; the annual wear is 0.198 inch, and if reduced to the traffic standard it amounts to 0.254 inch.

The Paddington Vestry have laid 125,232 square yards in various streets. Praed Street was paved in July, 1879, with 6-inch plain yellow deal blocks, at a cost of 10*s*. 7*d*. per yard, and the surface is in a fairly good state. At the eastern end the blocks have been repaired on several occasions, the present depth of wood in the centre part being 4 to 4⅞ inches.

Generally speaking, the plain system appears to have given satisfaction, but the mode of jointing with lias-lime grout, or a wider joint than ⅜-inch, cannot be recommended. Upon inspecting the lime-joint after a few years' wear, it may be ascertained that it wears below the surface, that dirt accumulates in the joints, and that the blocks have either become rounded or the top edges "burred," to such an extent that the surface has become bumpy. The lime-joint gave trouble when newly laid in Notting Hill in

December, 1878, as after a sharp frost the grout, so to speak, "spewed" up, the rain filled the joint, and considerable sections of the pavement were literally afloat until the defects were remedied. A system of blocking, which the author considers objectionable, is that by which the blocks of a new pavement are laid upon fresh unset cement floating, and, as the grouting is proceeded with, the blocks rammed with a "pavior's" rammer, so as to obtain a smooth surface. Under this process there is a probability of the blocks being injured or split, apart from which it is found that when the time arrives for renewing the wood, the surface of the concrete contains a series of indentations instead of being smooth and even. The difficulty may of course be surmounted by chipping off the projecting parts and refloating the surface, but the repaving cannot be so expeditiously or economically carried out. The result of the author's experience and investigation induces him to submit (1) that, the surface of the concrete foundation should be perfectly smooth and fully set before the blocking is proceeded with, and (2) that a carefully-made cement-joint ⅜-inch wide will not only be found simple and water-tight, but will prove as durable as the wood itself.

Cost.—With the exception of a small area, the whole of the wood pavement in Chelsea, about 50,000 square yards, has been laid by the Board's own staff. The estimated cost of the pavements in King's Road and Sloane Street was 11s. 3d. per square yard,* but, as previously stated, the actual cost amounted to 10s. 6d. per square yard, exclusive of £120 spent in the before-mentioned experiments. The pavement in Fulham Road cost 10s. 3d. per square yard, the difference being partially attributable to the fact that a portion of the old broken granite was used in the concrete foundation in lieu of ballast.

The details of the cost per square yard, are as follows, viz.:

ITEM.	31,333 yds. in Sloane Street and King's Road in 1879.	10,573 yards in Fulham Road in 1881.
	d.	d.
Labor in breaking up macadam surface and excavating	11.00	11.00
Cartage of old materials, including shoot for rubbish...........................	9.04	9.80
Portland cement for concrete and grout....	20.02	1.17
Thames ballast and sand for concrete, grout, and top dressing	8.56	5.80
Blocks.................................	58.66	60.82
Studs..................................	1.58	1.48
Labor in bottoming up and leveling, preparing and laying concrete, fixing stud in, wheeling and laying blocks, grouting, top dressing, watching and sundries	13.45	14.12
Labor and materials in permanently filling in margins.............................	0.86	0.90
Sundries — plant, tools, superintendence, testing cement, oil, repairs, etc.........	2.92	1.85
Total	126.09	122.94

NOTE—No allowance is made for value of the paving stones and broken granite taken up and re-used in other parts of the district, the minimum value of which amounted to £2,050, or about 1s. per square yard.

* Minutes of Proceedings Inst. C. E., vol. lviii, p. 75.

The variation in the prices paid for wood pavement in various parts of the metropolis has been somewhat remarkable, the maximum cost per square yard for laying a pavement and concrete foundation with entirely new materials having amounted to 18s. 6d., and the minimum to 10s. 6d. Owing presumably to competition and to the experience which has been gained, together with increased facilities, the cost has gradually been reduced to reasonable limits, as compared with the charges made eight or nine years since.

Maintenance.—However excellently a street carriageway pavement may have been constructed, its condition will soon become unsatisfactory unless its maintenance receives proper supervision. Good management implies not only that repairs shall be promptly and efficiently executed, but that the services of cleansing, watering, and sanding must be properly carried out ; in short, the essentials of proper management are to be found in the judicious application of the scraper and broom, of water, and of grit, and in the immediate removal of defective blocks. The reinstatement of gas, water, and drainage-trenches must be classed under the first heading ; and although an apparently small matter, yet, from the frequency of such openings, so serious an interference with the street surface is created that in the course of a few years surface uniformity cannot be maintained unless this work is very carefully executed, and ample time allowed to elapse before the traffic is allowed to pass over the work. After a pavement has been laid for three years the existence of defective blocks becomes apparent, as by this time, the first effect of compression having ceased, the fibres of such blocks begin to yield under traffic pressure, with the result that slight surface depressions are formed. When this happens a bumping motion is created, and as the wheels then strike upon the edges of the adjoining blocks it is obvious that the surface must become irregular ; and depressions or hollows a foot square or more are soon formed, which, unless promptly remedied, materially spoil the surface.

To avoid slipperiness and to insure many of the advantages claimed for wood pavement, it is essential that a thorough and systematic service of cleansing must be carried out, especially where macadam pavements are contiguous to wood, as in damp weather a considerable amount of mud is imported from them. In connection with the wood pavement in Chelsea there is a regular street orderly service, by which horse-droppings are removed and deposited in bins. In addition thereto the wood pavements are washed once or twice a week, and are cleansed daily either by horse-sweeping machines or by hand labor. In the absence of heavy rains mere sweeping fails to keep wood pavement clean, and washing then becomes essential. To effect this water-vans are sent out before midnight, and the surface is so thoroughly soaked that, by the time the sweeping machines commence to work at 3 A. M., the dirt is easily removed, the entire operation being concluded in the forenoon. The ascertained cost of this service, including labor and horse hire in washing and sweeping, street orderly work, and collection and removal of the sweepings, amounts to 4½d. per square yard per annum, as against 11d. per square yard for macadam previous to the substitution of wood. It has been asserted upon good authority that the cost of cleansing wood pavement is very trivial ; this is slightly misleading, the proportions being approximately :

Macadam............ .... 1.00
Wood... .. 0.41

Theoretically, the amount of mud created upon the surface of wood, as also in the case of asphalt, should be almost nil ; but practically the author finds that some 2,700 cartloads are annually removed from a length of about three miles in Chelsea. Therefore, after making every allowance for the conversion into mud of 350 loads of sand placed on the wood when slippery, it is obvious that a great portion of the mud is imported from the adjacent macadam.

The plentiful application of water prior to the work of cleansing is most beneficial, both in preventing dust, and, from a sanitary point of view, in removing the cause of obnoxious smells ; but as the metropolitan water-supply is not yet in the hands of the ratepayers, its use for this purpose is materially restricted. The author ventures to assert that the system of cleansing thus described is amply sufficient to obviate slippery surfaces caused by the accumulation of greasy mud, and that the summer watering may be so carried out that a minimum wetting will suffice to keep down the dust. Letters have recently appeared in the *Times* with reference to the watering and cleansing of wood pavements, in which it has been strongly urged that such pavements should not be watered at all. When it is considered in what an unskillful manner street-watering is sometimes done, and that, owing to the stupidity or carelessness of carmen, considerable danger to locomotion is caused by overwatering a dirty pavement, there may be some justification for the contention. Doubtless horses travel better on dry wood pavement than on a watered surface, but in the absence of rain, watering is an absolute necessity for keeping down fine dust, more especially upon a hot windy day, when at least five or six wettings are required. Watering is also necessary for the preservation of the wood itself, as without water it would be materially injured by abrasion under such conditions. It is also questionable whether the very fine dust which must be given off under a non-watering system, would not become so serious as to be injurious to health and promote disease of the eye ; but apart from this, the nuisance from the heat and dust combined would become intolerable.

In continuous damp and foggy weather and on frosty nights wood pavement is especially liable to become slippery ; therefore, to insure a safe foot-hold for horses, its surface should be covered with a thin layer of Thames sand or grit. In Chelsea it has been found that this operation can be more expeditiously and evenly carried out by horse-machines known as "sand-distributors" than by manual labor alone. Night gangs have been organized, and, according to the conditions of the weather, the machines are sent out either at night or early in the morning ; in the latter case the whole of the wood is sanded and made fit for traffic by 8 o'clock. The operation is beneficial to the wood itself, and might be advantageously carried out at other times, because the grit becomes so well worked into the ends or fibres of the blocks that it not only affords protection to them, but insures a better foot-hold for horses. The ascertained cost of the sandings does not exceed ¼d. per square yard per annum.

Durability.—One of the most important factors in connection with durability is the amount of traffic to which the pavement is subjected. As the author has been unable to obtain complete information thereon he has had to rest content with the available figures, and has reluctantly omitted results and comparison of other important streets. Table III. shows (1) the daily traffic weight per yard width ; (2) the depth

of annual wear of wood ; and (3) the annual wear of wood as reduced to a
standard of traffic equal to 750 tons per yard width daily, or 235,000 tons per
annum, exclusive of Sundays. It is gratifying to remark that there is a
growing tendency to make observations and keep records of traffic weight,
wear and cost, and it is only by these means that reliable data can be
obtained. It is to be hoped, therefore, that both local authorities and wood-
pavement companies will institute the desired inquiries, by which means
much valuable experience and knowledge will be gained. The comparative
annual wear reveals several inconsistencies. For instance, the wear of the
asphaltic pavement in Fleet Street, with the maximum traffic weight, or 24th
on the list, is but 16th in point of wear, whereas the same system with the 13th
and 9th traffic weight positions, takes the high places of 24th and 23d in wear.
Lloyd's pavement in Regent Street has an 8th traffic-weight position, but
is 20th in wear. The wear of the Improved, of Henson's, and of the plain
pitch-pine pavements compares favorably in all cases with the traffic weight.
The lesson to be drawn from these figures would appear to be that the

III.—COMPARATIVE WEAR OF WOOD PAVEMENTS AS REDUCED TO A
TRAFFIC STANDARD.

SITUATION.	SYSTEM.	Weight per yard width per day of sixteen hours.	Depth of annual wear of wood.	Comparative annual wear of wood as reduced to a traffic standard of 750 tons per yard width per diem.
		Tons.	Inch.	Inch.
*Fleet Street... ..	Asphaltic.......... .	1,360	0.456	0.251
Ludgate Hill	Improved..............	1,236	0.428	0.259
*Oxford Street ...	Henson's(east section)..	1,191	0.191	0.120
*Fleet Street......	Henson's	1,165	0.269	0.173
*Oxford Street. ..	Plain	1,164	0.475	0.306
*Oxford Street....	Asphaltic.......... ..	1,137	0.484	0.319
*Parliament Street	Improved....	1,106	0.154	0 104
*Leadenhall Street	Henson's	1,000	0.264	0.198
*Oxford Street....	Henson's (west section).	985	0.329	0.250
*Oxford Street....	Henson's (central section)..	948	0.323	0.255
*Leadenhall Street	Improved..............	808	0.200	0.186
Brompton Road ..	Asphaltic...	648	0.373	0.431
King's Road......	Improved.............	603	0.157	0.195
Brompton Road ..	Henson's	584	0.184	0.236
*Edgeware Road .	Plain	584	0.198	0.254
*Regent Street ...	Asphaltic...	558	0.286	0.384
*Regent Street ...	Lloyd's	558	0.214	0.288
King's Road......	Improved (pitch-pine) ..	558	0.089	0.119
King's Road......	Plain	551	0.144	0.196
King's Road......	Plain (asphalt bed).....	498	0.139	0.209
King's Road......	Plain (pitch-pine)	468	0.055	0.088
King's Road......	Creosoted blocks (mastic joint)	434	0.139	0.240
King's Road......	Creosoted blocks (lime joint)	407	0.111	0.204
Sloane Street.....	Plain	279	0.065	0.175

* Weight of traffic taken from Mr. Howarth's paper in these instances.

asphaltic and Lloyd's systems are not successful ; and the author cannot help regretting that Regent Street should have been paved upon either of these particular systems. The pavement in Parliament Street is stated to have worn ¼ inch only in three and a quarter years, but it must not be forgotten that its real wear has scarcely begun, and that the traffic weight is high. Should its life be eight years (which is extremely doubtful), it would take a higher position, as the annual cost would probably be 1s. 6d. per square yard.

It is interesting to notice that some pavements have exhibited a considerable degree of durability and have had a tolerably fair life. In the author's table the pavements and periods under the headings "actual" life relate to accomplished facts, while several under the heading "estimated" have already nearly realized the life allotted to them. In other cases the estimate is given after inspection and measurement or inquiry. The number of the pavements is necessarily restricted in consequence of the absence of traffic-weight records.

The author inclines to the opinion that it is not desirable to lay blocks of a greater depth than will provide for a life of seven years, as very few pavements retain a good surface after about six years' wear. In the case of the pavements previously described which have attained a greater life than seven years, it is proper to explain that those periods were only insured by the execution of frequent and somewhat costly repairs. For instance, a considerable number of new blocks, of depths varying from 3 inches to 5 inches, according to the extent of wear, had to be inserted in most cases, while in others the old blocks were taken up, reversed and relaid. These operations, however, are very unsatisfactory, both in appearance and ultimate result. Experience consequently suggests that if 5-inch blocks were adopted instead of 6-inch, it would be preferable ; and the author favors the opinion that the smaller depth would be found not only sufficient, but more economical and suitable, and would obviate much patching. Five-inch blocks are cheaper by 10s. per thousand, and it is estimated that in the first cost and twice renewal of a pavement which has an annual traffic of 750 tons per yard width, there would be a reduction of 1s. 6d. per square yard in fifteen years, or 1¼d. per square yard per annum. Even if the average annual wear of 6-inch blocks should prove to be very little, after seven years' wear it will generally be found that the surface is irregular ; but considerable hesitation is always shown before local authorities order the wood to be renewed, for fear that they may be accused of waste or extravagance in removing blocks which still retain a good depth, although they show a considerably worn and bumpy surface. In short, therefore, 5-inch blocks would give as good a surface and pavement as 6-inch blocks : there would be less waste of timber in renewal, and on the whole there is little doubt that pavements would be kept in a better condition. Blocks having a depth of 5 inches only have been laid in Oxford Street, Leadenhall Street and Aldersgate Street, and in Kensington a large area has been laid on this system ; and the result so far appears to be satisfactory.

The author has obtained numerous specimens of blocks from various streets, and he submits nearly sixty, which have been taken from Regent Street, Pall Mall, Canon Street, Oxford Street, Ludgate Hill, Brompton Road, Praed Street, Sloane Street, King's Road, and Fulham Road. It will be observed that the maximum depth of wood is from King's Road (pitch-pine), which is 5⅝ inches after four and a half years' wear, and that the minimum is from Oxford Street (near Rathbone Place) which is 1⅜ inches after six years' wear.

Wood Pavements on Grades.—Obviously there are local matters to be considered in connection with wood pavement, for instance, the effect of traffic in a wood-paved street having a sharp gradient. This has not specially come under the author's notice, except at Ludgate Hill, where the blocks laid in 1877 were removed in February, 1884, having been, it is alleged " kicked out " by horses' shoes, and not fairly worn out by vehicles. The inclination of the carriageway of Ludgate Hill is 1 in 25. Similar results are noticeable at the Western approach to Hyde Park Corner, at which place the inclination is 1 in 37. The question of gradient in wood-paved streets is also an important factor in regard to the tractive action of horses and the limit of safety for foot-hold, and it is regrettable that so little experience thereon is extant. In the city the steepest gradient paved with wood is in Ludgate Hill ; in some parts of Piccadilly the inclination is 1 in 25. It might therefore be assumed that, so far as actual safety is concerned, a gradient of 1 in 20 would not be too steep. In Chelsea the main roads are tolerably level, but there is little doubt that the annual wear of the blocks in King's Road (0.144 inch) is greater in consequence of the increased amount of omnibus traffic which has recently taken place, as, owing to a keen competition, omnibuses have been very rapidly driven along the street at times. A considerable amount of light traffic has also seriously tried the wood. In Sloane Street the nature of the traffic calls for no particular observation, and the annual wear of the wood (0.065 inch), together with the traffic-standard wear, gives a result which compares favorably and satisfactorily with any other street in London.

To insure durability, it may briefly be asserted that next to sound construction it is highly important that the number of openings for gas and water services should be limited, and that undue wear and tear can be mitigated by efficient cleansing and sanding. Neglect of the latter not only creates slipperiness, but is followed by permanent injury to the pavement itself.

In support of the author's views, he submits a number of blocks as taken up from various streets in the metropolis. Some are remarkable as specimens of excellent durability, while others exhibit considerable wear, and the very thin ones have created great surprise that a pavement could be held together with blocks so much worn.

Cost of Maintenance.—The author has frequently been asked whether the wood pavement laid in Chelsea has proved economical as compared with macadam, the answer to which may be found in the following statement, which is based upon the assumption that the average life of the wood blocks will be seven years, and which shows that the first cost, repairs, renewals, and cleansing, if spread over a period of twenty years, amounts to 1s. 9d. per square yard, whereas the previous cost of repairing, renewing, and cleansing macadam, but exclusive of first cost, amounted to 2s. 10d. per yard.

Estimated Cost of Wood Pavement Per Square Yard in Chelsea, for a Period of Twenty Years.

	£.	s.	d.
First cost	0	10	6
Repairs	0	0	6
Renewal of blocks every seven years	0	12	8
Interest on loans (at 4 per cent.)	0	2	9
20)	1	6	5
Per annum	0	1	4
Add cleansing and sanding	0	0	5
Total	0	1	9

If the cost be spread over a period of fifteen years only, the figures will be increased to 1s. 8¼d. per yard per annum for the wood + 5d. for cleansing, or a total of 2s. 1¼d.

Under the above circumstances it may be fairly assumed that the annual cost of properly constructing, repairing, and renewing wood pavement, exclusive of cleansing, which is subjected to a traffic of 500 to 700 tons per yard width per day of sixteen hours, and leaving it in a thoroughly good condition at the expiration of fifteen years, does not exceed 1s. 9d. per square yard; whereas the average annual cost of repairing and renewing the macadamized carriageways in Sloane Street and King's Road, formerly amounted to 1s. 11d. per square yard, and in Westminster similar repairs cost:

	s.	d.
In Parliament Street	2	10
" Whitehall	2	10½
" Victoria Street	2	0
" Great George Street	1	8

No doubt many similar instances might be adduced, in support of the assertion that, as a paving material, wood possesses the advantages of economy, independently of the saving in cleansing.

The annual cost per square yard for laying and maintaining wood pavements in various localities, including interest on loans, is given in Table VI. In most instances the actual cost has been supplied, and in the remainder it has been carefully estimated, after making due allowance for efficient and creditable maintenance.

The author regrets that he has been unsuccessful in obtaining fuller information as to the cost per square yard relatively to the traffic weight per yard width, but such information as he has obtained he has classified in the following table, and from which certain deductions are drawn. Until local authorities, or those persons directly interested in the question, adopt measures for ascertaining the latter, a considerable amount of theory must necessarily be exercised in deciding the question of cost according to the weight of traffic.

DAILY TRAFFIC WEIGHT PER YARD WIDTH OF PAVEMENT.

SYSTEM.	400 tons.	500 tons.	750 tons.	1,000 tons.	1,250 tons.
	s. d.	*s. d.*	*s. d.*	*s. d.*	*s. d.*
Plain yellow deal........	1 4	....	1 9	....	1 10½
Plain pitch pine.........	....	1 6	....	....	....
Creosoted yellow deal...	1 6½	1 10¾	....	....	....
Henson's.	....	....'	1 9½	1 9½	2 0
Improved........	....	....	... .	1 11	2 1½
Asphaltic....	....	....	2 0	....	2 0¼
Lloyd's.......... ...	... .	... .	2 2	.. ..	....

The figures undoubtedly give the plain system of pavement the highest position in an economical point of view, and show the comparative cost of other systems in a manner not hitherto attainable.

It is apparent that as the minimum net cost of a soundly constructed and properly maintained wood pavement amounts to 1s. 9d. per square yard for a traffic of 500, and not exceeding 750, tons per yard width, the absurdity of some of the maintenance contracts which have been entered into is remarkable. On the other hand, a good bargain was made by the Improved Wood Pavement Co. in 1876, when they undertook to lay and maintain a large area in Piccadilly, also upward of 2,000 square yards in King's Road, upon the "deferred payment" system, the rate for Piccadilly being 3s. per square yard per annum for a period of fifteen years. The result will consequently be that no less than 45s. per square yard will eventually be paid for an expenditure which in all probability will not greatly exceed 30s. In justice to the authorities, who entered into so costly a contract, it should be stated that in the year 1876 the modern system of wood pavement was in its infancy, and that public bodies were somewhat timid in incurring large outlays thereon ; and as the contract stipulated that payment would cease immediately the contractors failed to efficiently maintain the pavement, it was considered that the risk would be small, as a proportionate amount only would have been paid.

Much trouble has been caused by public boards accepting low tenders for first cost, and ridiculously low terms for continuous maintenance, and it has been truly stated on a former occasion by Mr. Burt, that some persons "were running a race to see which could get ruined the fastest." This prophecy has been literally fulfilled, but, unfortunately, as in the case of asphalt and other systems of pavement when improperly undertaken, the consequences had seriously damaged the reputation of wood pavement. Considering that persons enter into contracts to efficiently maintain large areas of pavements with a daily traffic of 600 or 700 tons per yard width, for a period of fifteen years, for the sum of 9s. per yard, whereas the net cost in all probability will amount to 13s., it is obvious that either the pavements will be insufficiently repaired or renewed, and the reputation of wood injured, or that "a day of reckoning" must come. With the experience already gained, it cannot be too strongly urged that public authorities should look ahead, and not accept a tender merely because it happens to be the lowest. Another matter ought perhaps to be mentioned, although, perhaps, a somewhat invidious one—namely, that ample and competent supervision should be provided, so that every detail in the execution of the works, especially the rejection of unsound blocks, may receive attention. The operation of inspecting every block is undoubtedly tedious; for example, in the Chelsea works, where the timber supplied was of fair quality, it was found necessary to separately sort out, reject, and mark about one block in twenty, the rejections being 96,000 out of a total delivery of nearly 2,000,000; and there is little doubt that, owing to the hurry of the work, many blocks

VI.—Annual Cost of Various Wood Pavements.

System.	Situation.	Traffic weight per y'd width per diem in tons.	Annual cost per sq. y'd for first cost, renewals, repairs, and interest on loans (exclusive of cleansing) if spread over a per'd of 15 y'rs. s.	d.
Plain..............	Sloane Street....	279	1	3¾
Henson's	Euston Road....	700	1	5¼
Plain (pitch pine....	King's Road....	468	1	6
Creosoted blocks lime joint).......	"	407	1	6½
Henson's..........	Oxford Street (C)	948	1	8¼
Plain..............	King's Road.....	551	1	8¾
Improved	Leadenhall Stre't	808	1	9¾
Asphaltic	Strand..........	1,100	1	10
Plain..............	Edgeware Road.	584	1	10¼
"	Oxford Street....	1,164	1	10¼
" (asphaltic bed).	King's Road....	498	1	10½
Henson's..........	Oxford Street (E)	1,191	1	10¾
"	Oxford Stre't (W)	985	1	10¾
Creosoted blocks (mastic joint).....	King's Road....	434	1	10¾
Asphaltic	Regent Street...	558	2	0
Improved..........	Aldersgate Stre't	....	2	0
"	N'th'mberl'd Av.	...	2	0
Asphaltic	Fleet Street.....	1,359	3	0¼
Henson's..........	"	1,165	2	1
Improved..........	Oxford Street...	985	2	1
Asphaltic	Brompton Road.	648	2	1
Henson's..........	" " .	584	2	1½
Improved..........	Parliament Stre't	1,106	2	1½
Lloyd's	Regent Street...	558	2	2
Asphaltic..........	Oxford Street...	1,137	2	2¼
Carey's..........	Canon Street....	...	2	4
Henson's..........	Leadenhall Stre't	1,000	2	8½
Improved	Ludgate Hill....	1,236	2	9½
"	Piccadilly.	800	3	0
"	Knightsbridge...	780	3	0
" (pitch pine)	King's Road....	558	3	2
"	" "......	603	3	2

were laid which escaped rejection, but assuredly these will be the first to fail. In works of considerable extent it would undoubtedly be a prudent expenditure to employ a competent inspector to look after the blocks alone.

To sum up, the author ventures to assert that a properly constructed and kept wood pavement meets with favor. He has personally ascertained that shopkeepers and residents like it, in consequence of the absence of noise, and the absence of the inconvenience usually experienced by the frequent closing of the carriageway for repairs, as in the case of macadam ; that cabmen give it their unqualified approval ; and that the Managing Director of the London General Omnibus Company prefers it, if properly kept, to either granite, asphalt, or macadam.   Of course exceptions have to be made in this, as in all other cases, but the public generally appear to be satisfied with it.   There are sections of the community, however, who do not view it with favor, especially carriage-builders, wheelwrights, saddlers and granite merchants ; but their disapproval will be readily understood.   The improved condition of the carriageways of some of the best thoroughfares has also created a favorable impression upon strangers which has not been lost, inasmuch as deputations have inspected the various systems and made numerous inquiries into their respective merits, not only on the part of the municipalities of the principal towns in Great Britain, but from France, Germany, and other countries.   At the present time a large amount of wood pavement is being laid in Paris and Berlin by the Improved Wood Pavement Company.

The author has described somewhat fully the details of what may be considered to be mere ordinary matters, but as the success or failure of wood pavements mainly depends upon a careful consideration of apparent trifles, he trusts that he may be pardoned for having taken this course. He is of opinion that local authorities should invariably adopt measures for ascertaining the traffic-weight per yard width in a street before deciding to lay down wood pavement, as such information might prevent contractors from submitting a maintenance-tender regardless of the duty the pavement has to perform, and that complete records of annual cost and wear should be kept.   He again ventures to remark that it cannot be too strongly urged that the greatest discretion should be exercised in the acceptance of tenders for construction and maintenance, and that no reasonable expense should be spared in providing ample and competent supervision.   Under these circumstances he is of opinion that a close-jointed plain system of pavement, judiciously and faithfully carried out and improved upon from time to time, will give good economic results, as well as insure a sound and suitable carriageway pavement.

, Lastly, the author submits :

(1.) That where the ascertained annual cost of maintaining and cleansing a macadamized carriageway exceeds 2s. 2d. per square yard, or where the traffic is so considerable that a quieter and cleaner pavement is deemed essential, the substitution of wood is desirable.

(2.) That experience has proved wood pavement to be an economical and a convenient carriageway pavement for the streets of the metropolis.

(3.) That, notwithstanding many former instances of failure, the modern system has achieved a fair amount of success, and there is no apparent reason why its use should not be extended.

THE DISCUSSION.*

Obnoxious Smells.—Mr. G. H. Stayton said he had been told that his paper did not deal with the obnoxious smells sometimes experienced in connection with wood pavement.   He thought, how-

---

* [The Italics when used in the text are ours.]—Ed.

ever, that he had dealt with the subject fully under the head of management ; he had certainly intended to do so by drawing attention to the absolute necessity for a plentiful supply of water for the purposes of cleansing. Of some samples of wood pavement which he exhibited, one block had been taken up from Oxford Street a month or six weeks ago, and it showed that wood pavement could be worn down to the thickness of a mere crust, and still retain a fair surface. It was originally about six inches in depth, and was laid in 1878. The wear was simply owing to the traffic upon it.

Traction and Foot-hold.—Sir Joseph Bazalgette, C. B., president, asked if the author could give any further opinion on the subject of traction and foot-hold with wood pavement as compared with other kinds of pavement.

Mr. Stayton said he could give no further opinion on the subject. It had been exhaustively treated a few years ago by Mr. W. Haywood, the City Engineer ; he had not, therefore, thought it necessary to deal with it again. The average thickness of the pavement at present in Oxford Street was 3.30 inches ; but at certain parts it was not thicker than the block to which he had drawn attention (1$\frac{5}{8}$ inch), which was a fair specimen of the pavement at those parts where the traffic was concentrated.

Rankin's Pavement.—Mr. W. Lawford remembered when, in 1841, a part of Whitehall and a part of St. Giles were laid with Rankin's patent wood pavement. It was down only six or eight months when it failed, owing, as he believed, to insufficient drainage and a bad foundation ; besides which the cost came to £2 10s. or £3 per square yard. The success of the present wood pavements was no doubt largely owing to the good foundations on which they were laid. He remembered seeing the pavement put down between the Chapel Royal, Whitehall, and the Horse Guards. There was no concrete, but only a wood framing as a sub-structure. A full description of it, with illustrations, would be found in the *Civil Engineers' and Architects' Journal,* for September, 1841. From the same publication he found that in 1838 many streets in Philadelphia were paved with wood, and he believed that the pavements remained to the present day, but he did not know the results.

Mr. W. Weaver considered that some apology was due from him as Surveyor of the parish of Kensington, where by far the largest area of wood pavement had been laid, for not having presented a paper on the subject to the Institution ; but as the terms under which he held his appointment involved the devotion of his whole time to the duties of his office, he had come to the conclusion that he could not abstract from those duties the time necessary for the preparation of such a paper. He agreed with the author in the three propositions at the end of his paper, especially the third, that notwithstanding many former instances of failure, the modern system had achieved a fair amount of success, and that there was no apparent reason why its use should not be extended. So far from there being no apparent reason why its use should not be extended, he thought there was a good apparent reason why its use should be very much extended in consequence of an injunction that had lately been granted against his board with reference to the use of *steam-rollers*. If that decision was not upset (the parish authorities were about to appeal against it) he did not think the public would like to revert to the state of things existing twenty years ago, when the traffic had to ¦plow its way through the newly-laid macadam, and in such case wood pavement was likely to extend very much.

Wood Pavements in London.—There were some details on which he could not quite agree with the author, one of which was the statement that wood pavement was laid down better and cheaper in Chelsea than in any other part of London. In Kensington seven different kinds of pavement had been tried—he believed every known kind except Carey's keyed-joint—and the result of his experience of those systems had led him to believe that the *most economical wood pavement* that could be laid was the plain deal, if it was to be laid under *competitive tenders* where the wood had to be inspected to see that it was of a proper description ; but if the work was not tendered for, if a good price was given for a good article, and it was wished to have the work done expeditiously, he preferred the system of the Improved Wood Paving Company with pickled blocks and asphalt joint. The work was got through much more rapidly in that way, and from the results in Kensington, he was sure that the system would last as long, if not longer, than the plain system ; the cost, however, was about 1s. 6d. per square yard more. The assumed superior merits of the wood pavement in Chelsea appeared to be due, according to the author, to the work having been executed by his own staff and the substitution of studs for asphalt or laths. But he could not see the advantage of the parish doing the work with their own staff. If they invited tenders for the supply of so many tons of Portland cement, so many cubic yards of Thames ballast, and for breaking up the surface, they had merely a series of competitive tenders for the work in detail, whereas in the other case they had one tender for the complete work ; and he thought that the one profit in the latter case was less than the various profits in the other. He might mention a practical illustration which had come under his own notice in the Fulham Road. It was the last extensive piece of work that the author had done, and at the same time Mr. Weaver was paving a portion of the same road under his charge extending from the Brompton Oratory to Thistle Grove. In his own case, the work was done by Messrs. Nowell and Robson under contract, at 9s. 5d. per yard, whereas in the author's case it was 10s. 3d. or 10d. a square yard more, and the time of execution was about forty per cent. longer. It would be apparent to every one practically acquainted with the subject that it was necessary to consider not only the question of cheapness in first cost, but the interests of those who were to find the money for the execution of the pavement — namely, the ratepayers, some of whom were abutting frontagers ; and it was a matter of considerable moment to a shopkeeper on a line of thoroughfare to have his receipts diminished £5 or £20 a week in consequence of the road being up. Time, therefore, was of great importance. Again, there was a great disadvantage in having to pick up the laborers almost as they came ; it might take several weeks to get them into working condition ; but if the order were given to an established firm they could at once send a number of men of experience in their several departments, so that they would get half-way through their work before the others had started. The question as to when it became economical to substitute wood-paving for macadamized roads was an important point for a board to consider when about to launch a large wood-paving loan. His figures were not quite the same as the author's, but they came to very much the same in the result. He was not able to separate the *scavenging* from the *maintenance* ; but he had always advised his board that if a macadamized road cost 1s. 6d. per square yard per annum for maintenance, it was cheaper to put down a wood pavement, and that was a very similar conclusion to the one at which the author had arrived. With regard to the question of *studs*, he had tried various kinds, single-pointed and double-pointed, but his experience was that they did not produce such regular jointing as asphalt or lath-joints, the lath being left in. If the lath was withdrawn, the brooms passing over the surface sweeping in the liquid grout disturbed the regularity of the surface. Studs were generally driven in by boys, who were not models of carefulness, hence some were driven in very hard (perhaps after the men had had their dinner), while in other cases, when the men were tired, a good deal of the stud projected. He would leave the practical point as to superiority to be decided by members who could examine and contrast the two pieces of pavement in the Fulham Road to which he had alluded ; that east of Thistle Grove, executed by Kensington, and the other portion west, carried out by Chelsea.

Mr. L. H. Isaacs regretted that he could not agree with either of the three proposals which the author had asked the institution to indorse. The paper appeared to be written

with optimist views, the author having charge of a suburban
or semi-suburban district, and apparently not being aware
of what *actual London traffic* was. The instances of wear and tear
which he had cited were confined to King's Road and Sloane
Street. Mr. Isaacs desired to set against them the experience he
had obtained in a central portion of London over which London
traffic, in the strict sense of the term, actually passed. The statement that wood pavement was calculated to last seven or eight years
was, he thought, misleading. He was ready to indorse all that had
been said as to the comfort and convenience of wood pavement,
but the question of cost ought also to be considered. The author's
first proposition stated that where the ascertained *cost of maintaining
and cleansing a macadamized carriageway* exceeded 2s. 2d. per square
yard per annum, or where the traffic was so considerable that a
quieter and cleaner pavement was deemed essential, the substitution
of wood was desirable. He entirely agreed with the second part of
the proposition, but the first was wrong. The author had admitted
that the statistics of his own office showed that the *cleansing*
amounted to 11d. per square yard per annum, leaving for *maintenance* 1s. 3d. per square yard per annum. If he had gone to London proper, he would have found that the cost of mere maintenance
was from 6d. to 1s. In his own district the cost in some streets was
6d., in others 9d., and in the majority 1s. Mr. Weaver had stated
that about 1s. 6d. was the proper sum due to maintenance. Which
was right? Or would engineers be justified in rejecting the advice
of both, and taking instead the evidence of a man like Mr. Haywood, the City Engineer, who had charge of streets over which
true London traffic passed? With regard to the *economy of wood
pavement*, the author had clearly failed to prove his proposition. It
was certainly convenient, indeed luxurious, and where the ratepayers were willing to pay for the luxury there was no reason why
they should not have it. When a rich banker drew a check for the
rates of his premises, rated perhaps at £10,000, it was of little consequence whether the amount of the check was £1,500 or £2,000.
In like manner it was a matter of indifference to a wealthy inhabitant of Prince's Gate whether in drawing a check for the rates of his
house the amount was £160 or £200. Such persons would rather
draw for a larger amount and have the comfort of a noiseless pavement than draw for the smaller amount and revert to the old state
of things. The author's third proposition, "that notwithstanding
many former instances of failure the modern system has achieved a
fair amount of success, and that there is no apparent reason why its
use should not be extended," had been drawn with great caution.
There had been no doubt "a fair amount" of success, and enormous

improvements had been made in the wooden pavements of the present day, as compared with those put down twenty-eight years ago ; but, after all, the question was very largely one of cost. He would invite the members to consider Chancery Lane, Southampton Buildings, High Holborn, Lamb's Conduit Street, Hatton Garden, and Great Ormond Street, which it would be admitted were fair representatives of streets with ordinary London traffic. They were all under the jurisdiction of the Holborn District Board of Works. *Chancery Lane* was first laid with wood in the Michaelmas quarter of 1876 by the Improved Wood Pavement Co., which he thought was one of the best wood-paving companies in London. The area was 1,960 square yards, and the first cost was £1,557 10s. or about 15s. per square yard. The complaints against it were numerous and grave. It was laid on the principle adopted by the company of transverse boards, with concrete as a foundation. The noise of the traffic passing over the granite pavement which previously existed was so great that the dwellers in Stone Buildings and Chancery Lane petitioned the board to lay down wood pavement, and even offered to contribute to the cost ; but after it had been down a few years they complained of the shaking of the windows and the general unpleasantness, and asked that it might be taken up again. The company, on being communicated with, stated that they had come to the conclusion that the system adopted was a mistake, and that they were prepared to alter it. To their credit it should be stated that they took up the whole of the pavement and relaid it at their own cost upon their modern improved system, with entirely new blocks and new materials. That was in 1881, and no complaints as to rumbling and vibration had been made since that time. If the pavement lasted till 1886 he should think it would have done its duty, and he would have no cause of complaint. The wooden pavement of *Southampton Buildings*, where the traffic was much lighter than in Chancery Lane, was laid down in the Christmas quarter of 1876. It contained 1,063 yards superficial, and the cost was £824, or 15s. per square yard. It was largely relaid in the year 1882. The first pavement that he took in hand when he was appointed Surveyor of the Holborn District was the wood pavement in *High Holborn*, which he removed, and for which he substituted granite pavement in 1857. In the Christmas quarter of 1877 a wooden pavement was laid down in High Holborn. The portion in the Holborn District contained 3,842 yards superficial, and its cost was £3,030 1s., which again was about 15s. per square yard. That pavement entirely failed to carry out the views, not only of himself and of the board, but of the company which supplied it, although they were paid the highest price for laying it down, and for subsequent maintenance. The pavement was continued through New Oxford Street, as far as Tottenham Court Road, and it was considered at the time as fine a sample of wood pavement as had been laid in London proper.

In the early part of 1882, when the pavement had been down less than five years, it became evident that its life had gone, and that it would have to be taken up. It had been his duty to take up the portion in Holborn in sections, in the years 1882-3 ; and not one block laid in 1877 was now to be found there. He might also mention, lest it should be thought that the circumstances had arisen from some want of care on the part of the officials of the Holborn District, that the portion laid down in the St. Giles's district as far as Tottenham Court Road had also been removed and *relaid with Val de Travers asphalt*. The wood pavement in *Hatton Garden*, which was considered a very suitable thoroughfare for the purpose, was laid in the Michaelmas quarter of 1878. It contained 4,679 yards superficial, and cost £3,743, or 15s. 6d. per square yard. It was laid in 1878, and yesterday, May 27, 1884, the Val de Travers Asphalt Company proceeded to take it up, and they were now laying down asphalt in its stead. With those facts before them he asked the members of the Institution to pause before they too readily indorsed the propositions of the author with reference to the economy of wood pavement.

Traction on Wood and Macadam.—Mr. W. E. Rich said, he thought the question asked by the President with reference to traction was most important. He believed that wood pavement was extremely favorable in regard to its low resistance to traction, and that was an important element which should encourage its extensive adoption. He hoped that some information would be forthcoming on the subject, obtained by means of the new dynamometer belonging to the Metropolitan Board of Works. He had himself had two hours' run with it a few months ago in some preliminary trials, and he had been surprised at the immense reduction in traction on going from a macadamized road to a wood pavement. He believed that the traction over a wood pavement did not exceed one half of that over a macadamized road, and it was much less than that over a stone pavement.

Merits of Asphalt and Wood.—Mr. E. Matheson inquired why the author had omitted all reference to asphalt. The paper seemed to imply that wood was the only alternative to macadam. Mr. William Haywood had exhausted the subject a few years ago, but he thought that later experience might induce him slightly to modify his views. In many respects asphalt was better than wood. The question of cost was not the only one to be considered. In regard to traction he had no doubt that asphalt was much superior to wood ; but its alleged slipperiness had at first condemned it, and hindered its adoption. He believed it had been found that the difficulties connected with *slipperiness* had arisen from the inexperience of the drivers, and the strangeness of the new pavement to the horses. In Holborn, in the St. Giles's District, the wood pavement was about to be replaced with asphalt ; and Cheapside also had an asphalt pavement. The low cost of cleaning, and the little delay in laying (an asphalt road being ready for use in twelve hours, while the author of the paper stated that wood pavement required a week), were important points in favor of asphalt over wood.

Mr. Hugh McIntosh said that he had not had sufficient experience to give a decided opinion upon the question in dispute. In his district the authorities were only just beginning the use of wood. There were certainly some disadvantages connected with asphalt, with which he had made himself acquainted by observation and inquiry, and from which wood pavement was entirely free. The little experience he had had inclined him to favor wood in preference to asphalt.

Mr. J. Lovegrove remarked that only a very small area of wood pavement had been laid in his district. It was put down two years ago by the Improved Wood Pavement Company, and it had proved a successful piece of work. There were two rails passing through the centre of its width, and blocks of Guernsey granite, well dressed, were placed on each side of the rails. That was a very successful way of dealing with the difficulty of obtaining a comparatively *noiseless pavement* opposite a public building. They had had Val de Travers asphalt at first, but the rails and the asphalt did not wear well together. Some years ago he had made some experiments with a view of testing the cost of the *maintenance of macadamized roads*, and the result was that he advised his board that when the cost was found to reach 2s. per square yard, it was time to get rid of macadam and lay down wood and stone paving. There was a good deal of saving with wood or stone pavement in the matter of cartage, by far less mud being made than on macadamized roads. It was very important that all the information obtainable with reference to asphalt paving, pitching and wood, should be laid before the members, and their thanks were therefore due to the author for his efforts in that direction.

Creosoting.—Mr. A. Giles, M. P., considered that the author had not done justice to the subject of creosote in his statement when he said that creosote to a certain extent closed the fibres of the wood, and tended to produce premature internal decay. Mr. Giles had had great experience in creosoting, and this was the first time that he had ever heard of creosote tending to promote premature decay. It was quite true that the pickling process was worse than useless, but if the blocks, before being laid, were properly creosoted they would last much longer than they did at present. It appeared from the statistics in the paper that the actual wear was only o.144 or ⅛ inch per annum. The average duration of wood pavement was six or seven years, and as the blocks were six inches deep, they only lost about one-sixth of their depth in the whole period of their life. The *destruction of wood pavement* was caused, not so much by wear as decay, for every one must have noticed that when wood pavement was being taken up, a great deal of it was as rotten as touchwood. That would not occur if the wood were properly creosoted before being laid. He was sorry the author had not gone back to the early days of wood pavement in 1842, of which Mr. Giles had a lively recollection. It was nonsense to say that the *slipperiness* of wood pavement was only felt by inexperienced drivers, for there were certain states of the atmosphere when the pavement became so moist and greasy as to render it absolutely unsafe for horses. If something could be devised to prevent that slipperiness, a great benefit would be conferred upon the traveling public.

Mr. S. B. Boulton said that he had had considerable experience in *creosoting*, but not much in the application of that process to wood pavement. It would appear that usually the paving blocks were merely dipped in creosote, and perhaps sometimes in a mixture of that and other more doubtful substances. Was this a wise course to pursue? It was well known how much the creosoting process had been improved by the abandonment, for ordinary engineering purposes, many years ago, of the mere steeping tank, in favor of the cylinder process by vacuum and pressure. In one instance referred to by the author, that of the King's Road, the blocks appeared to have been prepared by the latter process. In this case only seven pounds of creosote per cubic foot had been injected. The "resources of civilization were not exhausted," however, by the injection of so small a quantity. The author had said that these blocks were, when taken up, moist internally, although it did not appear that they were unsound. The internal moisture could only have resulted from one of two causes, either that the wood was not dry enough at the time of creosoting, or that the quantity of creosote employed was insufficient to prevent the subsequent infiltration of water. Many engineers caused to be injected for railway sleepers and other timber, 10 pounds and 12 pounds of creosote per cubic foot. Even with these quantities the injection was partially superficial, but the creosote completely saturated the sap wood and softer parts of the timber, filling up all cracks and fissures, whilst the ends of the logs, for some inches up, were usually gorged with the creosote. Small pieces of timber like paving blocks, should, if prepared, be saturated with creosote like the extremities of a sleeper. When timber was unprepared the ends of a log absorbed moisture freely ; and this must be the case more especially with paving blocks. Moreover, it was not pure water to which the paving block was exposed. Ammoniacal products, the very class of substances which were used in experimental putrefying pits for hastening the decay of timber, were largely present in the moisture of the London streets. It spoke well for the paving blocks, and for the selection of the wood of which they had been made, that, unprepared, or slightly prepared as they were, they had lasted as well as they had done. He was surprised to hear the author of the paper express the opinion that creosoting, by closing the fibres of the wood, tended to produce premature internal decay. General experience in this and other countries during the last forty-five years was entirely to the contrary. There had recently been exhibited at the institution a large collection of specimens of croesoted wood, sent by various railway administrations and from other sources, and which had been placed in many different kinds of soils. These specimens of ordinary creosoted fir timber had remained sound for periods varying from 10 to 32 years, and showed conclusively that the creosote, which had filled up the outer portion of the fibres, had completely protected the inner portion of the wood from decay. To produce such results it was, of course, necessary that the wood should be deprived of moisture, and that the creosote should be of a suitable kind. The wearing away of the top surface of the blocks appeared to be rapid ; any decay of the woody fibre would doubtless accelerate this abrasion. Creosote had not only a preservative, but also a hardening effect upon timber, and if thoroughly injected could scarcely fail to prolong the duration of paving blocks. As regarded the kind of timber to be used, he thought that a greater variety might be tried. Gothenburgh had been spoken of, and excellent timber could be procured from that port, but no better than from various other Swedish ports, or from Memel, Danzic, or Riga. Beech he thought would do good service; it absorbed creosote remarkably well, and evidence had recently been brought forward of the very long duration of creosoted beech sleepers on the Chemins de Fer de l'Ouest in France. English elm also absorbed creosote readily, which was not the case with the American rock elm. He had recently taken three pieces of ordinary fir sleeper wood, three of English elm, and three of American elm, and had subjected them to the creosoting process under nine hours' pressure. The pieces of timber were all cut to the size of ordinary paving blocks, 6 inches by 6 inches by 3 inches.

The results were :

Average Quantity of Creosote Absorbed per Cubic Foot.

	Lbs.	oz.
Three pieces of fir	22	1
" " English elm	27	13
" " American elm	4	10

In forests through which he had traveled in Canada, the United States, Russia and elsewhere, he had noticed a great waste of timber in cutting the lengths for the ordinary purposes of the market. If some uniform standard of size for paving blocks were adopted in this country, and if it became known that there was a constant demand, much of this waste timber might be utilized by being cut into blocks, either in the forest or at the shipping port, whilst they could be brought here as convenient stowage for vessels at a very low rate of freight. He could remember some years ago when his offices were in King William Street, near London Bridge, the effect produced by taking up the granite pavement and substituting wood. No words can describe the sense of relief in the mitigation of the roar of a mighty traffic, which was at once experienced by all the busy toilers whose work had to be carried on amidst such surroundings.

Mr. E. A. Cowper agreed that if blocks were creosoted at all they should be thoroughly saturated with creosote. It had been stated that the author had not dealt with London streets where there was a very heavy traffic, but he had actually given the first cost and maintenance of the pavement in Ludgate Hill, Aldersgate Street, Leadenhall Street, Fleet Street, and Oxford Street.

Asphalt Superior to Wood.—Mr. G. Allan did not think that wood pavement was a material to be recommended for further extension in the metropolis. About fifteen years ago his attention was first directed to the question when the Val de Travers asphalt was brought before him. He first had it experimented upon in Bombay, where it was used to pave the foot-paths of a number of leading thoroughfares, and it was in consequence of its success that he became the founder of the Val de Travers Co., and had the material introduced for the first time in London, their first contract being for Cheapside. When that contract was being carried out, Portland cement was used for the first time on an extensive scale. Lime was proposed, but upon his recommendation Portland cement was substituted in the contract, at an increased cost to the company of £4,000. No doubt it was to the excellence of that concrete *foundation* that the success of the asphalt was due. Wood should be regarded as simply a temporary material for paving ; so perishable an article should not be thought of for a moment for permanent use.

As granite sets had given place to wood, so wood would have to give place to an impervious material, as asphalt, or some future improvement upon it. Both frost and sunshine had a very destructive action upon wood, or any material of an absorbent nature, but asphalt had no absorbent qualities, so that whatever wet might fall upon it was rapidly dried up by the atmosphere. The reason why the use of asphalt had not rapidly extended was owing to the inefficient arrangements of the metropolis. These had no doubt been improved of late years, but not sufficiently to justify Vestries and other authorities in a further extension of asphalt. Asphalt pavement should be washed every morning as regularly as a stable or kitchen floor, and sanded if necessary, according to the state of the weather. If that were done, he believed asphalt would be everywhere demanded. In the case of wood sets there was nearly double the joint area of granite sets, and the joints were simply receptacles for dirt, mud, and horse droppings, which, in dry weather, were discharged in the form of dust. In asphalt there were no joints, and nothing to receive the droppings. All the dirt lay upon the surface, ready to be washed or swept away.

Mr. G. H. Stayton, in reply, said he had been glad to hear the various points discussed, particularly the criticisms of Mr. Isaacs ; but he regretted the course he had taken, feeling convinced that his conclusions were wrong, especially as to cost, economy, and further extension. With regard to the question of *absorption*, he would only call attention to a block which had been four or five years in actual use in King's Road. It had been taken up from the centre of the road, and it showed clearly that the absorption had been practically nil. His experience had been that if the wood was sound there was no fear whatever of absorption up to a certain point. He had no personal knowledge of the system of wood pavement referred to by Mr. Lawford as having been laid in 1841, neither had he ascertained the cause of its failure. The chief object of the paper was to draw attention to the modern system, as from its unprecedented extension within the last few years more advantage might be gained by a consideration of its merits than by going back a period of forty-three years. The remarks of Mr. Weaver were valuable from his great practical experience in the question, and considerable weight might be attached to his opinions thereon. It was satisfactory to note that he concurred with the author as to the advantages of the plain system, and the necessity for great care in the use of wood, but a few points demanded correction. In the first place it was to be regretted that Mr. Weaver should have fallen into the error of remarking that one of the author's statements was "that wood pavement was laid down better and cheaper in Chelsea than in any other part of London." The statements advanced in the paper did not in the least justify any such conclusion. He merely asserted that the *Chelsea pavements* comprised all the essentials of a sound and economical pavement, and that the result had been eminently satisfactory ; but he readily admitted that the Improved Wood Pavement Company, Henson's Company, Messrs. Mowlem & Company, and Messrs. Nowell & Robson, had also carried out extremely good and creditable work. The net cost of the Chelsea pavement in Fulham Road was rightly stated to have been 10*d.* per square yard more than the Kensington part, but Mr. Weaver was in error in asserting that the former took forty per cent. longer time to execute. The Chelsea work had been commenced on the 5th of September, 1881, and was completed, and the road reopened, on the 15th of November, thus giving 149 square yards per diem as the result, whilst the Kensington work was carried on in two sections between the 28th of August and the 21st of October, giving 143 square yards per diem for each section. The *Kensington pavement* had a large proportion of old macadam in the concrete, and the system of blocking, which he considered objectionable, had been adopted ; and there was little question that the cost would eventually be as great as that of the Chelsea pavement. The wood-paving works, executed in Chelsea in 1879 by the board's own staff, saved the ratepayers £3,160, which fact proved that system might be attended with substantial advantages. Mr. Weaver's remark as to the difficulty of organizing an efficient staff was, in his experience, purely imaginary. The conclusion that when the *cost of macadam* exceeded 1*s.* 6*d.* per square yard per annum it was time to adopt wood, fully bore out the author's view, inasmuch as the 2*s.* 2*d.* mentioned in the first proposition included 8*d.* for cleansing, thus leaving 1*s.* 6*d.* for repairs only. Notwithstanding what Mr. Weaver had urged against the use of *studs*, the author's practical experience satisfied him that they were preferable to laths, as he had repeatedly seen considerable displacement of blocks, and irregular width of joints, where laths had been adopted. The statements of Mr. Isaacs were fallacious and misleading, He challenged the accuracy of the author's conclusions, but failed to refute them by adducing a single reliable fact or figure as to annual cost of former or present maintenance, or records of traffic-weight, in justification of the position he had assumed. In a previous discussion in 1879, on street carriageway pavements, he also made certain statements at which Mr. Howarth expressed considerable surprise when correcting them. Mr. Isaacs must either have paid no attention to the paper and tables attached thereto, or

he must have neglected to formulate statistics as to weight of traffic
and cost in his district, otherwise he could scarcely have made state-
ments which had no practical value. Mr. Isaacs contended that
London traffic, in the strict sense of the term, might be seen in the
Holborn district, but not in the districts of Chelsea or Kensington.
To a certain extent this was correct, but subject to qualification. In
Holborn itself the daily maximum traffic-weight per yard width was
approximately 1,100 tons, in Brompton Road it was 648 tons, and in
King's Road 603 tons. The author made due allowance for the
variation in the traffic-weight in preparing the traffic table, from
which it might be assumed that the annual cost of the wood-paving
in Holborn, if spread over a period of fifteen years, would be 2*s.*
per yard, besides cleansing and sanding 5*d.*, or a total of 2*s.* 5*d.* Mr.
Isaacs has made no attempt to supply figures upon this point, and
without such information it could only be assumed that he had not
given full attention to the subject. He had apparently intended to
show that the experience of Mr. Weaver and the author was such
that their opinions upon wood pavements with "London traffic"
ought not to be indorsed, but he failed to adduce any serviceable
information in opposition to those opinions. It might, therefore, be
assumed that he had taken no pains to make himself acquainted
with the statistics relating to his experience, which would certainly
have been an easy matter in so small a district as Holborn,
which had but fifteen miles of streets (of which three-fourths of a
mile were macadamized, and 16,000 square yards were wood),
whereas in Chelsea and Kensington combined there were 108 miles
of streets, of which forty-one miles were macadamized, and 217,000
square yards were wood. But apart from this, the paper dealt with
pavements subjected to "London traffic" in its broadest sense.
Several instances were given in which the actual life of block had exceeded seven years, notwithstanding that the daily traffic-weight was
upwards of 1,000 tons per yard width. Mr. Isaacs was evidently
under a misapprehension as to the meaning of the item of 2*s.* 2*d.*
per yard in the first proposition, as he deducted the 11*d.* for cleansing from 2*s.* 2*d.*, instead of 2*s.* 10*d.*, as explained previously, and in
the foregoing reply to Mr. Weaver's remarks. The first proposition
was, "That where the ascertained annual cost of maintaining and
cleansing a macadamized carriageway exceeds 2*s.* 2*d.* per square
yard   *   *   *   the substitution of wood is desirable."
The figures (2*s.* 2*d.*) were arrived at by working out the cost of
wood pavement, including first cost, interest on loans, repairs,
renewals, and cleansing, for a period of fifteen years, in the following manner—viz.:

	s.	d.
First cost 10	10	6
Repairs..	0	4
Renewals of blocks every seven years.................. ...	12	8
Interest on loans	2	5
	25	11
Per annum..	1	8¾
Add cleansing and sanding..............................	0	5
Total	2	1¼

If, therefore, the author's experience was of any practical value, he was satisfied that wood was undoubtedly a desirable and economical pavement, as compared with macadam, under the above-named circumstances. The observations of Mr. Rich respecting traction confirmed the statement in the paper as to the low resistance of wood pavement to traction. The author had not used the new dynamometer belonging to the Metropolitan Board of Works, but as regarded the average distance that a horse might be expected to travel without falling, Mr. W. Haywood, M. Inst. C. E., reported in 1874 that on granite it was 132 miles, on asphalt 191 miles, and on wood 446 miles, whilst the injuries to horses and obstructions to traffic were greatest on asphalt and least on wood. Mr. Matheson's remark that in many respects *asphalt* was better than *wood* coincided with the author's views which also explained why the consideration of asphalt had been excluded from the paper. Within the last month the entire carriageway of a street under the author's charge had been paved by the French Asphalt Company, this system of pavement having been adopted principally on sanitary grounds. The author did not agree with Mr. Allan that wood was likely to be superseded by asphalt, at any rate in the principal West End thoroughfares, unless the reputation of the former became damaged in consequence of their neglect, as described under the head of " Management." Obviously, both *wood and asphalt needed constant attention* in the matter of washing, cleansing and sanding ; but however suitable asphalt might be for footways, the author could only repeat, that nothing but downright neglect in the management of wood was likely to lead to its extension, on account of slipperiness and the cruelty to horses which its adoption entailed. Mr. Giles and Mr. Boulton made some remarks pertinent to *creosoting*. The author in no way desired to do injustice to the system, his experience of creosoting piles having been favorable ; but however desirable in theory it might be to creosote wood-pavement blocks, his experience did not convince him that, it was good in practice. Some few years since he had seen wood pavement taken up at Westminster, where creosoted blocks had been used. Many of the blocks were internally as rotten as touchwood, and the sappy blocks had worn down greatly under the pressure of traffic. It was difficult to see that any advantage or economy could be gained by the adoption of creosote, and the author was strongly of opinion that the surface of blocks gorged at the rate of 10 or 12 pounds per cubic foot, would become so slippery, and the jointing so unsatisfactory, that the system would create dissatisfaction.

CORRESPONDENCE.

Mr. G. P. Culverwell approved of the *contour* of the carriageway in Fig. 2, but thought that, if anything, the slopes might be reduced. It was ordinarily admitted that wood was a most *slippery roadway*, and thus every precaution should be taken. The sharp camber of the old macadam or stone pavement was too often followed, but it was unsuitable, and placed vehicles at a disadvantage, especially when starting from the curb. The cross-section should be represented by two gradients, say of 1 in 40, from the channels to the crown, the latter being slightly eased off. In level roads the cross-section necessarily varied somewhat in order to drain to the gullies.

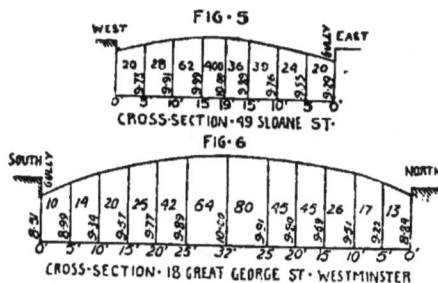

Figures 5 and 6 were typical sections, and had been carefully taken recently. The east and south ends respectively were at gullies, whilst the west and north ends were midway between gullies, both streets being level. It was to be noted that *steep gradients* transversely were especially objectionable, as in this direction the blocks were made to abut close one to the other without any joint, and this gave little foot-hold. He did not comprehend the remarks as to *pitch-pine* producing a "jarring, bumping motion." He had examined the portion of that pavement in King's Road, and traveled over it in various vehicles without noticing any increase of bumping motion, but there was somewhat greater resonance than where the wood was softer. He would expect to find the coefficient of elasticity of pitch-pine greater than of yellow wood. It was misleading to say that *creosoting* "tends to produce premature internal decay," although, indeed, perhaps the large amount of inferior work done in the market might form a sufficient excuse. Creosoting consisted in driving out the sap, and then substituting oil. If the first portion of the operation was not effected, the oil formed merely an external coating, preventing the sap from escaping, and "dry rot" set up. He, however, thought that creosoting was unnecessary and even in some respects undesirable for wood paving. Creosote was chiefly of use in the case of porous woods exposed on all sides to the weather, in prolonging the ordinary life of the sapwood and wood next the sap. It did not render the sapwood harder, or better able to withstand abrasion under traffic ; and it was almost useless to creosote the heart-wood of pitch-pine, the wood being already so full of resin that little oil was taken in. He had no hesitation in saying that the money expended in creosoting would be better laid out in the careful selection of heart-wood timber of uniform quality ; and he thought there were data to show that the life of pitch-pine pavement might, without undue maintenance, easily reach ten to twelve years as a first-class roadway under the traffic standard of 750 tons per yard width per day. The depth of block should be regulated to some extent by the weight of traffic, the aim being that the block should be fairly worn out by abrasion just before decay from weather or other causes set in. With pitch-pine, even 5 inches appeared unnecessarily deep, considering that the maximum annual wear in King's Road was given as under 0.1 inch, and in Oxford Street blocks (presumably deal), worn as thin as $1\frac{5}{8}$ inch, had not failed under the traffic.

" *Top-dressing* " was important, but the material used was often unsuitable and too much was put on, with a result that, upon the first rain, the roadway was almost impassable for pedestrians, as in the case of the Strand at Charing Cross when last repaired. The material should be fine-screened sharp gravel, deposited in successive thin layers. Stones of the size of walnuts only rolled about, the wood not presenting a sufficiently hard anvil to crush them, and the material did not work its way into the wood so well when in a thick layer.

The opening up of streets for repairs to gas and water mains, etc., was most objectionable, and peculiarly so in the case of wood pavement, owing to the time necessary for the proper restoration of the latter. It was to be regretted that a comprehensive scheme had not been carried out whereby to provide subway accommodations in the metropolis for gas and water pipes, telegraph and telephone wires etc. Many of the streets were much too narrow, whilst the evil daily increased, and thus every means should be taken to supplement their deficiencies.

More information was desirable so as to be able to determine the description of wood, size of block, and width of joint best adapted to varying conditions of traffic and gradient. He thought 1 to 20 too steep for any of the forms of wood-paving at present employed ; and whenever the gradient exceeded 1 in 40, extra provision should be made for cleansing and sanding, especially in time of frost.

Sloane Street at present afforded a good opportunity for examination, as the paving and concrete foundation were being taken up throughout its whole length for a width of about 16 feet, in connection with new sewerage works. Such examination showed that the concrete *foundation* was clean and sound, and there was no appearance of moisture or objectionable matter having infiltrated. The *blocks* were not perceptibly rounded on top, and were in good condition, showing little soakage, and the cement grout bonded to them remarkably well. The cement joints were of uniform width, and were very little below the level of the blocks. The blocks were burred about ⅛ inch upon the sides next the cement, and not at all at the ends where block bore against block. This pointed to the conclusion that blocks of uniform thickness, laid touching each other upon all four sides, and without any cement or other joint, would form a most durable pavement, whilst upon level roads it might give a sufficient foot-hold. The compression set up would effectually keep the blocks in place, and prevent infiltration of moisture. *Expansion* in the transverse direction of the street appeared to have been sufficiently allowed for, as the blocks bedded fairly upon the concrete, and no buckling away from it or disturbance of the foot-pavements, had been noticed. That the *compression* in the longitudinal direction of street was great was evidenced by the line of blocks continually cambering toward the free end. This camber averaged about one inch in the excavation width of sixteen feet. This showed the advisability of laying wood pavements in long even gradients ; and where the vertical curves at the summit levels were sharp, expansion should be provided for, otherwise buckling, followed by disintegration, might take place. Compression was most desirable for wood, and directly added to the life of the blocks, whilst further preventing objectionable soakage. Hence it appeared why a rigid joint, such as cement, was preferable to a mastic one. In the latter case, in place of the wood being compressed, the substance of the joint itself was expelled, and part, no doubt (especially if the transverse expansion had not been allowed for), found its way underneath the blocks, and uneven bedding, followed by disintegration, resulted. In connection with this matter of compression appeared a direct objection to the use of *croesoted blocks*, as these did not expand, and compression was not set up, and also the cement grout did not bond to them as well as to plain ones. This explained the several instances noted by the author of the case with which croesoted blocks were taken up, and of the moisture having been found between them.

He was informed of one case—that of the long-girder bridge over the River Foyle, at Londonderry—in which the cement grout had never set. This the engineer in charge attributed, no doubt correctly, to the unremitting vibration night and day, as every care had been taken with the materials. Henson's patent felt bed and joint appeared peculiarly applicable to such cases.

In conclusion, he congratulated the author upon having at an early stage adopted an excellent and economical system of pavement, singularly free from faults or objections, and one that bid fair, with minor modifications only, to be largely employed in the future.

Mr. W. H. Delano regretted that the author should have perpetuated confusion by using the words "*asphalt*," "asphaltic," when the substances he referred to were evidently *gas-tar* and gas-tar-mastic—namely, gas-tar mixed with chalk and gravel. Asphalt was a natural bituminous limestone ; asphalt-mastic was the same, mixed with natural bitumen, and when mixed with grit was called gritted asphaltic mastic.* He understood that pine sets cut from Swedish yellow deals six inches deep, three inches wide, and eight to eleven inches long, laid on a Portland cement foundation six inches thick, *cost* 10s. 6d. per square yard, exclusive of demolition of old roads, carting away old materials, and regulating the subsoil to a proper contour. This price was low and probably misleading as a guide to contractors' prices for similar work, for it seemed to include no general expenses. A contractor paid rates and taxes, interest on money disbursed, cost of plant and tools, rent, employees, which made his general expenses from fifteen to twenty per cent. Local boards, who were their own contractors, had none of these to count, but it was probable that their general expenses really exceeded those of a contractor, just as the management of railways by the State cost more than that of private companies. The *standard traffic* of 750 tons per yard width per diem seemed a misleading formula, for the wear and tear of a road did not depend so much upon the weight passing over it as upon the speed at which that weight traveled. There was also the important element of width of tire, the state of

---

* Minutes of Proceedings Inst. C. E., vol. ix, p. 249.

the atmosphere, the camber of the road, and the mode of traffic, cross, heavy, one side empty, the other loaded, etc., etc. Take the new three-horse Paris omnibuses carrying, when full, forty-six people including driver and conductor, weighing, when full, 5½ tons, the width of the wheel tires being only 3½ inches, running at a speed of seven to nine miles an hour, and compare the same weight on a 4½-feet wide smooth cast-iron roller dragged at 1½ miles an hour. The latter improved a road, the former rapidly destroyed it.

The author having alluded to the *wood pavement* laid down by the Improved Wood Company *in Paris*, he would make some remarks on that interesting work, having watched it daily from its beginning. It had been carried out in September, 1881, on a portion of the Boulevard Poissonniere, of the Rue Montmatre, and a cross-road formed by the two. The gradient of the former was nearly four in 100. It had been laid by skilled English workmen, with the thickness of concrete and wood, as above stated, and had resisted, up to the present, some of the heaviest traffic in the world. It was fair to state, however, that the winters of 1881–2–3 had been remarkably mild, and the summers of 1882–3 remarkably cool. In March, 1884, the first repairs had been effected, and now (July, 1884), although the road was in good order, some of the sets had become spongy, and those in the lines of traffic were rounded at the edges. On a wet day the wheels of the omnibuses, by expressing the surface moisture, left seemingly white tracks behind them. Now, large surfaces of wood pavement were being laid all over Paris, chiefly in substitution of macadam, and four rival contractors were in the field. The *price* for all is as near alike as possible, say 23 francs the square meter, or 14*s*. 10*d*. the square yard; but it must be noted that the Paris octroi on cement was 12 francs the ton, and on wood 7 francs 50 cents the cubic meter, which increased the cost price by about 1 franc 50 cents the square meter, or say 1*s*. the square yard as compared with that of London. *The system of deferred payments* had been adopted all round; the prime cost of 23 francs was spread over a period of eighteen years, interest and compound interest being allowed on both sides at the rate of 7 per cent. per annum. This gave 2 francs 35 cents per annum for first cost, to which was added for maintenance 2 francs 50 cents per square meter per annum, equal to 4 francs 85 cents per square meter per annum for first cost and maintaining, for a period of eighteen years, or 3*s*. 1*d*. per square yard. This price had been subsequently increased so as to include demolition of old roads, but to a slight extent only. When M. Léon Say was Prefect of the Seine, in 1872, he repudiated the system of deferred payments, stating that it was a disguised loan, and that a municipality could always borrow money for work at a lower rate of interest than from a contractor. The above price was about what the author considered a good contractor's bargain, as made by the Improved Wood Company in 1876, for the maintenance of Piccadilly roadway for fifteen years. It had often been remarked that the *materials for making roads* were few, say stone sets or flags, wood, macadam, asphalt, the tesselated pavement laid in concrete by the Romans, and lava flags, as in Naples and Catania. *Macadam* was certainly condemned for heavy traffic in all large towns as being too costly in maintenance and too disagreeable owing to its dust and mud, choking of drains, making mud-banks in rivers, etc., and the tendency of the day was to substitute for it a noiseless pavement. The macadamized road of the Champs Elysées, Paris, now nearly all replaced by wood, cost for maintenance 17 francs per square meter per annum, say 13s. 8d. per square yard ; that of the Boulevards, 6s. 8d. per square yard. The *proper way to lay wood* was upon a smooth cement concrete. Concrete was being used in Paris now as a foundation for granite sets, as in Berlin and Vienna. Wood might answer for wide, well-ventilated thoroughfares, but to use it for narrow streets was *anti-hygienic*. Wood absorbed the urine of horses, and the diluted filth of the streets ; horse-dung clung to it, and in dry weather it gave rise to horse-dung dust.

For traffic, *wood* was excellent for the first two or three years, but as soon as it became fibrous and worn, like an old tooth-brush, it would certainly give off *poisonous emanations* under a hot sun, and remain damp in winter. It lacked that first quality for a hygienic roadway of impermeability; and was far *less durable than asphalt;* for Cheapside, laid in asphalt in 1878, had never been renewed, and its repairs had never stopped the traffic for one minute since it had been laid, whereas wood was entirely renewed in six to seven years. It might be safely predicted that a reaction would set in against wood within the next few years. The Paris engineers stated in article 19 of their specification for wood pavements: "The administration reserves to itself the faculty of suppressing at any time any part whatever or even the whole of the roads paved with wood." He might add that the cost of the compressed *asphalt pavement* now being laid round and inside the new *Hotel des Postes,* Paris (about 10,000 square meters), consisting of a 6-inch Portland cement foundation and asphalt 2 inches thick was 19 francs 50 cents the square meter, say 12s. 5d. the square yard, and the yearly maintenance for ten years 2 francs per square meter, or 1s. 4d. per square yard. In the Rue de Richelieu the two systems of noiseless roadways, asphalt and wood, had been laid this year side by side ; a few years would prove which of the two possessed most durability.

Mr. H. Faija said that as it seemed to be admitted that the life of *wood pavement* was *dependent*, to a rather important extent, *on the foundation* on which the wood-blocks were laid, he agreed with the author that it was better to have good, clean ballast for the formation of the concrete, than to use the broken granite which existed in the old roadway, from which it would be hardly possible to make a good concrete. Even after the granite had been screened it was very dirty, and all the holes and interstices, which for the production of concrete should be filled with cement, were filled with dirt. Then, again, the screening removed not only the dirt, but also the smaller pieces of stone, leaving only the large pieces, so that to secure a sound concrete a much larger quantity of cement would have to be used than was necessary with the ballast. But as this extra amount of cement was not used, the concrete was very rough, very open, and consequently neither strong nor sufficiently smooth to receive the blocks; it was therefore necessary to lay a surfacing of sand and cement on which to set the blocks. This surfacing must, in time, break up and become reduced to powder, for it had not a thickness sufficient to resist hard wear, nor was it homogeneous with the concrete underneath. The author had stated the cost of the concrete at $2s$. $3\frac{1}{2}d$. per square yard, and it was therefore evident that if the concrete had to be relaid or materially repaired every seven years when the blocks were renewed, the maintenance would be considerably increased beyond the figures given in the paper; and might probably account for the high cost of maintenance which exists in the Holborn District. In fact, the concrete should be regarded as the permanent road which might from time to time be covered with wood or other material more suitable than itself to the requirements of the traffic. The concrete to be thus permanent should be one homogeneous mass throughout, without surfacing of any kind, such a concrete as resulted from a well-proportioned aggregate properly manipulated, and he therefore was certain that in putting a well-made concrete under the wood blocks, the author had acted to the best advantage for his Vestry.

With reference to the number of experiments of the *cement* which he had made for the author, and which had been published in their entirety, he should like to say that in every case he sent his report to the author on the completion of the seven days' test, and that therefore the extracts from these reports, which were given in the paper, were not to be considered as defining the nature of the cement as shown by future experiments, but simply as his opinion of the cement at that date. No doubt a finer-ground cement would have been preferable, and would probably have enabled the author to obtain the strength he required with the larger proportion of

aggregate—viz., 7 to 1 instead of the 5¼ to 1 which he found it
necessary to use, and would therefore have resulted in economy ;
otherwise the cement was of good quality and well suited to the
purpose.

Mr. A. Southam observed that his experience as Surveyor for
the Wandsworth Board of Works at Clapham, enabled him to confirm the author's conclusions.   In October, 1880, the *High Street
Clapham was paved with wood* by the London Tramways Company
and the Wandsworth Board.   In the centre of the road were two
lines of tramway ; these had been previously paved with several
kinds of asphalt, and all had failed ; the margins of the carriageway
were macadam.   *The tramway was paved* under the direction of the
engineer to the company, with plain deal blocks grouted with cement;
the margins, 3,600 square yards, were laid by the Improved Wood
Paving Company for the Wandsworth Board with blocks dipped in
creosote on a bed of 6 inches of concrete, composed of Thames ballast and Portland cement, in the proportion of 6 to 1, at a cost of
11s. per square yard.   The excavation was undertaken by the
Wandsworth Board ; the value of the macadam somewhat exceeded
the cost of breaking up, carting, and sifting it for use elsewhere.
The *pavement* had been *maintained* by the Improved Wood Company free of cost for three years, when they expressed their willing-
ness to enter into a contract to maintain it for a further period of
fifteen years at 10d. per square yard per annum, and to leave it in
good order at the end of that term ; but although that was considered a small sum the Board had thought it desirable to maintain it
themselves.   Both the pavement laid by the Improved Wood Paving
Company and the Tramway Company were in good order, no repairs
having yet been required, and the work had been done nearly four
years.   In any extension of wood-paving he would *use* plain well-
seasoned *yellow deal* wood blocks grouted with Portland cement,
laid on concrete, formed of the old macadam, sifted and mixed with
Portland cement.   He considered that it was expedient to have
large *works executed by contract*, but the *maintenance* should be undertaken by the parish authorities who had the control of the road.

Mr. G. F. White expressed his satisfaction with the paper,
which he thought was a valuable supplement to those of Mr. Deacon
and Mr. Howarth.   Considerable experience had been gained in
the intervening five years as to the endurance of wood as a pavement.   The statistics collected by the author should help, he
thought, to set at rest many questions on which opinions were
divided some years back.   There were two points on which he
desired to offer a few remarks.   The first related to the character of
the *foundation* to be used under the wood.   The second, to the

mode in which the blocks should be laid with reference to one another.
It seemed now to be quite agreed that the indispensable condition
of securing a good and lasting paving was a firm and unyielding
foundation, and it was also conceded that no material was so fit for
this purpose as concrete made of Portland cement and gravel, in the
proportion of 1 part of cement to 5 or 6 parts of gravel. The author
seemed to be of opinion that a thickness of 6 inches was sufficient for all purposes. Mr. White concurred in this view where the
soil under the concrete was hard and undisturbed ; but where the
ground had to be picked up to provide for gas or water pipes, as
was so generally the case, he thought the layer of concrete should
not be less than 8 to 9 inches, especially in thoroughfares where the
traffic was heavy. He had noticed in certain cases, and especially
in Pall Mall, where the wood paving was probably in a worse condition than in any street in London, that the excavation for the concrete base had been very unequal in depth, the sub-stratum having
in some cases been hardly removed at all, while in other parts there
were holes 12 and 13 inches deep, which had been filled up with hard
rubbish ; and over this surface had been laid a bed of concrete,
averaging hardly more than 3 inches in thickness. Such a mode of
proceeding could only have one result, which had been predicted
while the work was in progress—namely, the speedy breaking up of
the pavement, with the consequent necessity of replacing it within a
year or two. As a matter of fact that was what happened. The
pavement showed signs of subsidence almost immediately ; and,
though the middle of the roadway had been in part re-laid, the gen-
erality of the work was in as bad condition as ever. The author
had mentioned two cases in which the concrete had been laid 12
inches thick, and though in the one case (Regent Street) the advantage had been nullified by the inferior quality of the wooden road
laid upon it, it was to be inferred from the description of another
example in Parliament Street that the life of the wood road would,
in the author's estimation, be considerably increased by the extra
solidity of the deep foundation. From a careful consideration of
the whole question, Mr. White had come to the conclusion (1) that
the concrete should in all cases have a thickness sufficient to make
it act as a beam in bridging over these inequalities of excavation.
(2) That the life of the wood was in direct proportion to the immo-
bility of the foundation, which must be deep enough to resist from
the first the hammering action of horses' hoofs and the heavy pressure of wheels. The second consideration to which he would advert,
was the *way in which the blocks should be laid* together in the roadway, and which resolved itself practically into a question of joints,
or no joints. Now in the various pavings reported on by the author

every sort of jointing seemed to have been tried, and though he
had judiciously abstained from dogmatically asserting his opinion,
lest he should perchance be regarded as the partisan of any particular system, there seemed to be little doubt that he had a strong
preference for laying the blocks together as closely as they could be
put, and filling in the interstices with Portland cement grout. In
this view Mr. White heartily concurred, and was glad to find it was
also the opinion of Mr. Howarth. He believed that the use of
grooves or wide joints, as affording foot-hold for horses, was of very
doubtful advantage, since, when the paving was perfectly dry or
thoroughly wet, the foot-hold was complete, even with a jointless
material like asphalt ; and when the pavement was in the intermediate state of slipperiness, caused by frost or London fog, the grooved
paving gave no more support to horses than did the close joint, the
reason being that the mud filling of the joints, taken in connection
with the rounded edges of the blocks, was rather a cause than a preventive of slipperiness, which was altogether absent in the case of
the continuous paving. Another argument against setting the blocks
apart was the fact that their being so set helped the abrasion and
rounding of their arrises, which was not only in itself an element of
deterioration, but tended after a time to create a sort of corduroy
road, on which the wheels bumped from one block to another, producing thereby a jar very detrimental to comfort in driving over it.
If any proof were needed of the correctness of this statement, one
had only to drive in Oxford Street, over the road running from the
Marble Arch eastward, to feel the difference in smoothness and comfort of the close-jointed pavement in comparison with the grooved
and bumpy road which succeeded it further east.

Mr. White had been told by omnibus drivers, who were excellent judges, that if the wide-jointed pavements were universal in
London streets, there would not be a driver without a spinal complaint at the end of a twelvemonth. As regards the *direction in
which the blocks should be laid* in relation to the street, there would
seem to be no doubt that it should be transversely to its length, like
ordinary stone paving. In the paving before referred to in Pall
Mall, the blocks had for some inscrutable reason been laid diagonally, than which nothing could be imagined more unsafe for horses,
or more prejudicial to the comfort of those who drove over them,
especially when the joints were, as in the case of one exhibited, $1\frac{1}{2}$
inches in thickness. This particular pavement appeared from the
author's statement to have been constructed at a cost of about 8*s.* per
square yard, about one-half the price at which most other roads had
been laid. It was in his knowledge that the expense of making it
had been largely subsidized by the clubs and the War Office, who

had not taken the ordinary precaution of employing an inspector to watch its construction, but had left it entirely to the parochial authorities, of whose parsimony and ignorance it remained unhappily a convincing proof. It was to be hoped such a state of things would not easily recur, though he was not without fear that the neighboring roadway of Cockspur Street and Charing Cross, quite recently laid, would manifest before long the consequences of insufficient foundation. This led him to inquire whether it was impossible to institute some more *authoritative control* than at present existed over the construction of the wood roads of London. The information gathered by Mr. Howarth and by Mr. Stayton was as ample and specific as would have been collected by a select committee sitting on the question, and must surely therefore be sufficient as a guide to some uniform plan for the execution of such works. The question then became whether such uniformity could be enforced on the different parochial bodies of the metropolis. The roadways were for the comfort of the whole community—not of individual parishes ; and he could heartily wish that such a body as the Metropolitan Board of Works, which seemed to have most things under its management, could actively intervene to give the ratepayers the benefit of the investigations which had been made on the subject, instead of leaving them any longer to be the victims, both in pocket and in comfort, of every experimenter who might have a nostrum to recommend or a patent to push.

Mr. Stayton, in reply to the correspondence, remarked that such observations as those made by Mr. Culverwell were extremely practical and valuable. He, however, felt compelled to reassert his opinion that the *pitch-pine blocks* in King's Road *created an unpleasant "jarring" motion* when driving over them. He had many times experienced it, and on a recent occasion the effect was very apparent. Possibly this evil could be mitigated by the application of Henson's felt bed and joint. He was glad that Mr. Culverwell had so clearly expressed his reasons for declining to support the *creosote theory*, feeling convinced that the extra cost of creosoting could be better expended in the selection of the timber. He concurred in Mr. Delano's objection to the terms "*asphalt*" and "*asphaltic*" as applied to the wood pavement laid by the Asphaltic Wood Pavement Company ; obviously the British asphalt used by them was a manufactured article, and the name might lead to confusion. Where the word "asphaltic" appeared in tables, it merely referred to the "Asphaltic" Company's system, in the same way that "Henson's" or the "Improved" systems had been referred to. He did not find that the wood in Chelsea "gave off *poisonous emanations* under a hot sun," although it had been laid 5½ years ; this

fact, however, could be accounted for, because its surface was thoroughly watered and machine-swept twice a week during the summer, independently of the attention described under the heading of
management.   Any neglect of this service, however, would soon
create the unpleasant condition described.   He assumed that it
would be taken for granted that the *thickness of concrete foundation*
referred to by Mr. White (six inches) would only be adopted where
the soil was hard and undisturbed, and he could refer to numerous
instances in the Chelsea pavements where the thickness was from
8 to 10 inches; in fact, a depth of 12 inches had been laid for a long
distance over the site of a suspicious gas trench in Sloane Street.

WOOD AND ASPHALT PAVEMENT FOR CITIES AND TOWNS.*

From the experience of the city of Liverpool we may glean
some interesting facts respecting the use of wood for pavements.

Durability of Wood Pavement.—In special reports by the engineer
of Liverpool, Mr. Clement Dunscombe, it is stated that it would not
be prudent to assume a longer life than ten years for wood, under
the best conditions, and of the superior class laid down in that city.
In streets of minor traffic this may be exceeded, but in those of
heavy traffic it will be reduced.

(The life of natural-rock asphalt pavement may be taken as at
least twelve years).   In one street paved with wood, with a traffic of
94,000 tons per yard of width per annum, the wear was at the rate
of ¼-inch per annum.   In another street, with 302,000 tons per
annum, the wear was 0.58 inches.   The wear is found to be greater
in the latter years of the life of wood than in the first years.   Mr.
Dunscombe estimates the wear within tramway tracks at about one-
third more than the remainder of a roadway.

Gravel on Wood Pavement.—He states that to keep a wood
pavement in good condition it should be graveled.   " The fibres of
the wood ought never to come under direct wear, but the surface
should be kept indurated with sharp gravel."

Objection to Wood in Tramway Streets.—The unequal wear of
the blocks between rails of tramways he considers to be a serious
drawback to the use of wood in tramway streets.   "Properly laid
tramways ought not to present the slightest impediment to even the
narrowest-wheeled traffic, and in order to maintain them and the
track in proper order it is necessary to use materials for paving of
the most durable kind—such as the toughest syenite sets—which
shall approach most closely in wear to that of steel rails.   By such
a selection of materials only can the repairs to the pavement due

---

* XV, 292.

to the unequal wear of sets and rails be kept at a minimum." The life of steel rails in Liverpool tramways is about sixteen to twenty years, whilst the wear of the hardest wood is considerable. When the wood has worn down to the extent of half an inch at or near the rails, it should be relaid—at least that portion adjacent to and between the tracks, as it endangers the wheels of light vehicles if it is not done. This first repair will come where traffic is heavy in about two to four years with wood pavements, and the frequent tearing up of the street for repairs shows the necessity of using a more durable material.

Cost of Maintenance.—The cost of maintenance of wood pavement is placed at about 45 cents per yard per annum, and of graveling, watering, and scavenging at 22 cents. The cost of maintenance of syenite sets is placed on the contrary at 4 cents, and of watering and scavenging 14 cents per square yard only.

From a paper by Mr. Park Neville, M. Inst. C. E.,[*] we gather additional information as to foreign practice. It is well known that wood pavements have met with considerable favor abroad, on account of their freedom from noise and greater safety to animals against falling. The number of patents on wood pavement is considerable, eight different ones being described, and among them some of those with which we are familiar, but under different names.

Asphalt Wood Pavement.—The "asphalt wood pavement" seems to be the one most recently adopted. In this a thorough concrete foundation is made and accurately shaped. On this is placed a $\frac{1}{4}$-inch of asphalt mastic, on which croesoted wood blocks are placed, with spaces of half an inch between rows, and the blocks carefully breaking joint in the rows. The lower portion of the spaces for 2 to $2\frac{1}{2}$ inches up is filled with melted asphalt, and the remainder with cement grout with gravel. In London this costs $4 per square yard.

Cleanliness and Durability.—Colonel Haywood, of London, says wood can be kept cleaner than asphalt and at less expense. He estimates the life there at six to nineteen years, or ten years without repairs, but those who have ridden over the Piccadilly will have learned that in less time than this they become anything but a smooth pavement, owing to the unequal wear of the blocks.

Mr. Strachan estimates the cost of scavenging wood at one-sixth that of macadam.

While one-third of a load of mud is being taken on an average from wood pavements, there will be two loads taken from granite and four from macadam.

---

[*] A Description of Wood and Asphalt Pavement for Cities and Towns. Read before the Institution of Civil Engineers of Ireland, April, 1886.

Foundation.—Mr. O. H. Howard argues that the arch form of a street cannot be relied upon for distributing the pressure, and he, with others, claims that the wood should be considered merely as a surface or veneer; the *sine qua non* for wood paving is a thoroughly good concrete foundation, and this is the real basis of the pavement. Great stress is laid upon creosoting and upon the use of only hard, tough wood; and one wood pavement on a bridge is mentioned as having outlasted some granite blocks on the same bridge.

In reference to the concrete foundation, Mr. Deacon is quoted as follows : " The ground on which the foundation is to be laid is first well watered ; upon this is then scattered a layer of wet broken stones, on which is spread a thin stratum of cement mortar, and then a further layer of broken stones. This last layer is then beaten into the lower layer with beaters like large spades. This is followed by another layer of mortar, and a third of stones, until the required thickness is attained, when the surface is well beaten and finished by rubbing with the beaters to the proper curvature to receive the pavement. The cost of such a foundation is about 94 cents a square yard, and bituminous mastic foundations, six inches thick, cost in Liverpool 88 cents per yard.

Mr. James Newlands consolidated a foundation by excavating eighteen inches, then filling up to the grade needed for placing granite sets upon with old macadam material and allowing the traffic to come upon it several months. The surface was then leveled, an inch of pebbles placed on it, and the sets placed and grouted.

Mr. Neville gives finally his own conclusions :

He is in favor of wood pavement, on the understanding that it has a thoroughly good concrete foundation, but this should be provided for stone or asphalt as well.

Stone Pavement for Heavy Traffic.—Where there is very heavy or very much light traffic at rapid speed, he considers stone blocks " from the igneous, plutonic and metamorphic rocks or the syenite granites (pure granites are unfit)" as best. These should be thoroughly dressed so as to enable close jointing, and should be laid on a bed of Portland cement concrete mixed six to one with its top surfaced with cement mortar, and allowed to set ten days before the paving is done. It should be protected during this time by tarpaulins from heavy rains or frost. The sets should be wheeled to place on planks so as not to disturb the bed.

The sets he prefers to lay close or with not more than $\frac{3}{4}$-inch joints, which after careful ramming are to be filled with fine pebbles and asphalt. Such work he estimates to cost $3.12 per yard, and the best quality of wood pavement at $2.94.

Asphalt mastic pavements and compressed asphalt he favors except as to slipperiness.

Cost of Maintenance of Wood Pavement in London.—The annual maintenance of wood pavement is given on the authority of Colonel Haywood as costing in London from 18 to 30 cents per yard, of asphalt the same range, and of granite from 6 to 19 cents. The average cost per year, including first cost and maintenance, for wood 40 to 61 cents, for asphalt 33 to 59 cents, and granite 25 to 69 cents per yard.

The broken stone placed annually upon the macadam streets in Birmingham varies from 150 to 450 cubic yards per mile each year. This shows how unfit this class of pavements is for streets used for heavy traffic.

NOTES ON KING'S ROAD WOOD PAVEMENT, LONDON.*

These works were executed as relief works in the early spring of 1886, and done without intervention of a contractor. The *site* is between Limerston Street and Stanley Bridge ; the area paved being 6,666 yards.

Removal of Macadam.—The road was previously covered with macadam consisting of Guernsey granite. It was excavated by means of steel wedges driven through the crust by sledge-hammers. Then, when an opening was made, wedges were driven under the crust, and it was lifted by levers in pieces twenty feet square. These were broken to pieces by hammers, and screened through an inch sieve and then through a ½-inch sieve ; 1,440 cubic yards were carted to the wharves for use on macadam roads, and the remainder (about 2,000 yards) was waste. The bottom of the excavation was shaped so as to give the curve of road.

Foundation.—The surface thus prepared was covered with six inches of concrete, consisting of 5¼ parts of Thames ballast to one part of carefully tested Portland cement. On this was placed ½ inch of fine concrete, three parts Thames sand to one of cement, laid and floated to a perfectly smooth surface. The method of laying was as follows : At gulleys (or catch-basins), which are 6½ inches below the curb, the centre of the road was made at such a height that ¼ inch to the foot fall was given in the cross section. The channels (or gutters) had a fall of 1 in 150. The centre line of the crown of the road was made level, thus giving a fall in the cross section of ¼ inch to the foot at the *summits* of the channels. A narrow strip of concrete was placed in each channel and strips of wood laid on them and nailed to pegs at the exact grade. The crown of road had a similar strip of concrete and screeds also nailed. The templates were very accurately made to the curve of road, faced with iron, and

---

* Prepared for *The Engineering and Building Record* by George E. Strachan, A. M. I. C. E., Surveyor of Parish of Chelsea. xvii, 36.

when drawn along longitudinally by resting on the screeds they molded the fine concrete into a perfect curve.

The concrete was allowed to dry for seven days.

Blocks.—The blocks were yellow deal, 9x5x3 inches. Great care was exercised in rejecting bad blocks. The method used was as follows : An intelligent wood-pavior receives the blocks and rejects those which he is satisfied are bad, dividing those he passes into two lots : (*a*) Those that are without flaw or defect ; (*b*) those which have only slight defect. Blocks *rejected* by him are not examined again. The lot (*b*) are examined by the Clerk of Works, who rejects such as he pleases, and sorts those passed into (*c*) those he considers fit, and (*d*) those which are doubtful. The lot (*d*) are examined by the surveyor, and the good ones passed. The paviors have instructions to throw out any defective blocks which have escaped, and after the blocks are laid, but before they are grouted, they are again looked over. Even *then* bad blocks get in. A fine of 10s. a 1,000 for rejected blocks is enforced. At first the contractor has about 30 per cent. of his blocks rejected, but he soon learns to sort over the blocks before sending them. On this work five per cent. of the blocks delivered were rejected (1 in 20).

The blocks are separated by iron studs, as described in previous articles. They are laid with their lengths at right angles to the traffic, and the studs give a joint transversely of three-eighths of an inch.

The blocks are placed on the concrete, and the spaces at the joints filled in with cement grout (composed of 3 sand to 1 cement). Seven days are allowed for drying. The surface is then covered with very fine pebbles to a thickness of a quarter of an inch, and the traffic is turned on. The pebbles are crushed and the pieces are driven into the surface.

One inch in ten feet is allowed next the curbs for expansion.

Cost.—The actual cost was—

	Total.			Per square yard.
	£.	s.	d.	s.
Labor..	1,011	7	2	3.03
Blocks......................................	1,185	17	4	3.55
Cement......................................	406	3	5	1.21
Cartage.....................................	266	18	2	.80
Ballast, sand and pebbles.................	251	14	5	.75
Use and repair of tools...................	108	9	9	.32
Studs.......................................	34	13	6	.14
Incidental expenses........................	24	15	1	.07
Totals.,....................................	£3,289	18	10	9.87

As to Life of Wood Pavements.—Eight years under traffics of 500 tons per yard in width every sixteen hours. (Blocks not creosoted.)

First three years no cost for maintenance. Next five average cost of 2*d.* per square yard.

The wood gets bumpy towards the seventh year.

Scavenging wood pavements costs in King's Road one-sixteenth of what macadam does.

After washing and after rain wood pavements have a perceptible, but not unpleasant, odor.

Wood is practically noiseless.

Creosoting adds to the life of wood, but its cost is 1*s.* 6*d.* per square yard in addition to prices given. For streets with traffic weights under 500 tons per yard width in sixteen hours it does not pay, as creosoted and plain woods become bumpy after a certain wear, so that additional life is then an annoyance.

Hard woods are not good for wood pavements. The surface gets bumpy sooner than with soft woods.

Cement joints are water-tight and wear well.

Openings by water and gas companies are repaired by the Vestry. The excavation is rammed and watered, the concrete is carefully restored, and the contour preserved. The concrete is then allowed to dry ; then the space is repaved, and not until the grout is dried is the traffic allowed over it. The actual cost incurred are refunded by the companies for whom the work is done.

SPECIFICATIONS FOR PAVING MATERIALS, PARISH OF CHELSEA.*

Specifications for the Supply and Delivery of Wood Blocks for Paving Portions of the Streets known as King's Road and Pont Street, in the Parish of Chelsea.

1. *Blocks.*—The wood blocks are to be the best improved Swedish yellow deal, cut from Gothenburgh thirds, and are to be 9 inches long, 3 inches wide, and 5 inches deep. Notwithstanding any custom of trade, or any meaning usually attached to the description of blocks as "9 inches long and 5 inches deep," the length of the blocks to be supplied shall be such that when any twelve blocks are placed end to end in a straight line, so as to have their lengths in the same direction, the total length of the twelve blocks shall not be less than 8 feet 9 inches, and when any 8 blocks are placed in an upright position, with their sawn faces in contact, the height of the eight blocks shall not be less than 3 feet 3 inches. With the approval of the Surveyor to the Vestry, blocks of

* xvii, 68.

a less length than will comply with the above requirement may be supplied, but in such a case the contractor is to supply such a number of blocks to each thousand, at his own cost and without payment from the Vestry, as will make a total length at least equal to the length of the blocks as specified ; but the number of such blocks of a less length shall not be more than one-third of the total number supplied, nor shall any blocks be of a less length than eight inches. The blocks are to be sound, square, properly and uniformly cut, free from sap, shakes, warps, large or dead knots, and other defects. A sample of blocks can be seen at the office of the Surveyor to the Vestry.

2. *Surveyor may Inspect Works where Blocks are Cut, etc.*—The Surveyor to the Vestry, or other person authorized by him, may, at all times during working hours enter the saw-mills, wharf, or other place of the contractor where the blocks are being cut, to examine the same, and may reject any deals or timber from which the blocks are about to be cut which, in his opinion, are unfit for the purposes of this contract.

3. *Number of Blocks and Delivery.*—The contractor is to supply 640,000 blocks, of which 340,000 are to be supplied to King's Road, from Limerston Street to Lot's Road, and 300,000 to Pont Street, but the Vestry reserve the right to take a less number, or to require the supply of a greater number. The blocks are to be delivered on the streets named at any part thereof, or on any of the side streets within thirty yards of such streets, and are to be stacked in such a manner and at such times and places as the Surveyor to the Vestry shall direct. The contractor will not be required to deliver a greater number of blocks than 20,000 per day.

4. *Vestry may Obtain other Blocks if Contractor fails to Deliver.*— The contractor is to commence the delivery of the blocks within ten days after the receipt of an order in writing from the Surveyor to the Vestry, and is to proceed with such delivery in such quantities and at such times as the said surveyor shall direct. The said surveyor shall have power to suspend the delivery of the blocks, as he thinks fit, on giving twenty-four hours' notice to the contractor, without any charges which may arise in consequence, being chargeable to the Vestry.

5. *Blocks to be Approved by Surveyor.*—'I he blocks supplied are to be subject to the approval of the Surveyor to the Vestry, and any blocks which are not approved by him are to be removed from the works within six hours. The blocks are not to be considered as approved until they are laid in the streets as part of the pavement

---

* xvii, 68.

and grouted in. A sum of ten shillings per thousand will be deducted from the payments to be made under this contract to the contractor by the Vestry, for every thousand of the blocks so delivered on the works and not approved. If the contractor fails to remove the blocks not approved in the time allowed, the Vestry may do so at his expense.

6. *Penalties if Works Delayed for Want of Blocks.*—If the works are delayed or hindered by reason of the non-delivery of a sufficient number of blocks, the contractor shall pay to the Vestry the sum of £10 per day or part of a day they are so delayed, and the Vestry or their surveyor may obtain blocks from any person they think fit, and may deduct the extra cost of such blocks (if any) from the payments to be made to the contract or under this contract, together with the sums to be paid by him to the Vestry for the delay of the works, or may recover such extra cost and sums from him or his sureties as and for liquidated damages, as they deem best.

7. *Payment.*—Payment will be made to the contractor in monthly installments for the blocks laid in the street as part of the pavement on the certificate of the Surveyor to the Vestry.

8. *Surveyor's Decision to be Final.*—The decision of the Surveyor to the Vestry as to the meaning of this specification, or as to the quality or number of blocks supplied is to be final and binding.

GEORGE R. STRACHAN,

February 24, 1886.   Surveyor to the Vestry.

Specifications for the Supply and Delivery of Thames Ballast, Sand, and Pea Gravel for Paving Portions of the Streets Known as King's Road and Pont Street, in the Parish of Chelsea.

1. *Thames Ballast.* — The Thames ballast is to be taken from the River Thames, the stones are to be regular in size, and not to exceed three inches in any dimension, the sand clean and free from clay, mud or loam, and the proportion of sand to stones is to be to the satisfaction of the Surveyor to the Vestry.

2. *Sand.*—The sand is to be pit sand or Thames sand of the best quality, clean and sharp, free from loam or clay, and when screened through a sieve of 400 meshes to the square inch no residue must be left.

3. *Pea Gravel.*—The pea gravel is to be of clean water-worn stones, not exceeding three-eighths of an inch in any dimension.

4. *Quantity.*—The contractor is to supply the quantities of

materials set forth in the form of tender of this specification ; but the Vestry reserve the right to take a less quantity or to require the supply of a greater quantity of any or all of them. They are to be delivered on the streets named, at any part thereof, or on any of the side streets within thirty yards of such streets, at such times and places as the Surveyor to the Vestry shall direct. The contractor will not be required to deliver more than 100 cubic yards of Thames ballast, 20 yards of sand, or 10 yards of pea gravel per day.

5. *Delivery.*—The contractor is to commence delivery within seven days after the receipt of an order in writing from the Surveyor, to the Vestry, and is to proceed with such delivery in such quantities, and at such times as the said surveyor shall direct. The first order will be given on the 10th March instant. The said surveyor shall have power to suspend the delivery as he thinks fit, on giving twenty-four hours' notice to the contractor, without any charges which may arise in consequence, being chargeable to the Vestry.

6. *Materials to be Approved by Surveyor.*—The materials supplied are to be subject to the approval of the Surveyor to the Vestry, and any which are not approved by him are to be removed from the works within six hours. If the contractor fails to remove the materials not approved in the time allowed, the Vestry may do so at his expense.

7. *Measurement and Approval.*—The contractor is to bring an invoice with each delivery, stating the quantity and nature of the materials supplied, which is to be left with the person appointed by the Surveyor to the Vestry to receive such materials. If the contractor provides such invoice in duplicate, the person so appointed will sign and return it to the carter, if the quantity and nature of the materials are correctly stated thereon, and the production of such duplicate invoice, so signed, will be accepted by the Surveyor to the Vestry as proof of delivery of the quantity of the materials signed for. The materials supplied may be measured by the Surveyor to the Vestry, or any person whom he may appoint, in the cart in which it is brought on to the works, or on the ground, as he thinks fit. If the measurement is found to be short of that stated on the invoice, the cost of measurement shall be paid to the Vestry by the contractor. The materials are not to be considered as approved until they are used on the works.

Here follow the same clauses as to delay, payment, etc., as in the specification for blocks.

GEORGE R. STRACHAN, Surveyor to the Vestry.

March 3, 1886.

Specification for the Supply and Delivery of Portland Cement, for Paving Portions of the Streets Known as King's Road and Pont Street, in the Parish of Chelsea.

1. *Cement.*—The cement is to be the best Portland cement, very finely ground, weighing 112 pounds to an imperial striked bushel, capable of sustaining a breaking weight of 420 pounds per square inch of sectional breaking area after seven days immersion in water. It is to be delivered in sacks containing two bushels, each bushel weighing 112 pounds net.

2. *Quantity and Delivery.*—The contractor is to supply 5,000 sacks of cement, of which 2,660 are to be supplied to King's Road, from Limerston Street to Lot's Road, and 2,340 to Pont Street, but the Vestry reserve the right to take a less number or to require the supply of a greater number. The sacks of cement are to be delivered on the streets named at any part thereof, or on any of the side streets within thirty yards of such streets, and are to be stacked in such manner and at such times and places as the Surveyor to the Vestry shall direct. The contractor will be required to deliver the cement in lots of 500 sacks at a time, and to complete the delivery of each lot within seven days after the order is given. The order for the first lot will be given on the 10th March inst., for the second lot on the 11th March inst., and for the remaining lots at intervals of about seven days from the 11th March inst. The Surveyor to the Vestry shall have power to increase the length of these intervals, or to suspend the delivery of the cement on giving twenty-four hours' notice to the contractor, as he thinks fit, without any charges which may arise in consequence, being chargeable to the Vestry.

3. *Testing and Approval.*—When each lot is delivered, the Vestry will watch and cover it at their own expense until it has been tested. The Surveyor to the Vestry will cause samples of each lot to be taken and tested as to its breaking weight, and will cause other samples of each lot to be weighed and measured. If the samples so taken comply with clause 1 of this specification, the lot from which they were taken will be approved by the Vestry and their surveyor, and the expense of the tests and measurements will be borne by the Vestry. If, however, they fail to comply with clause 1 of this specification, the lot will be rejected and the contractor shall, at his own expense, remove the cement within forty-eight hours after notice of its rejection from the surveyor, and shall pay to the Vestry the cost incurred by them in testing and measuring the lots so rejected. If the contractor fails to remove such cement the Vestry may do so at his cost.

4. *Sacks.*—The contractor is to bring an invoice with each delivery of cement stating the number of sacks supplied in such

delivery, which is to be left with the person appointed by the Surveyor to the Vestry to receive the cement. If the contractor provides such invoice in duplicate, the'person so appointed will sign and return it if the number of sacks delivered is correctly stated thereon, and the production of such duplicate invoice so signed will be accepted by the Surveyor to the Vestry as proof of delivery of the number of sacks of cement and of the sacks stated thereon. The empty sacks will be returned to the contractor on the works as the cement is used, and are to be removed and signed for by him at his own cost within twenty-four hours after notice from the Surveyor to the Vestry. The Vestry will not make any payment for the use of the sacks, but will pay the sum of one shilling for every sack not returned within two months after its delivery on the works.

Here follow the same clauses as to delay, payment, etc., as in the specification for blocks.

GEORGE R. STRACHAN,

March 3, 1886.   ,   Surveyor to the Vestry.

SPECIFICATION FOR WOOD PAVEMENT, PARISH OF ST. GEORGE, LONDON.*

The specification given below is one of the best recently written for wood pavements in London. Among the things to be noted are : the requirement of a space next each curb to allow for expansion ; the requirement that the contractor give his own levels, etc., and making him responsible for their accuracy ; the six months' maintenance without charge ; the repair by the contractor of all openings in the street, at a fixed price per yard ; the sweeping and cleaning of footways during progress of the work ; and, lastly, that the contractor shall be present " whenever required, for the purpose of measuring and ascertaining the quantity of work performed," or else that he shall accept the measurement then made by the surveyor.

PARISH OF ST. GEORGE, HANOVER SQUARE, LONDON.

Specification for Paving with Wood the Carriageways of Buckingham Palace Road (from the Grosvenor Hotel to Ebury Bridge), Conduit Street, Half-Moon Street, Pimlico Road, Queen Street, and Wilton Road.—George Livingston, Surveyor.

1. *Description of Pavement and Mode of Laying.*—The pavement is to be composed of creosoted blocks, cut from the best yellow deal of first quality, 9 inches, 8 inches, and 7 inches long by 6 inches deep and 3 inches wide ; no block to be more than 9 inches nor less than 7 inches in length. Each block to be cut perfectly

---

*xvii, 53. This pavement mentioned in the specification was laid in two contracts at prices respectively of 8s. and 8s. 2d. per yard.

true in size and shape. The joints to be filled in with mastic asphalt of approved quality to a height of at least ¾ of an inch, measured from the bottom of the block, the rest to be filled in with Portland cement grouting, and no joint to be more than ⅛ of an inch in width. The blocks laid transversely are to be only laid to within 3 inches of the curb on each side of the street (to allow for expansion), the space so left to be filled in with Portland cement and sand, as shall be directed, but subsequently made good with blocks if necessary. No paving to be commenced on any portion of the foundation until six clear days after the concrete has been laid.

2. *Creosoted Blocks.*—The whole of the blocks used in the work to be creosoted * by a mixture of pure creosote, G. oil, and pure distilled tar, in proper proportions as shall be approved, but no blocks shall undergo this process until they have been first inspected and approved of by the surveyor, or other authorized officer ; and the surveyor shall have full power, notwithstanding such inspection, to split in two any reasonable number of blocks selected indiscriminately from those brought on to the works, and to reject any that he may consider unfit for use ; the same to be at once removed by the contractor.

3. *Excavation.*—The contractor to excavate the whole of the present macadamized roadways to the required depth to form foundation and pavement, such of the old macadam as the surveyor shall approve, if sifted and mixed with Thames ballast, to be used for concrete ; the surplus to be the property of the contractor, who shall at once remove the same at his own cost.

4. *Cement.*—The whole of the cement supplied for the works must be of the best quality, and from an approved manufacturer, and must conform to the usual tests to the satisfaction of the surveyor.

5. *Concrete.*—The concrete to be composed of Thames ballast (mixed with such of the old macadam and in such proportions as the surveyor shall approve) in the proportion of 7 parts by measure of hard core to 1 part by measure of cement ; the surface to be finished off to a smooth face with concrete 1 inch in depth, composed of Thames sand and Portland cement, in the proportions of 3 of sand to 1 of cement, and to be laid a depth of not less than 6 inches over the whole area of the street (from curb to curb), and to be composed of the best Portland cement, subject in every respect to the approval of the surveyor.

---

*The surveyor considers that all blocks should be creosoted under a pressure of 10 to 12 pounds per super foot.

6. *Old Paving Stone.*—All the paving stones in channels and crossings, etc., to remain the property of the Vestry, at whose expense they will be taken up and removed.

GENERAL CONDITION.

7. *Contractor to Provide all Materials, Labor, Tools, etc.*— The contractor shall provide all materials, labor, tools, tackle, implements, etc., for the proper execution of the works. All the materials used to be the best of their respective kinds, and applied in the most workmanlike and substantial manner possible, and to the entire satisfaction of the surveyor. The blocks shall be of the best yellow deal, of first quality, and the surveyor shall have full power to reject any materials which he may consider unfit to be used in the work.

8. *Setting Out.*—The contractor to set out and keep correct the works in every particular according to directions he may receive from time to time, and to be responsible for the correctness of the same throughout the whole term of the contract.

9. *Maintenance After Completion.*—The contractor is at his own expense, and without charge to the Vestry, to maintain the wood pavements, together with their foundations, in a state of perfect repair to the satisfaction of the Surveyor to the Vestry, for a period of six months from the date of completion of the entire work.

10. *Maintenance by the Vestry.*—Should the Vestry determine itself to maintain and repair the pavement of any or all of the streets at the expiration of the period during which the contractor is to keep it in repair free of charge, he is to be bound, if so required by the Vestry, to execute such repairs as he may be called upon to make at a price per superficial yard to be hereafter agreed upon. In all cases the old materials are to become the property of the contractor, and to be carted away by him without expense to the Vestry.

11. *Repairs Over Gas, Water and Other Trenches.*—The contractor is, during the term of his contract, and at any time after its expiration, if so required, to repair, within twenty-four hours after notice, all damage done to the pavement by, or in consequence of, the operations of gas or water companies, or other public or private bodies, or by the Vestry itself, and he is to do the work for such companies, or others, at a price per superficial yard (to be stated in his tender) measured as repaired. All old materials to become the property of the contractor, and to be removed by him at his own expense.

12. *Cleaning and Sanding Surface.*—The Vestry is to be at liberty to cleanse the pavement with water or by sweeping or scraping either by hand or by machines or in any other way it may be

deemed expedient, and may also strew the surface with fine sand, gravel or other materials with a view to prevent slipperiness, and the contractor is to have no claim for increased wear of the pavement should it result from the usage of such material, or from any mode adopted for cleaning the surface.

13. *Fencing, Watching and Lighting.*—The works are to be carefully fenced, watched and lighted, during their progress, both by day and night, by the contractor at his own expense.

14. *Footways to be Kept Clean.*—The contractor is to sweep and keep clean footways from ballast or any other material, so far as may be practicable during the execution of the works.

15. *Injury to Curbs and Footways.*—Should any portion of the curb stones or footway pavements be injured or displaced by the contractor's workmen, they are to be reinstated at the contractor's cost by the workmen of the Vestry, and such cost may be deducted from any sums then due or that may become due to the contractor.

16. *Injury to Sewer, Gas or Water Mains.*—Should any injury be done either to the sewers or their appliances, or to gas or water pipes or their appliances, by the contractor's workmen, the damage is to be made good at the contractor's expense, and the cost thereof may be deducted from any money then due, or that may become due to the contractor.

17. *Refixing Gas and Water Boxes at Altered Levels.*—The parish workmen will raise or lower as may be required all water or gas boxes, gully grates, man-hole covers, etc., to suit the new pavement at the expense of the Vestry.

18. *Dismissal of Workmen for Misconduct.*—The contractor is to dismiss from the work any agents, workmen, laborers or others in his employ for misconduct, if required to do so by the surveyor.

19. *Services of Notices Upon Contractors.*—All notices to the contractor, his foreman or agents, shall be deemed to be duly served upon him by their being delivered personally to any of his agents, or sent by post to his offices for the time being; and notice of such offices or the contractor's address is to be left with the Surveyor to the Vestry.

20. *Works to be Executed as the Surveyor Directs.*—The work is to be executed in such manner, lengthwise and widthwise of the streets, and at such times as the surveyor may direct, and be carried on in such a way as will least impede the business of the neighborhood and the public traffic, and so as not to obstruct or endanger passengers, animals, or vehicles more than may be absolutely necessary.

21. *Works to be Done to the Satisfaction of Surveyor.*—All work is to be executed and maintained throughout the contract to the satisfaction of the Surveyor to the Vestry for the time being, whose directions as to the work required and manner of performing the work are to be abided by; and his opinion on all points, both as regards execution and maintenance, and as to the moneys due to the contractor, shall be binding final and conclusive.

22. *Clerk of Works, etc.*—The contractor is to obey the instructions given him by the Clerk of Works to the Vestry, the surveyor's assistants, or other competent persons who may be deputed by the surveyor to superintend any part of the works.

23. *Contractor to Re-execute Improper Work.*—If any part of the work shall at any time, in the opinion of the surveyor, be imperfectly executed, the contractor is, at his own expense, on the requisition in writing of the surveyor, to remove the same and replace it with good sound work to the satisfaction of the surveyor.

24. *Upon Neglect to Re-execute Work.*—Should the contractor, for the space of forty-eight hours after receiving instructions from the surveyor, neglect or refuse to execute or amend any of the work which he may have improperly executed, or should fail to complete the work in time, or to maintain the work properly during the contract term, the Vestry shall have full power to execute the work themselves, by their own workmen, or agents, and deduct all expenses incurred from any moneys that may be due or become due to the contractor, or may recover the same as the law directs.

25. *Contractor Responsible for Accidents and Losses to the Vestry.*—The contractor, during the whole term of his contract, shall be held responsible for all accidents which may take place by reason of his works, or for want of repair in the pavement which he may undertake to maintain, whether notice of need of repair shall have been given to him or not; or for delay in executing repairs over openings made by the Vestry, or others, after due notice has been given to him to make such repairs; and he shall indemnify and hold the Vestry, and its officers, harmless against all actions, claims, compensations, losses, costs, and charges whatsoever in respect of the paving works, or anything arising out of the contract. And the Vestry shall have full power to deduct the amount of such costs, losses, etc., accruing or arising out of such accidents and compensation, or from any defect or omission on the part of the contractor in maintaining the pavement in a suitable condition, or performing the repairs in a suitable manner, and the Vestry shall further have power to recover such amounts from the contractor as the law may direct.

26. *Time of Completion.*—The contractor is to complete and finish the work according to the provisions and true intent and meaning of this specification, fit and ready for public traffic, within the following time from receiving the order from the surveyor to commence the work—viz.: Buckingham Palace Road, within eight weeks; Conduit Street, within four weeks; Half-Moon Street, within three weeks; Pimlico Road, within four weeks; Queen Street, within two weeks; and Wilton Road, within three weeks. And the contractor shall forfeit the sum of £5 per day by way of liquidated damages for any delay beyond the specified time and the Vestry may deduct the same from any moneys that may be then due, or become due to the contractor; but should the contractor be unable, owing to bad weather or other unavoidable causes, to proceed with the works, then such extension of time may be granted as the surveyor may think fit and reasonable.

27. *Measurement of Work.*—Upon the completion of the work, the contractor, by his agent or foreman, is from time to time, whenever so required by the surveyor, to attend at such time and place as shall be named by him for the purpose of measuring and ascertaining the quantity of the work performed, and in default thereof, the surveyor shall be at liberty forthwith to measure and ascertain the quantity himself, and his decision as to the quantity shall be final, binding, and conclusive upon all parties.

28. *Payment.*—No payment shall be made by the Vestry to the contractor for any work done or materials brought upon the ground until the surveyor shall have certified in writing that the work is completed according to the provisions of this specification, and to his entire satisfaction.

29. *Mode of Payment.*—The following payments will be made upon the work when certified by the surveyor: 90 per cent. of the total amount one month after the completion of the work, the remaining 10 per cent. at the expiration of the contract—namely, six months from the date of completion of the work.

30. *Sureties.*—The contractor will have to provide sureties (in such amount as the Vestry may consider necessary) for the due fulfillment of the contract.

31. *Tender not Necessarily Accepted.*—The Vestry does not bind itself to accept the lowest or any tender, and reserves to itself the power to give the whole or any portion of the work to any one or more parties tendering, and to increase or diminish the quantity of pavement to be executed under this contract, without in any way invalidating the same.

32. The contractor may tender for any or all of the streets mentioned, and state in his tender a separate price per yard for each

street, or he may state one uniform price per yard at which he is willing to undertake any or all of them.

NOTE.—The works will be commenced and proceeded with at once. Each street to be completed within the limit of the time separately specified, but the whole of the streets included in the contract must be completed within a period of nine weeks from the date of the commencement of any one of them.

The contract to be drawn up by the Parish Solicitor at a cost to the contractor not exceeding £5.

February, 1886.

One guinea is charged for this specification and form of tender, which will be returned on receipt of a bona fide tender.

Wood vs. *Macadam Pavement.*—The pavement called for in these specifications is the "improved wood" pavement, and in October following their issue, Mr. Livingston reports on the comparative expense attending macadam pavement on Piccadilly, and proposed improved wood pavement on Knightsbridge. The average amount of broken granite used *per year* for five years on 8,577 square yards of surface was 1,090 tons, costing for freight, carting, labor, rolling, watchmen, etc., 2s. 8½d. per square yard; and for the future he estimates it would be on Knightsbridge 3s. per yard.*

Cleansing Macadam in Piccadilly costs an average of 10d. per square yard per annum. For Knightsbridge it would be two cartloads of sweepings a day in dry weather, and ten loads in wet weather; and as statistics show wet days for one-half the year, there would be 1,872 loads at 4s. per load, or a cost of 11d. per square yard per annum on 8,218 yards. For wood, the estimate is 3d. per square yard.

Including cleansing, water, maintenance, and a proportional part of first cost and interest divided over fifteen years, he estimates a saving in the use of wood over macadam of about 1s. per yard per year.

The wood pavements can be contracted for at 3s. per square yard for first cost and maintenance for fifteen years.

WOOD PAVEMENTS IN PARIS.†

In continuation of the series of pavement articles that we have published in previous issues, we have prepared a description of the present standard practice of paving with wood and asphalt in Paris.

---

* There seems to be a discrepancy in this estimate as compared with a subsequent report to be quoted from hereafter, showing a much greater quantity of mud removed from Piccadilly, and a much greater saving than is shown by the estimate here made for Knightsbridge, but it is given as reported.

† xvii, 281.

The regulations and specifications for the work are translated from the *Annales des Ponts et Chaussees*, and will follow this introductory description of the conditions and methods of work in 1883, when the first satisfactory trial of wooden blocks was made.

We abstract below the paper of M. A. Laurent, Ingenieur des Ponts et Chaussees, published in the *Genie Civil*, at that time.

Numerous failures in the use of wooden pavements on public roads has discredited them in Paris, but their rapid extension and success in other countries, particularly in London, under very heavy traffic, has compelled the attention of the municipal authorities.

Objections to Macadam.—A new study of this mode of surfacing was demanded, not only on account of the hindrance occasioned by the constant reballasting of the more frequented macadamized public ways, and by their frequent renewals, with the attendant mud, dust and increased resistance to traction and restriction of traffic, but also by the quantities of waste material stopping the drains and endangering public health. The problem of replacing these macadamized pavements in the great thoroughfares involved its effect upon the sewers.

First Contracts.—The Administration favorably received the proposals of the London Wood Pavement Company, who have successively obtained the contracts for more and more important work. In 1881 they paved their first section on a small portion of the Boulevard Poissonniere and the Rue Montmartre; in 1882, and early in 1883, all the roadway of the Avenue des Champs-Elysées, from the Place de la Concorde to and including the Rond-Point, had been relaid with wooden pavement, and the Municipal Council has now just accepted their proposition to pave a large portion of six of the most important boulevards, the company guaranteeing its work for a long period, and the city in turn permitting them great liberty in the execution of the work by English methods.

The Process.—The process of the Improved Wood Pavement Company essentially consists in the construction, on a perfectly rigid foundation, of an impermeable surface of wooden paving blocks (pavés) thoroughly solidified. The resistance of the pavement to the action of the wheels depends almost entirely on the foundation, the wood being intended solely for a covering to protect this foundation, and to secure, by its elasticity and the perfect uniformity of its upper surface, the smoothest possible rolling of vehicles.

Rigidity of foundation, solidity of blocks, and impermeability of the surface are the three points this system aims to secure.

(1) *Foundation.*—This consists of a bed of Portland cement beton 0.15 m. (6 inches) thick, with top coat of cement mortar about 0.01 m. (⅜-inch) thick. The beton is thus proportioned : A

mixture of about one-third sand and two-thirds gravel is put in a
bottomless box containing half a cubic meter (0.65) cubic yards),
and after the removal of the box 100 kilograms (220 pounds) of
cement are emptied on the heap.   This is in the proportion, by
volume, of about one-seventh as much cement as there is sand and
gravel, since 1,400 kilos is the mean weight of a cubic meter of good
Portland cement heaped loosely.

The sand was dredged from the bed of the Seine and the gravel
taken from pits on the seashore.   The cement was furnished by the
manufactory of Demarle & Lonquety, of Boulogne-Sur-Mer.

The paving-blocks should have a uniform thickness and not be
laid on the bed of beton until after it has set, in order to exactly preserve the curvature of the surface of the beton required for the convexity of the roadway.   In the Avenue des Champs Elysées the convexity was 0.42m. (16¼ inches) in a width of 27m. (87 feet 7 inches),
which represents a mean transverse slope of a little more than 3 in 100.
This convexity, though less than first proposed by the company,
appears to be a little excessive, and it seems that for a road under
satisfactory drainage conditions the convexity might be diminished;
0.42m. is only a mean convexity, for, on account of the small longitudinal slope of the avenue, the grade of the gutters is not parallel
to the grade of the street, but presents a series of short slopes from
the hydrants to the sewer openings, consequently the convexity
varies from 0.39m. (15¼ inches) at the hydrants to 0.45m. (17¾
inches) at the sewers.

To exactly regulate the surface of the beton a series of transverse profiles were defined by stakes leveled to the grade of the top
of the bed.   Along each profile a strip of stiff beton was laid.   The
top of this beton was carefully leveled and smoothed and received a
guide rule, laid flat, whose thickness exactly corresponded with that
of the beton coating.   This series of rules thus formed a set of
guides close together, between which it was easy, with large straight
edges, to level the beton to the required surface.   The first leveling
could never be more than approximate, the surface of the beton naturally remaining somewhat rough.   The exact level required, as fixed
by the tops of the rules, was secured by the top coat of cement-mortar
which filled the spaces between the pebbles and made an exact surface.   This mortar was first composed of 200 kilos of cement to a
cubic meter of sand (336 pounds to the cubic yard), but this proportion proving too small it was increased to 300 kilos.   It was always
mixed with a great excess of water so as to penetrate the interstices
of the gravel.

(2) *Paving*.—The covering is formed of small uniform blocks,
of red northern fir, 0.15m. (6 inches) high, 0.22m. (8¾ inches) long

and 0.08m. (3⅛ inches) wide. These are set close lengthwise, with joints, transverse to the street, of about one centimeter (⅜-inch). The blocks are sent, ready for use, from England, where they were cut from planks of the ordinary size, 0.08m. thick by 0.22m. wide. The third dimension, taken in the length of the plank, forms the height of the block, so that in position the fibres of the wood are placed upright. The blocks are superficially creosoted after being cut.

When the foundation has set, two or three days after being laid, the blocks are set by the pavers. Owing to the light weight of the blocks the work of paving is very rapid. Between crossings the blocks are set in rows perpendicular to the axis of the street, with their longitudinal joints staggered exactly half the length of a block. The methods used at crossings to avoid a continuous joint parallel to the traffic are analogous to those used in stone paving. Special precautions should be taken to insure exact spacing and regularity of the rows. Before commencing a new row a strip, whose thickness is exactly that of the required joint, is set edgewise in contact with the last row and the paver has only to set the adjacent blocks in contact with it.

The blocks do not at first adhere to the foundation and are easily displaced after the removal of the strips, and to maintain them in place, as soon as the strips are taken out a small quantity of bitumen is poured in the joints. This liquid material fills the small spaces that may exist under the blocks and partially fills the joints, and in solidifying effectually seals the blocks.

The joints are then filled by a thin grouting of neat Portland cement, distributed by the aid of a broom. This should be done at least twice to insure perfect filling and the essential impermeability.

The pavement cannot be opened for traffic until after the cement in the joints has completely set, for which a delay of 4 or 5 days is considered necessary. During this interval the last operation is performed—viz., spreading a thin layer of dry sharp sand over the surface. The company claims that this dressing, crushed under the action of the wheels, incrusts itself in the wood and lends resistance to the wearing surface. It seems more probable that this coating is simply to protect the fresh mortar from the direct action of the wheels, for it can be maintained but a very short time on a traveled road, and is soon transformed into a disagreeable greasy mud.

Avenue des Champs Elysées.—The paving of the Avenue des Champs Elysées was divided into two parts, in the first of which the city removed the old pavement, and in the second this work was done by the company under a special contract.

For the first division, the removal of the old pavement was commenced October 16, 1882 ; first beton laid October 20 ; paving commenced October 24, and the first portion of the roadway opened November 2 ; the removal of old paving completed December 10 ; the last beton laid December 12; the surfacing finished December 14; the paving finished December 18, and the road completely opened December 22.

The total duration of work for 14,000 sq. m. (18,340 square yards) of surface was 67 days, corresponding to an average progress of 210 sq. m. (252 square yards) per diem. Considerable time is necessary for the various successive operations, but only sufficient space should be maintained between them to prevent crowding and interference between the workmen. These operations in the Avenue des Champs Elysées have always required 15 days, and though this time might be somewhat reduced by decreasing the length of a section opened up, it is not safe to estimate for any shorter time ; 15 days therefore constitutes the period necessary to put the work in operation and complete one section, and this time should be deducted from the total time to find the true daily advance, which would become in the above instance 270 sq. m. (324 cubic yards).

The second part of the work commenced March 29, 1883, and required 62 days for a surface of about 13,000 sq. m. (17,030 square yards), giving the same mean daily rate of 210 sq. m. The actual advance was 275 sq. m. (330 square yards) per diem, and would have been much more if it had not been delayed by lack of materials. The conditions were much more favorable on account of the season, the absence of hindrances attending the commencement of work, and because the removal of the old pavement was done by the company, so that but for lack of materials an advance of 300 sq. m. (360 square yards) would have been made daily.

SPECIFICATIONS.*

The following is a translation of the specifications and the principal instructions and conditions embodied in the " Direction of work in the service of the public way," issued to contractors in 1886 by the Prefecture of the Department of the Seine for Roads and Bridges.

CHAPTER 1.—*Regarding Contracts.*—ART. 1. The plan comprises the transformation to a wood pavement of about 34,500 square meters, (378,165 square feet) of the existing road on the upper part of the Avenue Champs Elysées, and before the principal entrance of the Palace of Industry, the city reserving the right to construct the stone pavement under the carriage stations.

---

*xviii, 16.

ARTICLE 2. The contract will be an unrestricted one to which only joint stock companies having their headquarters in Paris will be admitted, and they must conduct all their operations, unless it be the purchase of new material, in French territory, and present satisfactory financial guarantees for the performance of their obligations.

At least one month before the award of the contract the competitors must file with the Administration the necessary securities, a copy of their charter, a detailed statement of the method proposed for the preparation of the paving blocks and treatment of the joints, and specimens of the paving blocks and asphalt intended for filling of the joints.

ARTICLE 3. During the continuation of his bonds the contractor cannot, without the express authorization of the Administration, in any wise modify the process described, nor use any materials not rigidly conformable to the specimens exhibited.

CHAPTER II.—*Foundation and Paving.*—ART. 4. The demolition of existing roads and removal of old material shall be executed by the contractors for the new work.

ARTICLES 5, 6 and 7 provide for the removal and storage of the classified materials from the old roads.

ARTICLE 8. The contractor must do all necessary *grading* and prepare the road-bed in prescribed form, and must immediately remove the excavated material to the public dumping grounds.

ARTICLE 9. If the removal of the total thickness of the old road makes a *filling* necessary to attain grade for the bottom of the new road, no other material can be used than the ballast of the old road, which must be free from earth, turf, etc., and carefully laid and arranged.

ARTICLE 10.—*Foundation.*—The road-bed having been prepared and accurately surfaced, the contractor must lay a bed of Portland cement beton, at least 0.15m. (5⅞-inch) thick, which will cover all the space between the borders of the sidewalks. The Administration reserves the privilege of requiring, without extra payment, a thickness of 0.20m. (8 inches) where the soil is bad or the sewers have been recently placed.

All the *materials* used in the composition of the beton must be of the standard quality and inspected by the engineer before using. The beton will be composed of at least 200 kilos (441 pounds) of Portland cement to one cubic metre (35.3 cu. feet) of a mixture two-thirds pebbles and one-third sand.

The contractor will proportion the materials under control of the Administration by means of standard boxes approved by the engineer.

The contractor may mix the materials by any convenient process, but the mixture must be first intimately effected in a perfectl dry state, and the beton must be perfectly homogenous, so that no pebbles can be distinguished in the mass that are not completely enveloped in mortar. Sufficient water must be used for the employment of the beton in a fluid state. The material will not be emptied directly on the earth, but on a portion of beton already in place, whence it can gently flow to the place it is to occupy, where it will be lightly rammed with a shovel.

The surface of the beton must be perfectly smoothed by a straight edge following the prescribed curve. An immediate addition may be made, if necessary, of a thin coat of mortar to give a final surface that is absolutely uniform, without hollows, cavities, or any depressions whatever.

ARTICLE 11.—*Paving.*—When the above-described foundation is sufficiently dry, the contractor must establish upon it a surface of wooden paving blocks conformable to the given samples and laid by the prescribed method.

The *paving blocks* shall be of first quality Swedish red fir, made by sawing planks perpendicular to their length so as to give a length of 0.22m. to 0.23m. (8⅝ inches to 9 inches), a width of 0.76m. to 0.78m. (3 inches to 3⅛ inches), and a height of 0.15m. (5⅞ inches). The blocks must be homogeneous and free from knots for a distance of 0.5m. (1⅞ inch) from the upper surface. Sound sap-wood only will be accepted, of varieties approved by the Administration. The angles must be sharp and the faces perfectly square. No deficiencies will be tolerated in this respect nor in the sawed height, which must be precisely 0.15m. (5⅞ inches).

The blocks must receive in the contractor's shops a preparation protecting them as much as possible from dampness.

The Administration reserves the privilege of exercising in the shops a special surveillance to insure the original quality of the blocks and the proper application of the different preparations required in the process proposed and agreed to by the contractor. These applications must all be made in the shops and never in the yards, where the blocks must arrive ready for use in the work.

The blocks must be set on end in regular rows normal to the axis of the road. The joints in adjacent rows must be staggered and blocks of 0.11m. (4¾ inches) may be used to attain this requirement. The blocks in each row must be set in close contact, but joints of from 0.008m. to 0.010m. (⁴⁄₁₆ inch to ⁷⁄₁₆ inch) must be left between consecutive rows.

At the crossways the rows shall be arranged as directed by the engineer so that no continuous joint shall occur in the direction habitually followed by the carriages.

The surface of the pavement must be perfectly regular, and present the exact longitudinal slope and transverse curvature required.

The joints must be dressed with a mortar containing at least one part of Portland cement to three parts of sand.

The hold of the blocks before the joints are filled must be assured, either by a wash of some material like coal-tar, filling the bottoms of the joints, or by some other process previously agreed to by the Administration.

Whatever system may be adopted, all the joints must be exactly filled and present no cavities.

A space of a maximum width of 0.035m. (1⅜ inches) will, however, be left along the edges to allow play for the swelling of the wood.

After the completion of the pavement the contractor can, if he thinks it expedient, spread, at his own expense, a bed of fine gravel over the surface ; this gravel will be swept off at the expense of the Administration as soon as its engineer deems it requisite for the interest of the traffic, and may be replaced by the contractor at his own expense under the same conditions and as many times as he thinks useful.

ARTICLE 12.—*Time allowed.*—The work of paving will not be undertaken until after the completion of all sewers and the modification of the subterranean canals of every nature. The work will be commenced at a date fixed by the engineer, and must be completed within a maximum period of one day for every 450 square meters (4.842 square feet), and fifteen days more for the putting in train of operations —*i. e.*, the time necessary for removing a portion of the road from the moment it is occupied for demolition until it is re-opened for traffic. This period is rigidly fixed, and will only be extended in case of continuous freezing weather ; for every additional day consumed in the work 100 francs will be retained from the contract price, unless the contractor can present sufficient excuse.

The work shall be simultaneously executed over only about one-half of the width of the roadway in such a manner as to preserve the other half free for traffic.

The portions of the road simultaneously withdrawn from traffic shall not exceed 300 m. (9,842 feet) for each station. The number of stations will be fixed by the Administration.

At crossings the work must be conducted so as not to interrupt the traffic on the transverse streets, which must not be barricaded except in case of absolute necessity, and with the assent of the Administration.

CHAPTER III.   *Maintenance and Delivery.*—ART. 13.   The contractor will be charged with the duty of maintaining the wood pavement for eighteen years from the 1st of April following the completion of the entire work.

This *maintenance* will consist in preserving the surface and regularity of the profile, and in making all general or partial repairs necessary to keep the road in a perfect state, even if the dilapidations are the result of accidental causes, as fires, sinking of the subsoil, etc., excepting only the digging of pits. The maintenance also comprises the removal of the old material, and the tools, plant, etc., used on the new work.

ARTICLE 14.   The contractor will be required to make *general repairs* on all portions of the road where there is : 1. A reduction of the curve diminishing the original pitch by a least one-quarter.   2. Where the thickness of the paving blocks has been worn away 0.07 m. (2⅜ inches) or more.   3. Depressions or partial defects of the road numerous enough to make it rough, the Administration being judge of the time when it shall be required for this reason.

All the requirements of the foregoing Articles, 10, 11 and 12, are applicable to the work of general repairs.

The beton foundation will generally be preserved by simply adding Portland cement—beton on top if there is room for it—and its removal will not be obligatory except in case of its bad condition.

ARTICLE 15.   Besides the general repairs the contractor must insure the constant good state of the pavement by *partial repairs* that may be necessary.   He must immediately replace paving blocks that are decayed, crushed, broken or depressed by any cause whatever, also those which have become impregnated with urine or other offensive liquids, and emit a bad odor.

He must repair holes whose depth reaches 0.02 m. (¾ inch) for a length of 1 m. (3.28 feet) in any direction.

At the junction lines of the wooden pavement with the stone or asphalt pavement the Administration may require the paving blocks to be replaced when they shall have been worn away 0.01 m. ($\frac{7}{18}$ inch).

In all partial repairs the new pavement must have the same level as the adjacent ; no projections will be tolerated.

If any of the defects enumerated in this article are not repaired within three days after notification, a charge of 2 francs per square meter (10.76 square feet) will be deducted from the contract price for each day's delay.

ARTICLE 16.   *Renewals for trenches* opened by the Administration, for any cause, must be executed in the same time and under the same restrictions as above, under penalty, if delayed, of 50c. per square meter per day, unless this amount should be less than 2 francs per trench per day.

Always, except in urgent cases specified by the engineer, the contractor shall have the privilege of constructing, in not more than fifteen days, at his own expense, the gutters with pebbles or broken stone carefully leveled and rammed.

The methods employed must permanently prevent the scattering of rolling stones over the adjacent surfaces. The renewed portions will immediately pass into the maintenance of the contractor, who must preserve them according to the conditions of Article 15. No claims will be allowed for repairs required by sinking of the earth.

The contractor will only be paid for the area of the trenches measured when filled up.

ARTICLE 17. *The old material and rubbish* from the general repairs must be entirely removed to the work-yards at the expiration of the time allowed for the completion of the work, in default of which the contractor will be subjected to a penalty of 3 francs per day for each deposit not removed.

For partial repairs and trenching the removal must be made the same day the work is executed, under the same penalty for delay.

Finally, the materials of the wooden pavement, including its foundation, displaced by digging trenches, must be removed the day the trench is opened, or when notified by the engineer if this occurs subsequently. In default of this the contractor is subject to the same penalty, and is entitled to no compensation for loss of materials.

ARTICLE 18. *The cost of sprinkling* the streets will be sustained by the city, which will also pay for necessary scouring.

ARTICLE 19. At the expiration of eighteen years as defined in Article 13, *the roads must be delivered in perfect condition.*

Three months before the expiration of this time the engineers will make a statement accompanying that of the contractor, who shall furnish at his own expense a certain number of soundings. The road shall not be received unless it satisfies the following requirements :

(1.) There must be no holes having a depth of o.o15m. $\frac{5}{8}$-inch for 1 m. (5.28 feet) in any direction.

(2.) The transverse curve of the surface must not at any point be reduced so that the pitch is less than four-fifths of its original value.

(3.) The thickness of the blocks must at no place be less than o.12m. (4$\frac{3}{4}$ inches).

After the engineer's inspection and report the contractor will be allowed three months to place the work in required condition.

CHAPTER IV.—ART. 20.—*Payments* must be promptly made by the contractor for the wood pavement to the sub-contractors for the demolition of the old roads and removal of their materials, and if this is not voluntarily done the amounts will be paid from the contractor's annuity by the Administration upon the engineer's certificate.

The demolition and removal of old roads, the preparation of road bed and establishment of the wood pavement will be remunerated only by the fixed annuity, which will also include the interest on, and payment for, the expense of the original outlay and the required maintenance.

The total amount of this annuity is fixed in the contract and is subject to no variation during the period of the contract; it will be paid to the contractor for eighteen years beginning, for each street, on the first of April next after its completion.

ARTICLE 21. The contract price fixed for the renewal of the pavement will be paid for the repairs of the trenches, the demolition of the pavement being at the expense of the company opening the trench. The contractor must, if necessary, relay the pavement with entirely new materials, and can make no claim for damages to the work or its maintenance.

ARTICLE 22. The Administration reserves the privilege to suppress, at any time, a part or the whole of the wood pavement, and will, in that case, pay the contractor the amount per square meter agreed upon in the relative clause of the contract.

ARTICLE 23 relates to the non-fulfillment of the contractor's obligations and the penalties incurred by delays, etc.

ARTICLE 24. *The annuities* will be payable at the first of each quarter, subject to not more than two months' delay. Of each quarterly payment a sum will be withheld amounting for the first six years to 0.10 cent., for the second six years to 0.15 cent., and for the last six years to 0.25 cent. per square meter. These amounts cannot be reduced, and will be deposited in the municipal treasury, where they will form a total sum of 12f. per square meter, constituting a guarantee fund for the maintenance of repairs, etc.

ARTICLE 25. The above sum will be repaid, without interest to the contractor at the expiration of the eighteen years, if he delivers up the roads in satisfactory condition ; but in default of this a part or the whole of the amount may be used by the Administration to execute the repairs.

ARTICLE 26. If the contractor abandons the work he will forfeit all claim to any annuity and the securities deposited, and will receive no payment for materials furnished.

CHAPTER V.—*Obligations of Contractor.*—ART. 27. The contractor's carts or wheelbarrows will not be permitted on the sidewalks, crosswalks or promenades. If a pitchy material is used in the bottom of the joints between paving blocks, it must be brought, ready for use, from the shops, or prepared in closed furnaces arranged to produce no smoke.

ARTICLE 28. The contractor must inclose and properly light his shops, work-yards and store-yards, and provide watchmen when necessary.

He must light them only with lamps or closed lanterns.

He must, irrespective of police requirements, have at least one lantern for every 10 meters (33 feet) of the length of the work-yards.

ARTICLE 29. The contractor will be responsible to the city and to property owners for damages from fires originating in, or propagated by, the wood pavements.

ARTICLE 30. The contractor will be responsible to the city for the consequences of any thrusts which may result from the expansion of the paving blocks, or the action of ice in the open spaces left at the sides of the roadway. Such repairs will be executed by the city, and their cost deducted from the contractor's deposit or the payments due him.

ARTICLE 31 requires the surrender of any patent rights.

ARTICLE 32. The contractor is required to live in Paris, and must not absent himself without authorization from the Administration and duly providing for all the operations of his work.

The contractor or his agent must report at the engineer's office at a fixed daily hour, to receive instructions and report the progress of the work.

The engineer will give him a bill of the work to be executed the next day, and of the expenses, for his approval.

ARTICLE 33. The contractor must employ none but French workmen.

ARTICLE 34. One per cent. will be retained from all payments, for the benefit of the national asylums of Vincennes and Vesinet.

Schedule of Prices, Subject to Decrease by the Adjudication.— No. 1.—Annuity, comprising the interest and payments of all cost of changing the existing roads to wood pavements with a beton foundation, and the contract maintenance of these roads for 18 years, 4.80f. per square meter (10.8 sq. ft.)

No. 2.—Annuity to be paid in case of suppression, by the administration, of the wood pavement before the expiration of the 18 years, 2.40f. per square meter of surface of pavement suppressed.

No. 3.—Renewal of wood pavement at trenches, comprising the removal of old material, but not including the demolition, 23f. per square meter.

SPECIFICATIONS FOR WOOD PAVEMENT ON THE STRAND, LONDON.*

There is given below a copy of the form of contract and specifications used in 1887, under which wood pavement was laid on the
Strand, London. The italics are ours. It will be noticed that the
concrete foundation is extra thick where pipes are laid. Arthur
Ventris was surveyor to the Strand Vestry in 1887 :

Specification of works required to be done in forming and constructing
a wood carriage-way pavement upon concrete foundations in.... ....
for the Board of Works for the Strand District under the direction of the
surveyor for the time being of the said Board of Works.

The works comprise the formation of paved carriage-ways in the following sections, viz.:

....

which is in all about......yards, as otherwise hereinafter defined, and subject to the conditions specified, and the maintenance of the same in thorough
repair for the period hereinafter mentioned.

Contractor Responsible. Levels.—The contractors to set out and keep
correct the works in every particular according to this specification, or by
the directions they may receive from time to time, and to be held responsible for the correctness of the work throughout the whole term of this contract. The levels referred to in this specification are the finished levels of
the surface or distances therefrom, and shall form at the channels a true
and even incline towards the gullies or channels of existing streets, showing
a sufficient and proper depth of kerb, the surface of the wood paving and of
the concrete being formed transversely to a curved surface, having a rise
from side to center of not less than 1 inch in 5 feet, except where otherwise
directed by the surveyor.

Plant and Material.—The contractors to supply every kind of plant,
fencing, lights, watching and material that may be necessary to execute
the works, including all horses, carts, water carts, and rolling, if required,
and to leave the whole at its completion, in a sound and perfect condition,
to the satisfaction of the Surveyor to the Board.

Interference with Traffic.—The contractors are to give all notices that
may be legally or fairly demanded, to the owners or occupiers of the property, or to the gas and water companies, or to any other person or persons
who may be fairly entitled to consideration, and as far as practicable, to
study the public as well as private convenience during the progress of the
work, and to provide intelligent and responsible foremen to superintend the
same, and generally to perform the work with as little interference with the
traffic as is compatible with a workmanlike and sound completion of the
works. The contractors are not in any way themselves to interfere with the
gas or water mains, without the sanction of the various companies, except
the said mains or house services are accidentally injured, in which case the
contractor or contractors are to make good such injury at their own expense,
to the satisfaction of such company or other party interested.

Excavation and Foundation.—The roadway from time to time will be
handed over to the contractors free from the present granite pitching, and
the contractors, except as hereinafter provided, are required to excavate to a

depth of fifteen inches from the finished surface of the roadway, and to cart away at once such excavation. To form a Portland cement concrete foundation, except where otherwise mentioned, nine inches in thickness, mixed in the proportion of seven parts of ballast, to one of Portland cement, and the ballast to be clean Thames ballast and sharp river sand, free from loam, mixed together to the satisfaction of the surveyor. The surface of the concrete to be properly floated to a true surface, and formed to the required contour of the street. Should the present foundation, however, during the progress of the work be found sufficiently sound to require under the circumstances a less depth of concrete foundation, the same shall be measured and not disturbed, and shall be considered a substitute for concrete to be put in, and the amount shall be deducted from the tendered price at the rate indicated in this contract ; the decision of the surveyor to be final.

Extra Work.—Where a trench has been opened by either the water or gas company, or where an extra thickness of concrete shall be required by the surveyor, the same shall, if he so direct, be provided, including excavation, and shall be charged as extra work at per six inches in thickness of concrete, and the concrete so put in, shall be one foot three inches in thickness, measured from the underside of the wood pavement.

The Paving.—*The wood blocks* 6″×3″×9″ are to be of the best Baltic red wood or other equally good timber, prepared in creosote, and shall be placed directly upon the concrete in rows across the street, *the joints to be grouted with cement grout or run in with hot bituminous mastic* (the kerb joint excepted); the cement grout to be composed of one part of Portland cement to three parts of sharp river sand free from loam, the whole to be covered with a top dressing of improved ballast.

The works are to be executed in the following manner : Section No. 1 shall be commenced from the......end, and the whole of the excavation shall be made and the foundation completed, within......days from the day following the closing of the street, and the wood blocks shall be laid complete and the section opened for traffic in six days further, making in all......days including Sundays, from the day following the closing of the street. The removal of the granite pitching will be undertaken by the Board or their agents, but nothing shall exempt the parties to this contract from the terms of this clause, except as hereinafter provided.

During the excavation and the forming of the foundation, *the work shall proceed night and day* subject to an increase on the sum tendered per yard of 10 per cent., but this work shall be required to be done by notice in writing under the hand of the surveyor, and should no such notice be given the work shall proceed with due speed, and the excavation and foundations shall be completed within......and the blocks shall be laid and the whole thrown open for the traffic in a further period of six days, making in all......from the day following the closing of the street.

Section No. 2 *shall be executed in two halves longitudinally* or other approved method ; access being given as far as possible to......and to ......and these streets shall not be blocked at one and the same time, *neither shall the roadway of......be entirely blocked,* unless with the written consent of the surveyor. The removal of the granite pitching will be undertaken by the Board or their agents from time to time, but the joining of the concrete foundation shall be made in a manner as directed by the surveyor at the expense of the contractors, and the work shall proceed with due speed and be completed within......from the date of commencement of Section No. 1.

Section 3 comprises the taking up of the wood pavement where directed, and of making up the present concrete to the required levels, and floating to a true and even surface, the surface having previously been picked over and thoroughly cleansed by washing, etc., the laying of wood blocks as hereinbefore specified run in or grouted, and top dressing thrown on, and open for the traffic in......from the date of commencement of this section, and the price to be paid per yard for this work shall be the sum tendered for this section.

Extension of Time.—That the surveyor, if by reason of the Metropolitan Board of Works failing to give the necessary permission to close the street, or he sees just cause, shall grant an extension of time by writing under his hand, such extension of time to be either prospective or retrospective, and to assign such other day or days for completion as may to him seem reasonable, without thereby prejudicing or in any way interfering with the validity of this contract.

Inferior Materials, etc.—Should any materials be brought upon the works, or should there be any of the workmanship which, in the judgment of the surveyor shall be of an inferior description and improper to be used in the works, the said material shall be removed, and the workmanship amended forthwith, or within such period as the surveyor may direct. In case the contractor shall neglect or refuse to comply with the foregoing conditions, it shall be lawful for the surveyor on behalf of the Board and by their agents, servants and workmen, to remove the material and workmanship so objected to or any part thereof, and to replace the same with such other materials and workmanship as shall be satisfactory to him, and the Board, on the certificate of the surveyor, to deduct the expense thereby incurred or to which the Board may thereby be put or liable, or which may be incident thereto from the amount of any money which may be or may become due, or owing to the contractors, or to recover the same by action at law or otherwise from the contractor, as the Board may determine ; and in case the contractors shall fail to carry on the works with due diligence and as much expedition as the Board or their surveyor shall require, or neglect to provide proper and sufficient materials, or to employ a sufficient number of workmen to execute the works, the Board or their officers shall have full power, without vitiating this contract, and they are hereby authorized to take the works wholly or in part out of the hands of the contractor, and to engage or employ any other person or workmen, and procure all requisite materials and implements for the due execution and completion of the said works ; and the costs and charges incurred by them in so doing shall be ascertained by the surveyor, and paid for or allowed to the Board by the contractors, and it shall be competent to the Board to deduct the amount of such costs and charges out of any moneys due or to become due from them to the contractors under this or any other contract.

Fines.—That time shall especially be considered as the essence of this contract on the part of the contractors : and in case the contractors shall fail in the due performance of the works to be executed under this contract by and at the times herein mentioned or referred to, or other day or days to which the period of completion may have been extended, they shall be liable to forfeit to the Board the sum of £50 for each and every day which may elapse between the appointed and actual time of completion hereinbefore mentioned, or the Board may deduct the same from any moneys in their hands due or to become due to the contractors, and such payments or deductions shall not in any degree release the contractors from further obligations and forfeitures in respect to the fulfilment of the entire contract.

Tenders.—The contractors shall state in their tender the price per yard for wood paving laid on concrete, including excavation, in every respect according to this specification. Where, however, the surveyor may deem it necessary to order the contractors to execute extra work in connection with these works hereinbefore described, such extra work shall be paid for by the Board at prices to be fixed by the surveyor, and to be accepted by the contractors.

Maintenance.—The contractors shall keep and maintain the works in sound condition to the satisfaction of the surveyor, for two years next after the completion of this contract, free of cost to the Board, and shall, if required so to do in writing under the seal of the Board, keep and maintain the aforesaid works in whole or in part in sound condition to the satisfaction of the surveyor for a term of years (to be stated in the tender) for the tendered sum per square yard per annum, first payable on completion of the third year from the date of completion of the works.

Payments.—The contractors shall be paid one-third part of the contract on completion of one-half of the works, and one further third upon completion of the works, and one further third six months after such completion, but only upon the certificate of the surveyor of the satisfactory completion or condition of the said works according to the specification.

SANITARY ASPECTS OF WOOD PAVEMENTS.[*]

This is a report of a Board appointed to inquire into the alleged deleterious effects of wood paving upon the public health. The subject of pavements of different kinds in relation to cleanliness, freedom from noise, and surety of foot-hold is first discussed and the conclusions of the Board summed up in the table on page 106.

From this comparative statement it appears that no one pavement has all the qualities which are necessary to make a perfect pavement. Monolithic asphalt comes nearest to the model. It is the only material which can be kept perfectly clean, and it costs less for cleansing than any other. In point of hardness, smoothness, and noiselessness it is unmatched. But it is adapted only to low grades—grades so low that it is practically useless in this city except for sidewalks, and this failure depends upon the qualities of smoothness and hardness which, important as they are in other respects, render the foot-hold comparatively insecure. In a similar way each other kind of material may be criticised. Some objections, varying in character with the kind of road, appear in all; and these are only in part overcome by varying the material with the kind of country over which the road is taken, and with the kind of traffic for which it will be chiefly used. Whatever material is chosen from among those which, up to this date, are available, the pavement made with it must be a compromise. There is, at present, no pavement which can be called "best;" only some kinds are less objectionable than others.

---

[*]Legislative Assembly, New South Wales, Wood-Pavement Board Report, Sydney, 1884. xi, 331.

The Board examined some specimens of wood pavement as laid in the city of Sydney, taking up blocks at different points. In all cases the concrete bed underneath was moist ; in three cases a large amount of slimy mud was found, giving off an ammonical odor. In all these the joints and blocks appeared to be uninjured. The blocks were chemically examined to determine whether they had absorbed organic matter, with the result that some were found impregnated with filth to the very centre, while others were comparatively free from it.

A MODEL ROAD-PAVEMENT.	GRANITE BLOCKS.	SHEET ASPHALT.	WOOD BLOCKS.	MACADAM.
Should be impervious.	Joints pervious, equal to one-third of surface and retentive.	Impervious.	Absorbent and retentive.	Pervious.
Should afford good foot-hold.	Fairly good.	Good.	Variable.	Good.
Should be hard.	Too hard, injuring horses' feet and vehicles.	Hard and elastic	Not sufficiently hard.	Soft.
Should offer little resistance to traction.	Considerable.	Little.	Variable with age, etc.	Medium.
Should be noiseless.	The most noisy.	The least noise.	Less than granite; more than asphalt.	Less than granite; more than asphalt.
Should be cheap.	?	?	?	?
Should yield no detritus.	Less than macadam; more than wood or asphalt.	Yields less detritus.	Least, except as asphalt.	Yields most detritus.
The surface should be easily cleansed.	Less easily than asphalt.	Most easily cleaned.	Not easily cleaned	Not easily cleaned.
Should suit all traffic.	Practically suits all.	Suits all traffic.	Does not suit very heavy traffic.	Not heavy. traffic.
Should be adapted to every grade.	Not to very steep grades.	Grades must not be more than 1-50.	Not to very steep grades.	All grades.

The Board comes to the conclusion that wood is a material which cannot be safely used for paving unless it can be rendered absolutely impermeable to moisture, and so laid that while the entrance of water between the blocks is rendered impossible the separation of the fibres at the surface by the concussion of traffic is also effectually prevented. These conditions have nowhere, to the knowledge of your Board, been fulfilled.

So far as the careful researches of your Board go the porous, absorbent, and destructible nature of wood must, in its opinion, be declared to be irremediable by any process at present known ; nor were any such process discovered, would it be effectual unless it were supplemented by another, which should prevent fraying of the fibre. Still less can the defects of wood be considered to be of less consequence than the defects of other kinds of material.

In this city it may, perhaps, be considered that an amount of wood has not yet been laid sufficient to affect the public health, whatever its condition within reasonable limits may be ; and upon this ground your Board does not recommend that the present paving should be removed, but that the Board of Health should be empowered to examine it, and to report upon it, from time to time, with a view of ascertaining its behavior under longer exposure to weather and traffic than it has yet had, and that it should be no longer watered, but cleansed by sweeping at least twice a day (the sweeping to be done at right angles to the direction of the street or parallel to the courses, so that the latter may be cleared out by the broom) in order that destructive dampness and penetration of dissolved organic matter may be reduced as much as possible. But the presumption is upon the evidence here adduced, that in this climate the results alluded to would ensue if the extent of surface were sufficiently enlarged or fouling and decay sufficiently extensive. Your Board, therefore, recommends that the paving of the streets of this city with wood should be discontinued, and desires to add that this recommendation is intended to apply not to the particular mode of construction here adopted alone, but to the material itself, and to every known method of construction.

The appendix to the report contains some interesting data with regard to wood pavements, and the volume is one which municipal engineers will find interesting and valuable as a book of reference.

NECESSITY OF MAINTAINING A WOOD PAVEMENT.[*]

In the summer of 1883 a wood-block pavement was laid on Fifth Avenue, between Thirty-second and Thirty-third Streets. A good concrete foundation was carefully laid to the proper grade and transverse profiles ; on this blocks of pine which had been dipped in dead oil—not thoroughly creosoted—were laid in rows about one-eighth of an inch apart. Enough bitumen was poured into the interstices to hold the blocks in place, possibly a quarter of an inch in depth; the interstices were then filled with a Portland cement concrete, the whole covered with a top dressing of hard, clean gravel about the size of peas, and when the concrete had had time to set traffic was turned on to the pavement.

This pavement has stood since then, it is believed, without any other attention or repairs than sweeping. It has not even received a coating of gravel, which in European practice it would have received, on the theory that the gravel crushing in between the end fibres of the wood prevents wear to some extent.

-----

[*] Ed. xviii, 49.

All this time it has been almost entirely free from noise. It has been cleaner than any stone pavement in the city, and has been the pleasantest and safest piece of pavement in the city to drive over.

Now through inefficient creosoting or inherent weakness in some of the blocks they have commenced to fail, and between one and two per cent. of the pavement requires renewal. The practice in New York has been, with all wood pavements, in such a case to sit complacently and see the depression from a defective block, originally four inches by six, extend over a constantly increasing area, and these areas increase in number until the pavement was irretrievably ruined, when the wood pavements were condemned and the locality took to itself a stone pavement many times noisier and dirtier than before.   •

It is to be hoped that with the able engineer at the head of our Department of Public Works the old record will not be repeated, but that advantage will be taken of the small traffic on the avenue for the next two months to shut off half of the roadway at a time and replace all defective blocks, allowing the concrete full time to set; and we hope it is not too much to urge that more attention be given to that piece of pavement than it has received in the past.

WOOD PAVEMENTS IN LONDON AND NEW YORK.*

Construction and Maintenance.—The London *Financial Times* states that in 1883 the Strand was paved with wood, from Charing Cross to the Strand Post-Office, by the Vestry authorities, and from the Post-Office eastward to St. Mary's Church the work was done at the same time by the Improved Wood Paving Company. "As every Londoner knows, to his cost, the Strand was 'up' about a month ago, while the paving put down by the Vestry in 1883 was renewed. That laid down by the Improved Wood Paving Company is still good, and has not had to be touched during the six years of precisely the same wear and tear which has torn the other half of the Strand into holes and splinters." A part of the Strand, though it is not now possible to say with certainty which part, required renewal last spring, as the blocks were worn in places to a thickness of less than three inches.

The point which is of interest to Americans, and to New Yorkers in particular, is that two different samples of wood pavement lasted six years in the Strand, which has fully three times the traffic of Fifth Avenue, while on that avenue a wood pavement was not made to last five years before a barbarous return to stone was made.

The difference in the wearing of wood pavements in the two cities lies entirely in the fact that in London a pavement of any kind

---

* Ed. xx, 329.

is *maintained.* But the City of New York, for the past twenty years, has not been able to get authorities who understood the necessity of maintenance. The pavements have been patched, and they have been renewed, but not maintained. *The Engineering and Building Record* from time to time called attention to the necessity of re-placing the few blocks in the wood pavement formerly on Fifth Avenue which had proved defective, but we think nothing was done until the places started by single defective blocks were so far extended that the pavement was condemned and replaced by stone blocks.

Profits in Paving.—Moreover, the companies that have put down and maintained—for the maintaining part of the contract is gener-ally well enforced—London pavements for the local authorities seem to have themselves been very prosperous. The *Financial Times* says that the Improved Wood Paving Company paid 5 per cent. for four years ending with 1882, 8 per cent. in 1883, and has paid 10 per cent. since. The Val de Travers Asphaltic Company pays not less than 8 per cent., and the Compagnie Générale des Asphaltes de France, both doing business in London, yields a steady return to its shareholders of £1 on each £6 share ! The Limmer Company, also laying asphalt, met with heavy losses at the start in consequence of poor material, but is now paying satisfactory dividends. The necessity for having responsible, and therefore prosperous, com-panies to deal with is obvious, and the statements of the London contemporary certainly make a strong argument in favor of paving contracts with the maintenance feature, from all points of view.

RAWLINSON ON WOOD PAVEMENTS.*

Mr. Robert Rawlinson, writing to *The Builder*, on wooden pavements, which are now being laid extensively in London, says :

The character of wooden pavements may be known by the occupants of carriages in driving over them. Where the blocks are laid upon sand and boards, the vibrating and drumming effects are most distressing. When laid with open joints the surface becomes rapidly worn and uneven ; and when laid upon an imperfectly-formed or weak foundation, the surface also becomes uneven, alternately hills and holes, retaining dirt and wet, and so tending more and more to the destruction of material and the road. Blocks of wood, unexceptional in character, form, and dimensions of mate-rial, laid hard on an exceptionally good cement concrete foundation, close jointed, but without the felt bedding and jointing have a disagreeably jarring effect, though in a less degree than the examples previously described.

Some of the wood-paving companies must have been very stupid, and also very difficult to teach, or they would have learnt by their failures, sooner than they appear to have done, and we should not then have seen

---

*vi, 67.

most important main thoroughfares blocked for weeks at a time by the pulling up of the entire wood construction to begin again as from the beginning, and this process more than once over.

*   *   *   *   *   *   *   *   *

Cement foundations for wood-block paving ought to have ten or twelve days to "set," but in London this time can seldom, if ever, be given, and this may, in some cases, account for partial failures. Wooden pavements, in themselves, make no mud, but this is carried on from adjoining dirty macadam and "set" paving. When, therefore, wooden pavements are muddy, it shows neglect in street-scavenging.

PAVEMENTS OF COMPRESSED WOOD.[*]

The increased favor with which wood-pavements have of late come to be regarded in European cities, has led to the devising of numerous processes that have for their object the preparation of wood for use in pavements.

In a process of this kind, on the one hand, of course, strength and durability of the product must be aimed at ; on the other, it is necessary that the working expenses be as low as possible.

In this connection we note an article in the *Semaine des Constructeurs* on the process of Cyprien Mallet, which is said to fulfill these requirements.

The salts of the metals and various other substances having the property of preserving wood from decay, Mallet injects, hot, into the core of the pine-trees, an antiseptic fluid, and then compresses the logs about 1-10 their volume.

The antiseptic liquid is composed of : copper sulphate, 6 kilos ; juice sulphate, 6 kilos ; sodium chloride, 3 kilos, which are dissolved by boiling in 35 litres of water containing no lime salts.

Without interrupting the boiling, there is added then : resin oil, 40 kilos ; heavy oil, 40 kilos ; suet, 10 kilos. This solution is concentrated to a certain degree over a moderate fire. Ten litres of this liquid mixed with 90 litres of boiling water is the fluid used. The wood charged with this preparation becomes very hard, yet it retains sufficient elasticity ; it is not affected by the weather.

In Paris the cost of a pavement of this kind comes to about four dollars per square meter.

CEDAR BLOCKS VERSUS ASPHALT FOR STREET-PAVEMENTS.[†]

CEDAR RAPIDS, IOWA, March 16, 1887.

SIR: Do you think cedar blocks for street pavements as healthful as asphalt or other non-absorbent material? For like reasons, are the blocks when placed on boards and not treated with creosote as good as when treated and put on concrete ?   Respectfully yours,

JOSSELYN & TAYLOR.

---

*ix, 503. †xv, 431.

[Answering the questions of our correspondents in order, we would say:

First—That on general principles, that pavement from which water passes away with the least percolation to the soil below and which can be most thoroughly and easily cleansed is the most conducive to health.

Second—*No* pavement placed on boards or directly on the earth is as durable as when placed on concrete; this is the universal verdict of all municipal engineers who have had opportunity to observe the matter, and the latest practice in France and England, where wood has been used, is to make a carefully formed and substantial concrete foundation for the blocks.

Third—Creosoting undoubtedly adds to the life of the blocks, by making them wear out rather than decay irregularly. This question is complicated by the secondary ones of amount of traffic, kind of timber, care in selection, proper seasoning, etc.

Abroad a most rigid inspection is made, the blocks are thoroughly seasoned and thoroughly creosoted.

Fourth—The selection of a pavement should be guided among other things by the question of amount of traffic. On some streets in our cities nothing will stand the wear but the best granite blocks. Original cost oftentimes becomes also the deciding motive. We think it can scarcely be questioned that asphalt is superior to wooden blocks, provided other considerations do not essentially alter the conditions.]

CEDAR BLOCK PAVEMENT IN ST. PAUL.*

A recent number of the *Pioneer Press*, of St. Paul, Minn., contains the following.

Durability.—The cedar blocks taken up on East Seventh Street during the construction of the cable conduit show an average wear of less than 2 inches. This is very noticeably light, because that pavement was the first cedar block pavement put down in St. Paul. This was in 1882, seven years ago. The pavement is good for ten years of service yet, with occasional repairs, and that is a first-class record on a street where a heavy traffic is carried on. The blocks are in a perfect state of preservation. The planks beneath them are 1-inch planks, but show some effects of seven years' pounding. Two-inch planks are used now.

In regard to the above item L. W. Rundlett, City Engineer of St. Paul, says:

The statement of the *Pioneer Press* is not strictly correct; the paving was laid in 1882 and was among the first cedar block pavements put down in the city. The roadway is 40 feet wide, with street car tracks occupying a width of about 15 feet in the centre. It is a retail street and one of the principal thoroughfares of the city, although not subject to as heavy traffic as some other streets. On account of the street car tracks being in the centre the line of travel was thrown to a considerable extent on each side of these tracks, which made the wear somewhat unequal; the wear of the blocks, more particularly in the line of travel, being from 2 to 2½ inches; the blocks originally being 6 inches in depth. The blocks themselves show very little decay, some of the foundation boards, which were only an inch

---

* xx, 86.

thick (our present construction is 2-inch plank) were somewhat decayed, and in repairs probably 25 per cent. were renewed.

The Water Board have put down a larger main on the street this year; the gas company have put down a new gas main, and the cable line has been substituted for the horse line; a subway for putting wires underground is also in process of construction. This work, with the necessary changes in the house connections, has caused the street to be pretty well torn to pieces; we have, however, used considerable care in replacing the blocks, and I think it will probably be about three years before the street will require to be repaved.

WOOD PAVEMENT IN DULUTH. *

† Cedar blocks were used for paving and were laid upon a permanent concrete foundation, so that when the traffic on the street became heavier, stone blocks could be substituted in place of the cedar without relaying the foundation.

The method of laying the pavement was as follows: After the street was brought to the required sub-grade it was repeatedly rolled with the steam road-roller. A granite curb was then laid on either side, under which was a tile drain, to prevent water getting under the concrete. The concrete of the foundation was then laid six inches deep over the entire roadway, and was composed of five parts natural cement, twelve parts coarse sand and twenty-four parts broken stone, well mixed together and quickly laid in place. The surface of the concrete was then made smooth by a thin layer of mortar, consisting of one part cement and two parts sand, brought to the right form by means of templates. This was allowed to harden for about five days, when, outside the street car tracks, round cedar blocks were laid directly on the concrete and covered with tar and gravel in the usual manner. The street car company paved the space between the rails of their track with granite blocks set in sand.

Where the grade is 10 feet in 100, cedar block pavement would have been too slippery in wet or frosty weather, so rectangular pine blocks four inches wide were laid on end in parallel rows, three-quarters of an inch apart, on a 2-inch plank foundation, the space between being filled with gravel and tar.

A twenty-ton Aveling & Porter *steam road-roller* was purchased in 1887. During the first months it was constantly in mud holes, owing to the fact that there were but few street foundations in the city strong enough to hold it up. All contractors were required to use it, however, and the result has been far more faithful work on

---

* Second Annual Report of the Board of Public Works of Duluth, Minn., for the year ending February 28, 1889. William B. Fuller, City Engineer.

† xx, 53.

their part than could have been secured in any other manner. In the preparation of street foundations and for rolling macadam paving this roller has been invaluable.

CEDAR-BLOCK PAVEMENT AT LEAVENWORTH, KANSAS.*

Noticing in an exchange a discussion regarding proposed new pavements for Leavenworth, the statement that a *cedar-block pavement* on a certain street had proved *a failure*, we wrote to City Engineer George T. Nelles, asking for facts and details as to the manner in which the pavement was laid. The following is from his reply :

The pavements in question were put down on two of our principal business streets and on one resident street, and consisted of 7-inch sound white-cedar blocks 4 to 8 inches in diameter laid on a 6-inch concrete foundation. The work was all done in accordance with the specifications in general use for this class of work. The work was done during the summer and fall of 1887, so that the reliable test of use cannot be brought into play. So far as I am able to judge, from my knowledge of the manner in which the work was done, and from present indications, the work will prove as permanent and lasting as any pavement of this class. The cedar blocks were of unexceptionally good quality, live and sound, and free from the usual imperfections of this class of material. The assertion made in the clipping herewith to the contrary is absolutely without foundation. Owing to the inferior quality of the sand and broken stone available here for carrying on works of any magnitude, the concrete foundation may not be up to the standard. Still, I confidently believe it is ample for the purpose, and that it will outwear a great many sets of cedar blocks. I have recently had occasion to examine some of the concrete laid during freezing weather last fall, and found that, although not perfectly set, it has not been in the least disturbed by the frost or the traffic over it. I have not arrived at the true inwardness of the attack on cedar pavements, but think that the idea that the pavements laid last year are failures arose from the fact that several very ugly settlements have taken place where the pavement was laid over heavy fills, which, until repaired, greatly injures the appearance of the work ; and from the fact that the blocks raise along the street-car line, which was put down without cross-ties by laying the stringers directly on the concrete, and being dependent entirely on the blocks to hold it down, and, in consequence, has raised the blocks in places. There is nothing, so far as I can see, connected with this work that can be cited as an experience with cedar block pavements.

COST AND DURABILITY.†

A Report on Pavements, made to the Common Council of Topeka, Kan., says : Wood pavements of the old varieties are conceded to have proved failures. The white cedar blocks (more properly cypress) sawed from small trees are being laid in Chicago, Kansas City, Omaha, upon concrete foundations and also on boards.

---

* xvii, 323. † xv, 375.

The life of the latter they consider to be about three years. That on concrete would cost in Topeka about the same as granite block, or sheet asphalt, and if the cypress blocks were creosoted, as are the pine blocks now used in England and France, the cost would be still greater.

PLANING DOWN A WOOD PAVEMENT.*

Some recent experiments in Manchester, dealing with the question of repairing roadways laid with wood blocks, are of considerable interest to vestries and public boards. Hitherto, owing to the unequal wear of the blocks in any one road, resulting from the varying density of the wood, it has frequently been necessary to pull up and relay a road with new blocks, discarding all the old ones, of which, possibly, a great number may have had good "life" in them still. To obviate this necessity Mr. A. C. Bicknell, of the Sandycroft Foundry Company, Chester, has invented a machine of traction-engine type, self-propelling, having in front a revolving table fitted with cutters. A section of a road was laid with old blocks, which had been discarded as unfit for further service. The method of laying was the usual one, in cement and sand, and being composed of old blocks pitted with sand and gravel the experimental road presented the normal features of an old road requiring repair. When several days had elapsed, to allow the cement to set, the machine was brought to work, the cutting-head being gauged to cut one-half inch below the surface of the lowest face of the roadway. In the course of the work it was necessary to take off three inches from some of the blocks, and the result throughout was satisfactory. The system of planing down a roadway is certainly novel.

.

* xiv, 224.

CHAPTER III.

ASPHALT PAVEMENTS.

THE NATURE AND USES OF ASPHALT.*

ASPHALT is a variety of bitumen, found in a native condition and not manufactured, and in a solid form is commercially known as glance pitch. Glance pitch is found in limited quantities in various parts of the Rocky Mountains and in Texas. It is very pure and is used to make a high grade of varnish, but its brittleness makes it useless for paving or roofing compounds.

Occurrence.—The asphalt of Trinidad is found in a so-called "lake" about 130 feet above the sea-level, on the island of that name. The "lake" is a level tract, about 114 acres in area, of brownish material of an earthy appearance. It is sufficiently hard to bear the weight of carts and animals, and yet its consistency is such that excavations fifteen feet in depth are filled up by the flow of adjacent material in a few months. It is estimated that the amount of asphalt in the lake is upwards of six million tons. On partial analysis it yields approximately 40 per cent. of pure bitumen, 40 per cent. of earthy and vegetable matter, and 20 per cent. of water. The material is heated in large tanks at a temperature of about 300° Fahr., to drive off the water and let the larger portions of the earthy matter settle and the vegetable matter to be skimmed off the surface. This refined asphalt contains about 60 per cent. of pure bitumen and 40 per cent. of finely divided earthy matter invisible to the eye. This material is too brittle for commercial use, and it is therefore mixed with a heavy, dark oil, known as the residuum of petroleum, in the proportion of six parts of asphalt to one of residuum. This is the material so largely used in paving and roofing compositions.

On the coast of California, near Santa Barbara, and also in certain portions of Colorado, Utah and New Mexico, are found large beds of sandstone containing from 15 per cent. to 20 per cent. of bitumen. Recently this material has been used for paving in cities on the Pacific coast, but it has not yet been in use long enough to prove its desirability.

* A paper read by Captain F. V. Greene, Vice-President of the Barber Asphalt Paving Company, before the Society of Arts at the Massachusetts Institute of Technology, and published in the Boston *Transcript.*—xix, 258.

In the valley of the Rhone in France and Switzerland, in Sicily, in parts of Italy, in Spain, and in Hanover, there are very large beds of a fine amorphous limestone naturally impregnated with bitumen, and it is from these mines that the asphalt pavements of various cities in Europe have been obtained. Those most suitable for paving contain about 10 per cent. of bitumen and 90 per cent. of fine limestone.

Uses.—The uses of asphalt may be divided into five classes— viz.: 1st, as a varnish for paint; 2d, as an insulating material; 3d, as a water-proofing material; 4th, as a cement in ordinary construction; 5th, as a cement in roofing and paving compounds.

In its natural state it is too brittle for any of these purposes. *For varnish* it is mixed with oil of turpentine, linseed oil and shellac. Such varnishes are used on leather, producing the so-called patent leather, and on iron, the black japan varnish being well known.

For insulating compounds the exact mixture is not divulged. For water-proofing arches and similar construction it is sometimes used in the form of a layer of mastic, made from bituminous limestone spread over the arch, and sometimes as a cement between the joints of the bricks, the bricks being heated and dipped in hot asphalt, and the joints poured with similar material after the bricks are laid.

As a roofing material, asphalt is used in the form of asphalt cement, very similar to paving cement. The roof is covered with one or more layers of felt; on this a layer of the cement is poured, and before it has cooled fine gravel or pebbles are spread over it.

Asphalt Pavements.—The amount of asphalts used for paving is about 95 per cent. of the total consumption.

I shall endeavor to show that the asphalt pavement is the latest and, all things considered, the most satisfactory solution of the paving problem yet devised. It is not as durable as cast-steel, nor as noiseless as velvet, nor does it afford as firm a foot-hold as the loose earth of a race-track. But it is much smoother and less noisy than stone, much more durable than wood or macadam, is water-proof, contains no decaying vegetable matter, can be kept perfectly clean at comparatively small expense, is less slippery under ordinary conditions (as shown by careful observations in Europe and America) than either wood or stone, and it enables larger loads to be drawn by the same force and with less wear on vehicles than any other form of pavement ever used. It has thus many advantages and fewer defects than other pavements in common use.

The speaker next gave a brief history of pavements in general, so as to trace the origin and development of asphalt pavements.

Macadam.—He said that no improvement has ever been devised upon MacAdam's system as a road covering outside of cities. It has also been widely introduced within cities; London has about 600 miles and Paris 100 miles of it to-day. In America it has always been popular in New England cities—three-fourths of the streets of Boston are paved with it—but it has not found favor in other cities except on streets reserved for pleasure driving only.

Its advantages are a firm foot-hold for horses, and a reasonably smooth surface. Its defects are heavy resistance to traffic, great cost of maintenance, and the impossibility of keeping it clean. When sprinkled it is always muddy, and when dry it is invariably dusty. These defects are inherent and cannot be remedied. The Paris Budget shows that it costs over $900,000 per annum in that city for the single item of repairs to macadam pavements. This is equivalent to 45 cents per yard per year. It is estimated that the annual cost of repairing the macadam pavements of Beacon Street, Boston, is 50 cents per yard.

Stone Pavements.—Stone pavements were next taken up, and figures presented showing that the cost of repairs was from five to twelve cents per yard per year, while the annual cost for laying and maintaining the best quality of granite-block pavements on concrete foundations is twenty-six cents per yard.

Wood Pavements.—Wooden pavements were shown to cost annually for laying and maintaining about sixty-one cents per yard, but it is now claimed by its advocates in London to be forty-two cents per yard under moderately heavy traffic, or about two-thirds the cost of macadam under the same conditions. Captain Greene gave the experience of Washington as a warning against its use. Upwards of $4,000,000, derived from a loan not yet paid off, were expended in that city under the Shepherd government of 1871-74 in laying fifty miles of wood pavements. They proved a complete failure in a few years, and have all been since replaced with asphalt at a little more than half the original cost of the wood.

Rock Asphalt Pavement.—Bituminous limestone or rock asphalt began to be used for paving in Paris in 1854, in London in 1869, and in Berlin about 1880; and its use has been continued with success. The total area of this pavement in use now in Europe is about 1,500,000 square yards, covering a length of about ninety miles of roadway.

A uniform system has been used in laying in all the cities. A solid bed of concrete is used to give a foundation. On this the asphalt surface is laid about 2½ inches thick. The preparation of this surface requires great care and skill. The asphalt rock, as it is quarried from the mines, is crushed in a rock breaker to a size of about three inches. These are then passed through toothed rollers and again through smooth rollers until the rock is reduced to powder. This is then heated in revolving cylinders to a temperature of about 280° Fahr., and the heated powder is carried in carts to the street where it is to be used. There it is spread on the concrete and raked with hot iron rakes until it forms a uniform layer of loose powder, about twice its ultimate thickness. This is then quickly compressed by pounding with hot iron rammers, after which a small amount of hydraulic cement is swept over the surface and the pounding is continued until the pavement will no longer yield under the rammer. It is left until the next morning to cool, when the street is opened to traffic.

Cost and Maintenance.—In London the first cost has been about $3.75 per yard, and the maintenance eighteen cents a yard per year, the street to be delivered at the end of seventeen years as good as new. Including first cost, the total expense is forty cents per yard per year.

Tar Pavement.—The success of rock asphalt pavement in Europe gave rise to a demand for such pavements in America, but the expense of transportation of the rock from France was so great that inventors sought to find a substitute. They first tried the tar produced in large quantities at the gas works, which they erroneously supposed to possess the same qualities as the natural bitumen in the asphalt rock. This was combined with sand, limestone, sulphur, sawdust, etc. The material did not look unlike the real asphalt, and a craze for such pavement started in Washington in 1871 and spread all over the country.

The majority of these efforts were complete and costly failures, and as they all claimed to be asphalt, the result was to create a prejudice against all pavements of that character, which it required years of careful experiment and proof to overcome. The defect of them all lay in the tar, which contained volatile matters which evaporated under the influence of the sun, and left the pavement a mass of dry black powder.

Trinidad Asphalt Pavement.—A Belgian chemist conceived the idea of using the asphalt of Trinidad as the cementing material, knowing that it had been exposed for centuries to a tropical sun, and that the sun's rays could have no further effect on it. With this he combined clean sharp sand and a small amount of powdered limestone. The sand in it afforded a firmer foot-hold for horses. It was used on a part of Pennsylvania Avenue, in Washington, in 1876, the asphalt-rock pavement of Paris was used on the other part, and they have been in constant use ever since. The French pavement proved more slippery and more costly than the Trinidad, and no more of it was laid; but the Trinidad asphalt gave entire satisfaction, and has been constantly laid with succeeding years, until now its area in Washington alone is but little short of one million yards. After seven years' successful use in Washington, other cities began to use it, Buffalo being the first, and it now rivals Washington in extent of its use. It is now used in thirty-four cities, the total area being about four million yards.

These pavements are laid on a solid *foundation* of concrete six inches in thickness. The asphalt surface is 2½ inches thick, laid in a similar manner to those in Paris. But the asphalt is prepared quite differently. The refined asphalt is mixed with the residuum of petroleum to make the cement. This is heated at 300° Fahr., and the sand is also heated to the same degree; these are mixed in a large box in which agitators are constantly revolving, and there a complete mechanical mixture is formed. The hot powder is then taken to the street, spread the proper thickness, and immediately compressed by large steam rollers, weighing from five to ten tons.

The cost of maintenance for the Washington asphalt pavements is about two cents per yard per year. On many of them, now ten years old, no repairs of any kind have been made, and they are still in perfect order. These streets are subject to very light traffic, and are kept clean. In other cities where the pavements are not kept clean, and the traffic is heavier—in some cases the most destructive traffic in American cities—the expense is much larger. But under ordinary conditions the cost of maintenance does not exceed ten cents per yard for a long term of years. Including first cost, the total expense of maintenance for seventeen years would be about thirty cents per yard per year. From this it appears that asphalt is a little more expensive than stone, but it is the cheapest of the smooth pavements.

Asphalt Block Pavement.—Asphalt blocks have been tried, but they are deficient in durability under heavy traffic, but they have been very satisfactory on residence streets of light traffic. These can be laid as ordinary paving stones, thus doing away with an expensive plant in every city.

Traction and Cleanliness.—On asphalt pavements the same force will draw a load three times as heavy as on the ordinary stone pavement. The former can be kept perfectly clean at small expense; the latter has one-fifth of its service composed of joints filled with stable filth, which cannot be removed in cleaning.

If some one gives voice to the current belief that horses are constantly falling on the asphalt, I will show him the result of careful observations in ten different cities on 736,000 horses, of which 84 fell on stone pavements, and only 71 on asphalt. The proprietors

of the livery stables of Washington and Buffalo say that they invariably use the asphalt in preference to the stone pavement, and that
there is far less injury to horses, as well as to their vehicles, on the
asphalt.

ASPHALT PAVEMENTS IN THE UNITED STATES.*

At a recent meeting of the Board of Improvements of Cleveland,
O., City Engineer Walter P. Rice made the following report of his
investigations of asphalt pavements:

Material.—The best material for pavement purposes seems to be the
Trinidad asphalt, which is controlled by the following companies: Barber
Asphalt Paving Company, Warren-Scharf Asphalt Paving Company, and
National Vulcanite Company. The latter company is able to command
asphalt only for laying pavement other than the Barber, while the two
former lay pavement under the same specifications, but in separate territory. There can therefore be no competition and but one bid can be
received under the same specification, and to all practical intents cities or
individuals using the street pavements are subject to such dictation as may
be imposed as to cost of all repairs and resurfacing.

Washington and Buffalo Pavements.— I examined over twenty
streets in Washington, D. C., paved with asphalt, vulcanite, and coal-tar
distillate. The age of these pavements ranges from five to fourteen years.
In general they are in good condition, show frequent repairs, and are
subject to light traffic. Pennsylvania Avenue and K Street are typical
respectively of heavy and light travel. Pennsylvania Avenue, the pavement of which is part Trinidad and part Neufchatel rock asphalt, is about
105 feet between curb lines, age 12 years, never resurfaced, about 10
per cent. of wearing surface gone, largely patched in places, lack of
uniformity, some ruts. In K Street the pavement is vulcanite, 50 feet
between curb lines, age 14 years, never resurfaced or patched, condition
almost perfect. K Street is used as a standard of comparison in all
vulcanite specifications.

The Buffalo pavements, with a few exceptions, are not yet of a sufficient age to justify the formation of any opinion except in case of early
failure, as on Broadway, paved three years ago, resurfaced once and giving
out again in places. Franklin Avenue, paved with asphalt nine years ago,
is the oldest exception and without repairs, is in excellent condition.
Cottage and Bryant Streets, paved eight years ago, are in good condition.

Formula for Asphalt and Vulcanite Pavements. — The reduced
formula for asphalt is: Asphalt, 13.16 to 15.78; heavy petroleum oil, 1.84
to 2.21; fine sand, 70.00 to 65.00; pulverized carbonate lime, 15.00 to 17.00.
There is little or no bond between the base and wearing surface.

The reduced formula for vulcanite is: Asphalt, 4.4 per cent.; coal-tar
distillate, 13.4 per cent; clean sand, 53.6 per cent.; pulverized stone, 26.8
per cent.; hydraulic cement, 1.2 per cent.; flour sulphur, .14 per cent.; air-
slacked lime, .28 per cent.

The main differences are the percentages of asphalt and the value
of coal-tar distillate versus petroleum. The vulcanite is the cheaper as
regards first cost, although costing slightly more for maintenance.

---

* xvii, 361.

Defects of Asphalt and Vulcanite Pavements.—Asphalt pavements are liable to the following defects: First, the formation of "wave surface," especially on grades (probably due to lack of cohesion between the wearing surface and base). Second, the formation of transverse cracks, more apt to occur in wide roads and intersections and undoubtedly due to contraction and expansion caused by variations of temperature. Third, a rot or disintegration takes place in the gutters and necessitates the use of stone or a coating of coal-tar; this does not seem to be the case with vulcanite. Vulcanite or coal-tar distillate pavements: First, are affected at a lower temperature than asphalt, and sometimes present evidence of a flow of material toward the gutters during warm weather often rising nearly to the top of the curb and necessitating a cutting away of the material. This is not observable in the asphalt. Second, vulcanite is not so liable to transverse cracks (the forerunners of repairs), or to wave surfaces as the asphalt, on account of greater longitudinal strength due to a closer union of base and wearing surface. On the other hand, in making repairs the wearing surface of asphalt can be easily removed and renewed, while in the vulcanite no such separation can be made and in resurfacing it is necessary to overlay the whole surface. This slightly affects the grade of the street by raising the same above the curb grade. Third, vulcanite has a more granular surface than the asphalt.

Sheet pavements are sightly, pleasant to drive over, and easy to clean; but if not cleaned, as might be the case in this city, any accumulation of material would, from the nature of the surface, be ground to an impalpable powder, unpleasant to travelers and detrimental to lawns, the fine dust being blown into houses and stores. Such pavements would undoubtedly on business streets attract travel from parallel routes, thereby compelling the pavement of the latter with the same material, as a matter of self-defense, no matter what might be the ultimate value of the pavement. The mixture has to be nicely tempered, requires the satisfying of many conditions and expert manipulation. It is largely affected by temperature and climatic conditions. "In extreme cold the surface cracks and becomes friable; in extreme heat the surface rolls or creeps under traffic, presenting a wave surface uncomfortable to travel over." In this connection I would state that the temperature at Washington ranges from 150° above zero in summer to 10° below in winter, or 160° Fahr., and Captain Griffin, United States Corps of Engineers, of that city, is authority for the statement that the Washington pavements suffer more during the three months of winter than during the remaining nine months of the year.

The expense of laying foundations for street pavements is a large part of the total cost, and when the paving is worn out, or a city is tired of repairing the same, then the expense of removing the foundation will be large, as it is too near the established grade or curb grade to lay stone or other than sheet pavement on. It cannot be laid next to street-railway tracks, as it will not stand the shearing action of wheels. The science of laying sheet pavements is still to a large extent tentative. That good ones can be laid has been proven, but there is no certainty that the standard can be reached on any particular street; two separate batches of material mixed under the same formula and specification may be totally unlike as regards durability.

Standard of Comparison.—In discussing sheet pavements a standard of comparison might be outlined as follows: First, moderate first cost;

second, durability ; third, minimum resistance to traction ; fourth, secure foot-hold for horses ; fifth, healthfulness ; sixth, noiselessness.

Asphalt vs. Stone Pavement.—First, sheet pavements do not satisfy the first requirement much better than dressed block ; second, I should give dressed block the preference as regards durability and ultimate cost ; third, the resistance to traction is undoubtedly less on sheet pavements ; fourth, such few experiments as have been made indicate fewer accidents to horses on sheet pavements than granite, but do not think it would hold good for Medina ; fifth, both sheet pavements and dressed blocks are healthy, and the asphalt filling of the latter sufficiently fulfills all sanitary requirements ; sixth, sheet pavements perhaps fulfill the last requirement better than stone, but there is the sharp click and ring of the horse's hoof, although the vehicle itself makes little noise.   In making repairs in the case of a stone pavement 50 to 75 per cent. of the old material can be used over again ; it is worthless in the case of a sheet pavement.

Finally, I consider sheet pavements a luxury, rendering any city attractive where successfully laid and maintained.   Their introduction should be simultaneous with the adoption of a continual repair system, to secure the best results.   There is small choice between the asphalt and vulcanite pavements.   I would advise against laying asphalt or vulcanite pavements on any of the thoroughfares of this city, and would further advise that the city should only consider its introduction as an original pavement and not on the repaving of any street.   Further, if property-owners are desirous of obtaining these pavements, the city should, in my estimation, demand not less than ten years' guarantee, the pavement being turned over in good condition at the expiration of that period.

REGARDING W. P. RICE'S REPORT ON ASPHALT PAVEMENTS.*

NEW YORK, June 9, 1888.

SIR: I have been reading the report of Mr. Rice, City Civil Engineer of Cleveland, quoted in your issue of May 26, on asphalt pavements in the United States, and in following this series I infer it is your purpose to publish abstracts of official statements, whether you agree with the conclusions of the writer or not.   With regard to this report of Mr. Rice, I think it is well to comment on some of his inferences and conclusions.

Is not the following standard of comparison adopted by him for pavements—viz.: "First, moderate first cost; second, durability; third, minimum resistance to traction ; fourth, secure foot-hold for horses ; fifth, healthfulness ; sixth, noiselessness," as set forth in your abstract of his report, susceptible of a better arrangement?

Pavements are primarily intended for the convenience of the inhabitants of those towns or cities which possess them, and the health of those who live beside them seems a consideration of the first importance.   Putting moderate first cost and durability first would lead Mr. Rice, if a logical man, to clothe himself in a flannel shirt and stogey boots, articles of less first cost and greater economy in maintenance and repairs than those ordinarily worn by persons in his position.   Fortunately for the arguments of those who contend for good pavements, noiselessness and minimum resistance to traction are closely associated with healthfulness in pavements. For the nearest approach to a perfect pavement, in a sanitary view, seems accomplished by a pavement which is impervious to fluids, presents no inequalities in which solid and fluid fæcal or other matter can lodge, and admits of cheap and thorough cleaning. This evidently calls for a sheet pavement, for no block pavement is laid that is free from inequalities, and however well the joints are filled with concrete or a mastic of pitch or bitumen, the filling in the joints will wear out faster than the stone, or even wood blocks wear, leaving shallow but numerous cavities to hold dirt and give off, at least, unpleasant exhalations under the heat of the summer sun, while all sheet pavements, on the contrary, when made either from asphalt or bitumen, are virtually non-absorptive ; that is, the small amount of water absorbed by *asphalte comprime*, or compressed rock asphalt, and pavements in which bitumen is the base, like those of Washington, is a mere matter of maintenance, having no sanitary influence whatever ; while pavements of *asphalte coule*, or the mastic made from rock asphalt, and those in which coal-tar is the basis, are free from this objection. Now these sheet pavements, in addition to their healthfulness, offer the minimum resistance to traction, are on the whole less slippery than any other except wood (excluding macadam), and are undoubtedly less noisy than any other pavement except wood. And there is another point which should not be overlooked : horses, besides doing more work, last longer, and there is less expense in repairing running gear and harness on these pavements than on any other.

There is an apparent anomaly in this mater of pavements which your correspondent is able only to call attention to—not to explain. Statisticians who make such subjects their study assert that the average income per annum in the United States is $200, in Great Britain and Ireland $150, in France $120, in Germany $100, and in Italy $80. Italy has long had the best maintained and constructed roads. Berlin, the capital of Germany, has better pavements than Paris, while that city until less than a decade since was decidedly better paved than London. But in New York, which may be regarded as the Capitol city of this country, one sighs for the quiet and repose of an active boiler-shop when the garbage-carts of the city pass through its best residence streets, and Philadelphia, the home of our oldest and stateliest families, has "pavements fanged with murderous stones," and in most of our cities one can count, in the words of Coleridge,

"two-and-seventy stenches "—

"all well-defined and genuine stinks!"—any pleasant summer day ; while Washington, which has no productive or other business which is not parasitic, has, considering its traffic, possibly the largest area of luxuriously and economically paved streets of any city in the world, and Buffalo has over forty miles of noiseless pavements.

Washington, as is well known, aspires to have the position of the first residence city in our country, counting greatly on its pavements for help in reaching it.

It might be suggested that Mr. Rice commenced his study of pavements without previous expert knowledge of this branch of the subject, or he would never have classed the compressed asphalt on the upper end of Pennsylvania Avenue with the pavement made of Trinidad bitumen on the lower part, or have failed to notice the rather peculiar outlines of the areas repaired in the compressed asphalt, so peculiar that persons unacquainted with the circumstances under which it was laid might have thought that the areas were denuded as a source of material for making asphaltic mastic.

But a little inquiry should have shown him that the pavement had been laid on a foundation not sufficiently dry, and the hot powder had drawn steam from the damp concrete on which it was placed, which prevented the adhesion of the powder so affected, resulting in *macaroons*, a disease which was prevalent in compressed asphalt pavements, till its cause was ascertained ; or, in other words, through the action of the steam, separate masses about the size of almonds, were formed under the smooth surface, produced by *pillonage*, which had no adhesion to each other. After three or four years these places had to be renewed, and the darker material in general use in Washington, which was used, gives, what otherwise would have been one of the finest pavements in any city, a very patchy appearance.

It is rather surprising to see the statement that the shearing action of the wheels prevent its being laid next to street-railway tracks, as it is thought all specifications now provide for a toothing of granite blocks to be laid on each side of the track. But this toothing should extend so that the wheels of a wide-gauge truck should be supported by it. While omnibuses were running on Broadway, between Seventeenth and Twenty-third Streets, where one wheel was generally on the inside track, the outer wheel cut into the granite blocks at the rate of over half an inch a year, and it is absurd to suppose that a material, which must be more or less plastic, not to break up in the winter, can stand the concentrated action of heavily-loaded wheels without being pushed out of place. The frequent repairs now necessary on Chambers Street could be dispensed with if the toothing was carried over the outside tracks to cover the width of broad-gauge trucks.

E. P. NORTH.

ASPHALTE* PAVEMENTS IN PARIS.†

The following is an abstract translated from the specifications and instructions issued in 1884 by the Department of Bridges and Roads to contractors for roads and pavements in Paris :

CHAPTER I.—*Object, Duration and Extent of the Enterprise,*—ART. 1. Relates to the object and conditions of the work.

ARTICLE 2. The work is divided into five divisions, not more than three of which will be awarded to one contractor.

ARTICLE 3. Proposals for the first three divisions will be received only from contractors who have already successfully executed important work in compressed asphalt. The committee must be convinced that they possess sufficient capital to assure the progress of the work during the period of the contract, and that they can furnish the specified brands of asphalt.

The necessary brands and samples must be submitted to the committee one month before the award of the contract.

ARTICLE 4. The duration of the contract is fixed at ten years, ending March 15, 1894. The amount of work is undetermined, depending on the funds applicable for its payment. The total annual expense is estimated at about 1,100,000fr.

---

* All asphalte pavements in Paris are either *asphalte comprimé* or *asphalte coulé*, that is, the wheelways are paved with compressed asphalte, the rock being powdered, heated and pressed in place, and the sidewalks are paved with a bituminous mastic made of powdered rock asphalte which is melted by immersion in hot bitumen (often called "asphalt" in this country), sand and gravel being added to form a wearing surface. Crude bitumen is always refined by melting it in refined bitumen or shale oil, which acts as a flux. Heating the crude bitumen alone will burn it.—ED.

† xviii, 230.

ARTICLE 5. The contractors must deposit in the municipal treasury, for the different divisions, a total security of 370,000 fr. If the security is in money, interest will be paid at three per cent.; if bonds or deeds, the contractor may collect the income.

ARTICLE 6. When the owners of adjacent property pay the whole or part of the cost of these works, the contractors must execute them in the same manner as for the city and under supervision of the same officers.

CHAPTER II.—*Form and Dimensions of the Work.*—ART. 7. The width of the sidewalks for each locality will be determined by the Administration, its slope by the engineer. The curb between the sidewalk and the roadway will not be included in this contract.

ARTICLE 8. The pavements of *asphalte coule* will be formed of a bed of natural bituminous mastic, melted with sand, at least 0.6.of an inch thick, resting on a foundation of hydraulic lime concrete 4 inches thick which is covered with 0.4 of an inch of hydraulic lime mortar.

ARTICLE 9. The Administration reserves the privilege of making the asphalt roads on an old pavement whose surface will be redressed and coated with Portland cement to regulate the form.

CHAPTER III.—*Quality and Brand of Materials—Preparation of Bitumen and Asphalt.*—ART. 10. *The asphalt rock* must be a homogeneous carbonate of lime, brown, fine grained, of a close texture, and uniformly impregnated with bitumen so as not to present black or white spots ; it must be free from iron pyrites, and not contain more than 2 per cent. of clay, and any portions producing less than 5 per cent. of bitumen will be rejected.

The rock employed in the manufacture of compressed asphalt must be obtained from the mines of Val de Travers, Switzerland ; Volant, Upper Savoy ; Pyrimont, near Seyssel ; St. Jean de Maruéjols, Gard ; or from other sources approved by the Administration.

For the preparation of *asphalte coule* there will be accepted, besides the above, the products of Lovagny, Upper Savoy, of Dallay and Pont-du-Château, Puy-de-Dôme ; and such other mines as give analogous materials satisfactory to the Administration.

ARTICLE 11. *The bitumen* must be from the mines of Lussat and Malintrat, Puy-de-Dôme ; or Maestu, Spain ; or from others having a similar product acceptable to the Administration.

The bitumen must contain no foreign substance, neither water, clay nor light oils. When maintained at a temperature of 250° Fahr., for 48 hours it must not lose more than 3 per cent. of its weight.

It must be viscous at ordinary temperature, never brittle nor liquid ; when drawn out in threads it must elongate and only break in very fine points.

Trinidad bitumen will be accepted if refined by melting and decanting in the contractor's shops in Paris.

The necessary flux for this operation may be either fine natural bitumen from one of the above sources, or shale oil from the Autun slate, excluding gas-tar, the so-called fat bitumens and analogous products.

The mixture of Trinidad bitumen and shale oil must be heated for eight hours, during the first six hours it must be vigorously stirred, but remain undisturbed during the last two hours in order to permit the impurities to fall to the bottom of the boiler whence they must be removed after each removal of the supernatant refined bitumen.

The latter may be employed immediately for compounding the mastic or may be poured on suitable dry surfaces.

The refined bitumen must not contain more than 25 per cent. of its weight of clay, and must be free from dirt, roots, etc.

ARTICLE 12. *The bituminous mastic* employed for new pavements, and the repairs and renewal of old ones, will be composed of a mixture of powdered asphaltic rock and mineral bitumen.

The old compressed asphalte obtained by the removal of the old roads, may, if carefully separated from sand and all foreign matter, be substituted for the asphaltic rock; but artificial bitumen, gas-tar, fat bitumens and bitumens with a slaty base and all analogous products are rigidly proscribed.

The asphaltic rock must be reduced, cold, in the most perfect mechanical crushers to the finest possible powder, which must pass through a sieve having meshes not more than 0.1 of an inch square.

It will be melted and stirred for six hours with a suitable quantity of mineral bitumen, forming a mastic which, cooled, presents a homogeneous mass, slightly elastic and not softening at a temperature of 140° Fahr. This mastic must be molded in blocks bearing the maker's brand, its yield of bitumen must not be less than fifteen per cent. nor more than eighteen per cent. of the total weight.

ARTICLE 13. *The surfaces of asphalte coule* of the first class must be formed of the mastic described in Article 12, of mineral bitumen, and of sharp sand, clean and dry, in the proportion of 100 parts by weight of mastic, four parts of bitumen, and sixty parts of sand. The sand, perfectly dry, will be successively added in the boiler to the melted mixture of bitumen and crushed mastic blocks which must be stirred for at least eight hours. It is forbidden to make the bitumen in the movable boilers intended only for its transportation.

ARTICLE 14. The Administration will permit in certain cases the surfaces of *asphalte coule* to be composed of 100 parts by weight of new mastic, ten parts of bitumen, sixty parts of sand, and 170 parts of old bitumen from worn-out sidewalks, and carefully separated from all sand and foreign materials.

The classification must in all cases be precisely stated in the bill accompanying each load.

ARTICLE 15. *The powdered asphalte*, mechanically crushed according to Article 12, must, when intended for the composition of *asphalte comprime* contain not less than seven per cent. and not more than thirteen per cent. of its weight of bitumen, the Administration reserving the right to fix the exact proportion according to the circumstances of the case and under the following conditions :

1.   The rocks shall not differ except in their yield of bitumen, which must not be less than five per cent.

2.   The pieces of rock must be mixed before crushing, or if mixed in the powdered state it must pass through the crusher again.

The mixture of old asphalte from the removal of worn-out pavements with powdered new asphalte is forbidden.

ARTICLE 16. The powdered asphalte must be raised to a uniform temperature of 248° to 266° Fahr. in rotators mechanically turned in a uniform and continuous manner which shall prevent adherence and burning.

The apparatus must be of the most perfect type, and the powder must be maintained at the required temperature long enough to expel all watery vapor.

The use of machines called *decrepitoirs* is forbidden, either to reduce the rock to powder or warm the powder.

ARTICLE 17. One month before the award the competitors must deposit in the Chief Engineer's office for each division of the public roads, and in the testing laboratory: (1) samples of asphalte rock and natural bitumen which they wish to use; (2) blocks of mastic; (3) samples of refined Trinidad bitumen and shale oil; (4) a statement of the composition of the mastics and the brands, compositions and proportions of the asphaltic rocks which they propose to use.

The samples of rock and bitumen must bear the signature of the contractor and the maker's brand.

During the period of the contract all materials used must conform to the samples deposited and bear the same brands. If it is wished to use other material they must be submitted to and accepted by the Administration, and samples deposited.

ARTICLE 18. *The sand* will be dredged from the channel of the Seine above La Marne; sand dredged opposite or below Paris will be rejected.

Sand intended for the preparation of the surfaces of melted bitumen must be free from all earthy and foreign matter; it must be sharp and dry, and separated by successive screenings from all grains less than 0.08 or more than 0.16 of an inch in any dimension, and conform to the samples deposited.

Sand intended for mortar must be of the same kind and quality; no grains must be larger than $\frac{1}{4}$ inch in any dimension.

Sand from the seashore may be specially admitted if it is of first quality, perfectly pure and well washed, and has been dredged, but a less price will be allowed for it than for sand from the Seine.

ARTICLE 19. *Gravel for beton* must pass through a ring $2\frac{3}{8}$ inches in diameter, and must not be able to pass through a ring $\frac{3}{8}$ of an inch in diameter; it must be freed by abundant washing from earth and all foreign materials, and all round or smooth pebbles must be rejected or broken.

ARTICLE 20. *The lime* must be hydraulic, powdered and delivered at the yards in sealed sacks, marked with the manufacturer's name, and presenting no exterior seams or patches. It must be provided from the kilns of Moulineaux, Mancelliére, Echoisy, Ville-sous-la-Ferté Raincy, Bougival, Coucou, or from some other source recognized as at least equally good.

Vassy cement must come from the works of Grenan, Millot, Gariel, or Prévost. Portland cement must come from the works of Famchon, Boulogne-sur-mer; or Quillot Freres near Lézinnes. The cement must be furnished in sacks, bearing the brand and seal of the works, the seal of the city, and presenting no external seams or patches. Any sack not satisfying the above requirements that is found in the yards will be immediately rejected and the contractor fined.

CHAPTER IV.—*Execution of Work.*—ART. 21. The contractor will make the necessary *excavations* indicated by the grade marks placed for them. After the concrete is laid the iron or wooden grade stakes must be withdrawn and the holes immediately filled with concrete and mortar. When the excavation does not exceed two inches in depth, and the haul is not more than 100 feet, the excavations will be considered as simple surface dressing, whose payment is included in the price of the pavement above it. When the haul for such excavation is above 100 feet, payment will be allowed for transportation only.

All excavated material must be removed at the end of each day's work. If neglected by the contractor he will be charged with the cost of doing it, and fined for every cubic meter not promptly removed.

ARTICLE 22. Existing *sidewalk pavements* must be demolished, and the materials either removed every night to the city storehouses, delivered to the adjacent property owner, or piled up for his disposition.

Before tearing up the paving blocks they must be counted by the agent of the Administration and the contractor, who will be responsible for the recorded number, which he must deliver at the storehouse or be charged with the value of an equal number of new blocks of the same brand and pattern.

The curbstones, crosswalks, etc., must be removed every night and placed at the disposal of the owners, or transported to the city storehouses by the engineer's orders.

ARTICLE 23. *The mortars* must generally have the following proportions, by volume: Hydraulic lime mortar, 2 parts of powdered lime to 5 parts of sand; Portland cement mortar, 1 part of powdered cement to 3 parts of sand; Vassy cement mortar, 2 parts of powdered cement to 5 parts of sand.

The Administration reserves the privilege of changing these proportions.

For all important work, and wherever else required, the use of mixing machines will be obligatory for making all mortars not composed of quick-setting cement.

The use of sewer water is forbidden for mixing mortar for which suitable pure water is required, that the contractor must take and pay for in the usual way.

ARTICLE 24. *Hydraulic lime and Portland cement concretes* must be ordinarily composed of two parts by volume of mortar and three parts of gravel, rigidly measured in boxes whose capacity has been approved. The gravel will be brought in wheelbarrows to a grating, where it will be abundantly washed in clear water. The mixture must be thoroughly worked by a rake and shovel on a plank platform of sufficient size, until all the stones are completely enveloped in mortar. For Portland cement beton sufficient water must be added to make the mixture suitable for use in a fluid state.

ARTICLE 25. The ground over which an *asphalte coule pavement* is to be made must first be loosened, then sprinkled and carefully rammed, especially along the border. When the soil has thus been solidified and well dressed, the contractor must establish the foundation bed, formed of hydraulic lime concrete 4 inches thick, including a coat of ½ inch of lime mortar.

The concrete must be carefully rammed to close all open spaces and make the mortar overflow the surface, after which the surface will be immediately regulated to the required profile.

The bitumen pavement will not be made until this foundation has set and is well dried; meantime the surface of the foundation will be protected by a bed of fine dry sand, which will be carefully swept off before the application of the asphalte coulé. Watchmen and barriers against traffic must be maintained at the contractor's expense.

ARTICLE 26. The asphalte coulé, prepared as above required, must be quickly brought from the shops in closed locomotive furnaces, warmed and supplied with suitable apparatus to continue the mixing during transportation and up to the moment of application, so as to maintain intimate mixture and prevent direct action of the fire. It must not be prepared in fixed furnaces at the points where it is used, when not otherwise specified. The coat of bitumen will be 6 inches thick ; it will be poured in sections about 6 feet wide, perpendicular to the curb, and limited by iron rules of the required thickness.

The asphalte coule will be spread by means of a wooden float in such a manner as to form a perfectly dressed surface without lumps or hollows.

Each new section must be perfectly welded to the preceding in such a manner as to present no open joints, cracks or offsets.

The mastic must have no cavities, must be level and exactly match all curbs, gutters, paving, man-hole covers, tree boxes, hydrants, etc., and join the house walls without any cracks or cavities.

In constructing isolated sections the concrete foundation must extend one inch each side beyond the mastic, which will be limited by regular lines.

As soon as completed, the surface will be lightly sprinkled with fine dry sand and gravel, no grains exceeding one-eighth inch, nor being less than one-twelfth inch in diameter.

It is absolutely forbidden to throw water on the surface while it remains warm.

Before admitting traffic, the contractor must barricade and guard the pavement and protect it until it is able to sustain the passage of pedestrians.

The barrier posts and pickets must never be driven into the pavement, but must be supported by plaster bases.

Article 27.—*Asphalte Pavements.* The ground on which the asphalt is to be placed must always be first loosened, then sprinkled and carefully rammed, particularly along the curb. The Administration reserves the right to roll it at its own expense.

When the soil is thus sufficiently consolidated and well dressed, the contractor will make the foundation bed, generally formed of Portland cement concrete, six inches thick, but always subject to the requirements of the Administration, who may augment or diminish the thickness or prescribe the use of hydraulic lime in certain cases.

The *concrete foundation* must extend with the same thickness, six inches under the curb, and four inches behind its rear face. And the contractor must, if required, set the curbs in the concrete in a mortar bed at the same time that the road foundations are made.

Stone and other materials must be transported on planks over the concrete foundation.

Wheelbarrow loads of concrete must not be emptied directly on the soil, but on the work already finished, and afterward spread into the required position, and rammed until the mortar flows to the surface, which will then be carefully regulated to the required curve and *immediately* covered with a coat of Portland cement mortar about one-half inch thick, that will be smoothed and cannot be replaced by a coating made at a later time.

The concrete must be left to dry and set for the longest possible time, in every case, during at least three days in dry and five days in wet weather ; meantime the contractor must provide barriers and watchmen to protect the work.

Immediately after it is laid, the concrete foundation must be protected by a thin bed of very fine sand, which will be carefully swept off when the asphalt is applied.

The Administration reserves the right to have the *asphalte* roads built on rolled stone foundations, whose surface shall be cleaned and receive an evening coat of Portland cement mortar.

ARTICLE 28. *The powdered asphalte*, heated and prepared, must be brought to the required point in covered carts, of a pattern approved by the Administration, and designed to prevent, as much as possible, the cooling of the material during transportation. The powder must be spread about two-fifths thicker than the required two inches. It will be brought in special wheelbarrows, trundled on planks over the foundation bed. The powder will be leveled with a rake, and all foreign material carefully removed, and it will be carefully rammed, with increasing energy, by cast rammers, properly warmed in movable furnaces.

The ramming will be commenced at the edges, and must be conducted so as to assure complete union between adjacent bands, and exact junction with the edges of the pavement limiting the *asphalte* surface.

After being twice rammed, the pavement must be smoothed by a suitable curved hot iron tool, after which it must be vigorously rammed and rolled until thoroughly cold, by a roller of at least 1,100 pounds weight. The cooling must not be hastened by the application of cold water.

The operations must be conducted so as to secure a perfectly uniform bed of *asphalte*, with a surface rigidly conforming to the required profile and presenting no lumps or hollows.

The pavement will be sprinkled with sand and not opened for traffic until thoroughly cool, until which time the contractor must protect it by barriers and watchmen.

ARTICLE 29. The above requirements apply to the construction of carriageways, footways, and analogous works, and to the maintenance of repairs of junctions of drains, etc.

Carriageways constructed by special order must have, when made of *asphalte coule*, a foundation of hydraulic lime concrete 6 inches thick, including the mortar surface, and covered with two coats of *asphalte coule* 0.6 inches thick, successively poured, the second one being checkered.

When made of *asphalte* the carriageways must have a foundation of Portland cement concrete 6 inches thick, covered 1¼ inches thick with a bed of compressed *asphalte*.

ARTICLE 30. Within three months after the award of the contract the contractor must distribute furnaces for use in repairing the *asphalte coule*. These furnaces must be provided with boxes to contain the old materials removed from the road. Similarly, for *asphalte* repairs, the carriages bringing the hot powder must contain one closed compartment for the heated powder, another for the grate, fire, irons and the tools necessary for making the repairs, and a third to carry the *debris* and material resulting from the demolition of the old pavement.

After each operation of new work or repairing, before returning the furnaces to the shop their site must be carefully swept and cleaned of ashes, coal, *debris*, etc.

CHAPTER. V.—*Maintenance, Under Forfeit, of the Sidewalks and Areas.*—ART. 31. The contractor is required, 1st, to make all necessary repairs and removals, and furnish the labor and materials requisite to maintain the sidewalks and *asphalte* and *asphalte coulé* areas in perfect condition, and to maintain all paved gutters included in the given district.

2d. To repair yearly, equal to new, at least the fifteenth part of the *asphalte coulé* areas.

3d. To repair yearly, equal to new, such parts of the *asphalte* areas as are designated by the Administration. This comprises repairs due to any cause whatever, except those necessitated by trenches dug by the city or by private or public companies.

The contractor is not entitled to extra payments for loss or expense from any unforseen circumstance or condition.

ARTICLE 32. Each year, before March 15, the contractor must present to the Administration a statement of the surface to be repaired, this statement, subject to modification by the Administration, must be conformed to by the contractor; if it is not presented by him it will be prepared by the engineer. One-half of the *asphalt coule* repairs must be completed by July 1, two-thirds must be completed by August 15, and they must be entirely finished by October 1, and the contractor will be liable to a retention of payment if the work is delayed beyond these dates.

The repairs implicitly comprise the demolition and removal of the old *asphalte coule* which belongs to the contractor, and the rectification and repairs, if necessary, of the old foundation.

The *asphalte coule* surface must be perfectly plain and regular, without humps or hollows, and so that a straight edge, 1.0 m. long (3.20 feet), laid in any direction will not leave an opening more than 0.4 inch deep between its side and the pavement at any point.

There must be no holes nor cracks in the surface.

Throughout the $\frac{1}{2}$ ms. of *asphalte coule* surface not specially repaired, the contractor must renew all places where the surface is cracked, split, depressed, swelled, or in any way perforated, where it matches imperfectly with tree boxes, fountains, man-holes, etc., and especially where sunken near trenches and fountains.

It is formally stipulated that in repairs of the first fifteenth of the surface, only asphalt of the first class can be used; in other repairs asphalt of the second class may be used.

ARTICLE 33.—*Repairs of Asphalte Surfaces.* The requirements are substantially the same as those of Art. 32, excepting the last clause, and it is further specified that when the foundation is found defective, the Administration may require it to be replaced with Portlant cement concrete.

Defective spots must be carefully cut out with a sharp hatchet and at least 1.67 feet larger in every direction than the defective place; the sides must be cut on straight lines; there must be a perfect union of the old and new material, and the surface must show no irregularities.

On September 1, or sooner in case of bad weather, a general examination will be made with the contractor, who must immediately begin repairs on doubtful surfaces not likely to endure through the winter.

In rainy weather the bottoms of the patches should be sponged and dried as carefully as possible with fine hot ashes, and then be well brushed.

Special care must be taken to clean all sand, powder, etc., from the bottom of the patches.

If the concrete is reached it must be entirely renewed underneath the patch, or a Portland cement dressing be laid on top, flush with the bottom of the patch throughout.

During bad weather no repairs shall be made of *asphalte*, unless expressly authorized by the Administration.

Melted bitumen, which may be of the second class, will generally be used for a bed for the patch. The patch is only temporary, and must be replaced by the 15th of May, or sooner, if possible.

The contractor is absolutely forbidden to use pebbles for filling holes in the *asphalte*. When the contractor fails to make necessary repairs, and the Administration, exceptionally and in default of other available means, fills holes with broken stone, the contractor must pay for the work and materials, and cannot claim damages for injury to the pavement caused by the loose stones.

In winter, when the Administration recognizes the impossibility of making lime or cement concrete, holes in the foundation that is to be repaired may be filled by a mixture of three parts by volume of pebbles to one part of hot asphalt, but this provisional foundation must be removed as soon as possible and replaced in the standard manner, and this bituminous material cannot be re-employed. All repairs must be made with new powdered asphalt, the same as for new work.

ARTICLE 34. Relates to statements, estimates and payments for the work done.

ARTICLE 35. Relates to the measurement of surfaces, etc.

ARTICLE 36. Relates to the prices and payments for the different classes of work.

ARTICLE 37. The contractor will be paid for repairs at sewer trenches, water and gas conduits, etc. He can make no claim for settlement or other injury at these places, and must maintain the pavement there in the same condition as elsewhere.

To assure a perfect welding at the edges of the *asphalte coule* or the *asphalte* it will be paid for of a width 2 inches greater in every direction than the trench, and must not fall short of these dimensions.

To provide for earth settlements the contractor may temporarily maintain these areas during a period of not more than eight days with broken stone, 2½ inches in diameter, that is rammed, sprinkled, swept and maintained, so as to prevent injury from loose stones. After eight days, if final repairs are still impossible, the contractor must, at his own expense, make a provisional surface six inches thick of bituminous concrete, to be removed for final repairs.

All old materials taken from trenches, and all abandoned depots, must be removed the same evening.

CHAPTER VI.—*General Conditions and Regulations.* ART. 38. Relates to the inspection and acceptance of materials and work.

ARTICLE 39. Relates to extra payments.

ARTICLE 40. The contractor must remove at his own expense, and buy at schedule prices, all the asphalte provided by demolishing the old work. No allowance will be made for the nature or condition of the materials. The asphalte and asphalte coule will be bought by the square meter ; the price of the former will be according to the mean thickness, and for the latter will be uniform for all thicknesses.

ARTICLE 41. Relates to transportation of material.

ARTICLE 42. Relates to the inspection, testing, marking, etc., of materials and work.

ARTICLE 43. The contractor must provide at his office and shop, a telephone for the use of the engineers in special cases, and to give them notice of the exact hour when their presence is required ; the telephone must be connected with the general system and have an attendant capable of receiving and transmitting orders and messages.

ARTICLE 44. Relates to police regulations, numbering, marking furnaces, etc., and unloading materials.

ARTICLE 45. The contractor is forbidden to burn any material in his furnaces that will cause disagreeable fumes, discolor the paint on adjacent houses, or injure the neighboring vegetation, and the furnaces must be placed so that the smoke will be carried as little as possible by the wind to adjacent gardens or parks.

ARTICLE 46. On holidays or exceptional occasions, the Administration may require the contractor to cease digging or depositing materials in the street, and he shall be entitled to no damages therefor or extension of time on account of such delays.

ARTICLE 47. In an urgent case the contractor may be required to establish a night-turn, for whose labor he will receive 50 per cent. above the regular schedule price, during that time only which is specified by the Administration.

ARTICLE 48. Any employee not obeying the engineer's orders, or who practices any fraud or disobedience of the regulations, must be immediately discharged by the contractor at the engineer's request. Payments will be retained for any portion of the work badly or fraudulently executed by the contractor's agents, and such work must be renewed before it can be estimated.

If an employee is not immediately removed when required, all work done by him will be subject to a double penalty and will not be estimated.

The contractor must furnish a certificate showing cause for leaving, to every employee who leaves any department of his municipal work, and this certificate must be immediately submitted to the engineer, who may refuse to permit re-employment.

ARTICLE 49 provides that if the contractor is absent without authority, fails to perform all his duties, or to obey the injunctions of the Administration within ten days, he shall be notified that he is considered to have abandoned the undertaking, and after twenty-four hours the Administration may either prescribe a re-letting of the old contract or may entirely cancel it without returning the material ; or it may execute the work at the expense of the contractor, by workmen under competent supervision, or with contracts by private contractors, according to the urgency of the case. The expenses will be covered by the amounts due the contractor, and by the contractor's deposit in the city treasury.

ARTICLE 50. The contractor is forbidden to sublet any part of the work without the written authorization of the engineer, which authorization is always subject to revocation.

It is especially prohibited to sublet the digging and transportation to a different party from the one who furnishes and constructs the concrete foundations.

ARTICLE 51 requires that the contractor provides a responsible agent for each division of the work ; that the contractor and his agents must personally meet the engineers at any time they may appoint, and relates to the regular reports, receipt of instructions, etc.

ARTICLE 52 is a schedule of amounts to be retained by the Administration from the payments due the contractor, in cases of fraud, delay, imperfect work or materials, etc.

For the use of unauthorized or impure asphalte, each offence, 500f.; for the omission of new concrete where required under patches, 50f.; for delaying the repairs of $\frac{1}{15}$ of asphalte coulé, from 100f. to 300f.

For delaying repairs of asphalte, from 200f. to 500f.; for delaying to commence work ordered begun, 20f. per diem.

Other fines are specified for insufficient thickness in pavement and foundations thereof, for the use of improper and insufficient material, for poor or improper workmanship, for unsatisfactory illumination and protection, for various delays and insufficient progress, etc., etc.

ARTICLE 53 provides for the regular reports of the engineers, which must be countersigned by the contractor.

Provision is made for settling disagreements between the contractor and engineer and for paying the former, if necessary, from the deposit in the treasury.

ARTICLE 54. The schedule prices for the different parts of the work will be held invariable.

ARTICLE 55. The contractor must, at his own expense, properly light, barricade and guard the work and shops, conforming to police regulations and special instructions from the engineers. An inclosed lamp or lantern must be placed at least every 10 meters (33 feet) along the working place, which must have red lights toward the open road.

ARTICLE 56. One per cent. of all sums due the contractor will be reserved for the benefit of the national asylums of Vincennes and Vesinet.

ARTICLE 57. The contractor must pay the contract and registry fees, and at the expiration of the guaranteed period will receive his deposit, less any deductions that may have been incurred.

ASPHALT PAVEMENT CONTRACTS IN NEW YORK.*

During the past few days, the New York *Evening Post* has, in double column articles with conspicuous head lines, published attacks on Public Works Commissioner Gilroy, of New York, alleging improper conduct in the matter of letting contracts for asphalt pavements, and intimating this was with a view of favoring adherents of the political organization known as Tammany Hall, of which he is a conspicuous leader. The charges are, substantially, that a pavement, except in one instance, was called for that one corporation practically controlled ; that the price paid the contractors on this work was exorbitant, and that the bond exacted to guarantee the 15-year maintenance, though 25 per cent. of the contract price, was insufficient. Seventy per cent. of the contract price was payable on the completion of the work, and 3 per cent. of the remaining 30 per cent. being payable yearly for ten years after the first five years.

---

*xxi, 98.

We are satisfied that he honestly tried to get the pavement that he believed was the best, and that he acted on the advice of those whom he believed competent to advise and who had no interest in any particular pavement.

In regard to the cost, the *Post* quotes Mr. Gilroy's statement that the average price of the contracts is \$4.45 per square yard, with a 15-year guaranty, while the ruling price in Washington for several years has been \$2.25 per square yard, and urges that there is no good explanation for the difference. The pavements in the two cities are laid by the same concerns in the same way, and are substantially alike. Of course it might happen that New York paid too much, relatively, without dishonesty of its officers, but as a matter of fact there is obvious explanation for a considerable difference in price between the two places.

The cost of labor, including teams, is in the region of 30 per cent. less in Washington than in New York, and in the first named place the refining plants are within the city and in hauling and handling material there is a large advantage over New York. Again, Washington is a city of very light traffic and the cost of maintenance is small, while in New York it will form a heavy item. The general theory in the last named city is that asphalt pavement will need but slight repairs for the first five years in residential streets, but after that it will cost from 10 to 15 cents per square yard per year to keep it in order. This, with the fifteen year contracts, of course involves a charge of from \$1 to \$1.50 a square yard in the price. This is put in the contract price for paving instead of being made directly in the shape of future payments for maintenance, it may be remarked, because of a provision in the laws forbidding contracts for payment of money not already appropriated. Of course the proper difference in cost of the pavement at Washington and New York can be only approximated, but the facts and figures given at least show that it is considerable.

The *Post* contended for a bond of more than 25 per cent. of the contract price, yet the fact that it was put even so high effectually excluded many parties from bidding, simply because they could not find capitalists who were willing to wait fifteen years to be released from an obligation. It should be clear that if the bond was higher the competition would be more restricted and the price greater. It seems to us the bond is sufficient.

The people of New York certainly have manifested a desire for asphalt pavements on streets for which it is suited, and Mr. Gilroy should not be charged with improper conduct in asking for what he was advised and had satisfied himself had been satisfactorily tried elsewhere, especially since his specifications called for no pavement that one company has a monopoly of putting down. This is shown by the fact that contracts were awarded to the Barber Asphalt Paving Company and Matthew Taylor, rival concerns, to the amounts of $325,287 and $256,590, respectively. In our opinion the average price of $4.50 per yard, on a 6-inch concrete foundation, involving as it does a fifteen-year maintenance, was a low one to be made with responsible guarantees.

ASPHALT PAVEMENTS OF BUFFALO, N. Y.[*]

Buffalo has a large number of its principal streets paved with asphalt. We, therefore, obtained from the city engineer copies of the specifications under which these pavements were laid. Deputy City Engineer Edward B. Guthrie informs us that under the "Barber Specifications," which we reprint below, about 50 miles have been laid since 1882, and the experience has been satisfactory to the citizens and the municipal authorities. Besides this a contract has been made for five miles of vulcanite pavement this year under a five years' guarantee, which is being laid under the specification also reprinted below.

A. L. Barber's specification for laying his genuine Trinidad asphalt, pavements, 1888.

1. We propose to lay Trinidad asphalt pavements two and one-half (2½) inches in thickness when compressed, with a base of hydraulic cement-concrete six inches in depth.

Roadway.—2. All unnecessary material will be removed from the street; soft or spongy places, not affording a firm foundation, will be dug out and refilled with good earth, well rammed, and the entire road-bed will be thoroughly rolled.

Foundation.—3. Upon the road-bed thus prepared will be laid a bed of hydraulic cement-concrete six inches in thickness, to be made as follows:

One measure of American cement, equal to the best quality of freshly-burned Rosendale cement, and two of clean, sharp sand will be thoroughly mixed dry, and then made into a mortar with the least possible amount of water; broken stone or brick, thoroughly cleaned from dirt, drenched with water, but containing no loose water in the heap, will then be incorporated immediately with the mortar in such quantities as will give a surplus of mortar when rammed. This proportion, when ascertained, will be regulated by measure. Each batch of concrete will be thoroughly mixed. It will then be spread, and at once thoroughly compacted by ramming until free mortar appears upon the surface. The whole operation of mixing and laying each batch will be performed as expeditiously as possible. The upper surface will be made exactly parallel with the surface of the pavement to be laid. Upon this base will be laid the wearing surface, or pavement proper, the cementing material of which is a paving cement prepared from pure Trinidad asphaltum, unmixed with any of the products of coal tar.

---

[*] xix, 60.

Wearing Surface.—4. The wearing surface will be composed of:

Asphaltic cement............. ....from	12 to	15
Sand....... .... "	83 to	70
Pulverized carbonate of lime.............. "	5 to	15
	100	100

In order to make the pavement homogeneous, the proportion of asphaltic cement must be varied according to quality and character of the sand. The carbonate of lime may be reduced or omitted entirely when suitable sand can be obtained.

The sand and asphaltic cement are heated separately to about three hundred degrees Fahrenheit. The pulverized carbonate of lime, while cold, is mixed with the hot sand in the required proportions, and is then mixed with the asphaltic cement at the required temperature, and in the proper proportions, in a suitable apparatus, which will effect a perfect mixture.

The pavement mixture, prepared in the manner thus indicated, will be laid on the foundation in two coats. The first coat, called cushion coat, will contain from 2 to 4 per cent. more asphaltic cement than given above; it will be laid to such depth as will give a thickness of half an inch after being consolidated by a roller. The second coat, called surface coat, prepared as above specified, will be laid on the cushion coat; it will be brought to the ground in carts, at a temperature of about 250° Fahr.; it will then be carefully spread, by means of hot iron rakes, in such manner as to give a uniform and regular grade, and to such depth that after having received its ultimate compression, it will have a thickness of two inches. The surface will then be compressed by rollers, after which a small amount of hydraulic cement will be swept over it, and it will then be thoroughly compressed by a steam roller; the rolling being continued as long as it makes an impression on the surface.

Should there be a railroad track on the street, the edge of the asphalt pavement adjacent to the track will be protected by a line of stone paving blocks on each side of the rail.

SPECIFICATION

For.............paving...................................... ....
with "A. L. Barber's Trinidad Asphalt Pavement," including all necessary connections as per specifications and quantities given:

The carriage-way shall be graded two (2) feet wider than the width of the pavement ordered and to a depth of eight and one-half in ches below the surface of the street on the centre line, the cross-section to conform to a crown of....inches, in accordance with stakes set for line and grade by the engineer, or person appointed.

Roots of trees, when so directed by the engineer, to be saved as much as possible, and grading to be done carefully around said trees ; also the sodding on sidewalks to be protected as much as possible, if the street is to be graded only for the width of the paving.

The sidewalks are to be graded, if petition or the order for paving this street includes the grading of sidewalks, with a rise of one-half ($\frac{1}{2}$) inch per foot from the top of the curb-stone to the side of the street. Wooden or stone sidewalks must be removed carefully and must be replaced in front of each lot where they originally belonged, in as good a shape and manner as they were found by the contractor before removing them in order to grade the street.

All embankments shall be thoroughly settled before any sand or gravel or concrete for paving is laid on them. The contractor (if necessary) must flood the entire bed of the sub-grade with a sufficient quantity of water, provided the water-pipe lays on the street, after excavation to proper sub-grade, and must pound well all spots that show a depression by being flooded. The cost of the water the contractor must include in his contract price.

The width of paving between the curb-stones is to be....feet.

The curb-stones to be of the best quality of *hard* Medina sandstone, not less than four (4) inches thick and eighteen (18) inches deep, not less than two (2) feet six (6) inches long, cut with a proper bevel on the top to a true and uniform surface. The end joints are to be cut at right angles with one-quarter ($\frac{1}{4}$) inch joint for a depth of ten (10) inches. The face of the curbing from top to the under side of the asphalt coating must be neatly dressed. The curbing must be carefully set to straight line and a true grade in a full bed of sand or gravel, and backed up with eight (8) inches thickness of same material in the rear.

At crossings of paved streets, connections of curbs must be made at corners with a true circle of six (6) feet radius.

Protection curbing to be not less than three (3) inches thick on top, and dressed to uniform width on top, with end joints cut at right angles not to exceed $\frac{1}{4}$ inch for a depth of at least three (3) inches, to be not less than eighteen (18) inches long and fourteen (14) inches deep, and to be hard Medina sandstone.

The gutters shall be of a depth of... inches, or as shall be directed.

The asphalt pavement to be laid in accordance with the specification for A. L. Barber's Trinidad asphalt pavement, hereto attached.

Where the curbing comes above natural ground, it must be backed up by a bank....wide on top, rising $\frac{1}{4}$ inch per foot towards lines of street, with a slope to natural ground near lines of street of $1\frac{1}{2}$ to 1. Where such banks come on stone or wooden sidewalks, these must be first carefully taken up and properly relaid on top of bank in front of each property, where they originally belonged, as to be directed.

General Specifications.—All material found in the grading of the street, except sidewalks proper, hitching or other posts, step stones or other private property, *is the property of the contractor ;* also all chips, stones, rubbish, etc., that may have accumulated on the streets or sidewalks during the process of paving, and after completion of grading and paving the street, block after block, as the work progresses, must be removed therefrom by him.

All old material found in the street, that after being properly redressed corresponds in reference to size, shape and quality to these specifications, *may be used in the new work.*

All material furnished and all work done, which, in the opinion of the engineer, shall not be in accordance with these specifications, shall be immediately removed, and other material furnished and other work done that shall be in accordance therewith.

In preparing *concrete*, none but *coarse*, sharp, clean sand will be allowed.

The work shall be done *subject to the inspection* and approval of the engineer, or such person as appointed, both as to material and workmanship, and must be completed in.................working days, after a written notice from the engineer to begin the work has been served on the contractor. The work shall be deemed complete when the contractor is allowed the final twenty per cent. on his contract.

The contractor to be subject to a *penalty* of $........for every successive day after the above specified time until the work is completed and accepted, unless otherwise ordered by the Common Council.

The engineer, or such person as appointed, will set the stakes for paving, provided a block has been properly excavated, to sub-grade from curb line to curb line of cross street, for line and grade, which the contractor must protect and maintain and keep uncovered for examination.

The contractor shall pay all *damages, or losses*, or claims recovered, that the city may be made liable for, and save the city harmless in all things, from any accident which may happen or arise by reason of failure, or neglect, or refusal on his part to comply with the ordinances of the city.

In the event that the *contractor shall abandon the work* and refuse to commence it again within three (3) days after a written notice from the engineer, directing him to resume the work, has been served on him, then the sureties on the contract will be notified and directed to complete the work.

The contractor will be required to pile all material that may be necessary for the work neatly on the front of the sidewalk, and not within three (3) feet of any fire hydrant, and in such manner as will preserve sufficient passageway of not less than three (3) feet on the line of the sidewalk.

The contractor must accept orders for *extra work*, not included in this contract, which may be necessary ·and connected with the paving of the street, if so directed by the engineer, and execute them properly at reasonable prices. He must make sewer and water connections where ordered, in accordance with attached specifications for same, to be paid for as stated in said specification, at prices not to exceed the ones given below.

Each proposal must state a certain sum for furnishing materials, labor, and finishing the work complete ; it must be accompanied by bonds required by law ; it must include the sum of $....... per day for the purpose of paying an inspector, to be appointed by the engineer, to superintend the paving during the entire time the work is progressing ; said inspector to be paid on the recommendation of the engineer, by warrants drawn against the assessment roll for said work. The inspector shall not be employed in any manner by the contractor, and shall be discharged at the option of the engineer.

The orders of the inspector are to be obeyed by the contractor, or in absence of the latter by the superintendent.

In case the city finds it necessary to *lay pipes for gas or water*, or make other improvements in the street, or private parties or companies want to make connections with pipes laid in the street, or want to make repairs or adjustments of their pipes, stop-cocks, etc., before the concrete bed for paving is made in any one place, the contractor shall not interfere with the progress of such work, but any time lost by him on account of the city, private parties or companies making such improvements, etc., before paving, will be added to the time allowed to him to finish the work complete.

The contractor is obliged to be either himself at the work, or in his absence, to have a foreman present at the work, who will be responsible for the whole work under contract, and who must obey the order of the engineer or his assistant in relation to every part of work to be done under their contract.

The contractor agrees that he will pay, punctually, the workmen who shall be employed on the aforesaid work, in cash current, and not in what is denominated store pay,

And that all laborers to be employed on the said work shall be paid at the rate of not less than one dollar and fifty cents per day of ten hours' work.

(SECTION 39.) "No contract for the pavement of a street or avenue, or any part thereof, shall hereafter be let or made until the person or corporation to whom it is proposed to let the contract shall furnish *a bond to the city of Buffalo*, with two sufficient sureties, who shall each justify in full penalty of the bond. The condition of said bond to be that the contractor will, at his own cost and expense, keep and maintain the pavement in good condition for five (5) years from the completion of each pavement. Said bonds shall be approved by the Mayor, and filed with the Comptroller, and the amount of said bond to be equal to 33⅓ per cent. of the contract."

According to above section 39, chapter 3, the contractor binds himself to *repair the street* during said five (5) years whenever and wherever and in the manner as directed by the engineer within five (5) days after a written notice to that effect is served on him ; and in the event that the contractor refuses or fails to comply with such order of the engineer, then the sureties of said bonds will be notified and directed to proceed with the repairs as directed by the engineer. He further binds himself to repair the street only on the order of the engineer.

Payments will be made semi-monthly, according to an estimate of the engineer, of work completed and material delivered within twenty (20) per cent. to be drawn from the proper fund; said twenty (20) per cent. to be retained until the work is completed and accepted by the engineer in legal form.

In accordance with a resolution adopted February 7, 1887, all curb, crosswalk and other cut stone used in the paving of this street must be cut within the city limits.

Specifications of Connections.—Proper connections have to be made with all asphalt or stone pavements abutting to or crossing this street. The asphalt connections have to be made either according to "A. L. Barber's Asphalt Specification" or according to the kind of asphalt laid on these streets previously as to be directed.

The stone pavement connections must be made according to the .... class pavement specification, with the exception that excavation below is not required, but all necessary sand and stone is to be furnished by the contractor to fetch such repavement to proper grade and quality as to be directed.

The quantities for excavation and filling include all connections of unpaved cross streets, at a rise or fall of one foot in thirty-three, and for a width as to be directed in each case ; also all openings of gutters on said unpaved crossing streets, or other streets, where, and as far as necessary, as directed by the engineer.

The wooden or other crosswalks, where disturbed by grading said dirt street, must be carefully relaid in as good condition as they were before, at each crossing where they originally belonged.

Quantities.—Excavation, cub. yards ....; embankment, cub. yards ....; new curbing, lin. feet ....; protection curbing, lin. feet ....; old curbing reset, lin. feet....; length of street to be paved, lin. feet ....; new gutter, sq. feet ....; old gutter reset, sq. feet ...; new crosswalk, sq. feet ....; old crosswalks relaid, sq. feet ....; new paving, sq. yards ....; repaving with more or less old material, asphalt, sq. yards....; repaving with more or less old material, stone, sq. yards....; round corners....; connections with sidewalks....; inspector per day, $....

Engineer.

Specifications of the Filbert Vulcanite Asphaltic Pavement, laid by the National Vulcanite Company of New Jersey.

1. The vulcanite asphaltic pavement will be eight and one-half (8½) inches in thickness, as follows : The wearing surface will be one and one-half (1½) inches in thickness when compacted, with a bituminous base and binder seven inches in depth.

2. The space over which the pavement is to be laid will be excavated to the depth of eight and one-half (8½) inches below the top surface of the pavement when completed. Any objectionable or unsuitable material below the bed will be removed, and the space filled with clean gravel or sand well rammed. The bed will then be trimmed so as to be exactly parallel to the surface of the new pavement when completed, and the entire road-bed will be thoroughly rolled with a heavy steam-roller. Upon the foundation will be laid the base and binder, seven (7) inches in thickness, in the following manner :

3. The "base" will be composed of clean broken stone, that will pass through a three (3) inch ring, well rammed, and rolled with a steam-roller to the depth of five (5) inches, and thoroughly coated with hot paving cement, composed of No. 4 tar distillate in the proportion of about one (1) gallon to the square yard of pavement.

4. The second or "binder" course will be composed of clean broken stone, thoroughly screened, not exceeding one and one-quarter (1¼) inches in the largest dimensions, and No. 4 tar distillate. The stone will be heated by passing through revolving heaters, and thoroughly mixed by machinery with the distillate in the proportion of one gallon of distillate to one (1) cubic foot of stone.

5. The "binder" will be hauled to the work, spread upon the base course at least two (2) inches thick, and immediately rammed and rolled with hand and heavy steam-rollers while in a hot and plastic condition.

6. The wearing surface will be one and one-half (1½) inches thick when compacted, made of paving cement, composed of twenty-five (25) per cent. of asphalt and seventy-five (75) per cent. of distillate mixture, with other materials, as follows : Clean, sharp sand will be mixed with pulverized stone of such dimensions as to pass through a one-quarter (¼) inch screen in the proportion of two to one.

To twenty-one (21) cubic feet of the above named mixture will be added one peck of dry hydraulic cement, one quart of flour of sulphur, and two quarts of air-slacked lime. To this mixture will be added three hundred and twenty (320) pounds of paving cement, to compose the wearing surface.

7.   The material will be heated to about 250° Fahr.—the paving cement in kettles, the sand and stone, etc., in revolving heaters. They will be thoroughly mixed by improved machinery, and the mixture carried upon the work, when it will be spread upon the binder course two (2) inches thick with hot iron rakes and other suitable appliances, and immediately compacted with tamping-irons, hand and steam rollers, while in a hot and plastic state. The surface will be finished with a dusting of dry hydraulic cement rolled in.

8.   The pavement so constructed must be a solid mass, eight and one-half (8½) inches thick, and will be thoroughly rolled and cross-rolled until it has become hard and solid.

9.   The pavement shall be equal in every respect to that laid on K Street, between Ninth and Eighteenth Streets, N. W., Washington, D. C., in 1874 and 1875.

ASPHALT PAVEMENT ON STONE FOUNDATION IN NEW YORK.*

The following specifications for the particular kind of pavement required is of interest, since they are prepared by the Department of Public Works, presumably after careful investigation, it being remembered that for the expenditure of the $3,000,000 appropriation for new pavements, a special engineering department has been created, with Stevenson Towle, M. Am. Soc. C. E., as engineer in charge.

Specifications for Laying an Asphalt Pavement on the present Stone-block Pavement in Park Avenue, between Thirty-fourth and Fortieth Streets, New York City.

1. *Work and Materials must Agree with Specifications.*—All the materials furnished, and all the work done, which, in the opinion of the Commissioner of Public Works, shall not be in accordance with these specifications, shall be immediately removed and other materials furnished, and work done that will, in the opinion of said Commissioner, be in accordance therewith. Before any materials are placed upon the street or avenue, the Commissioner of Public Works shall approve of the quality and finish of samples of the same, which shall be furnished at his office.

2. *Inspectors on Subdivisions of Work.*—The work under this agreement is to be prosecuted at and from as many different points in such part or parts of the street or avenue on the line of the work as the said Commissioner may, from time to time, determine, and at each of said points inspectors may be placed on the day designated for the commencement of the work thereat. Whenever any work is in progress at or from one or more points at a time, an inspector may be appointed by said Commissioner to supervise each subdivision of the same, whether such subdivision be the culling of the bridge-stones, or the excavation for and preparation of the foundation, or the laying of the pavement, or the laying of the bridge-stones or otherwise. The aggregate time of all the inspectors so employed will be the time with which the time allowed for completion of the work under this agreement will be compared. The inspectors will be paid each at the rate of three and one-half dollars per day.

*xx, 256.

3. *Right to Construct Sewers, etc., prior to Laying of Pavement.*
—The right to construct any sewer or sewers, or receiving-basins or culverts, or to build up or adjust any man-holes, or to reset or renew any frames and heads for sewer man-holes, or for Croton water or gas stop-cocks, or to lay gas or water pipes, or to construct necessary appurtenances in connection therewith in said street, or to grant permits for house connections with sewers or with water or gas pipes, at any time prior to the laying of the new pavement over the line of the same, is expressly reserved by said Commissioner ; and said Commissioner of Public Works reserves the right of suspending the work on said pavement on any part of the line of said street or avenue at any time during the construction of the same, for the purposes above stated, without other compensation to the contractor for such suspension than extending the time for completing the work as much as it may, in the opinion of the said Commissioner, have been delayed by such suspension ; and said contractor shall not interfere with, or place any impediment in the way of any person or persons who may be engaged in the construction of such sewer or sewers, or in making connections therewith, or doing other work above specified, or in the construction of any receiving-basins and culverts, or in setting or resetting any curb or gutter-stones on the line of the street or avenue.

Contractor to Remove Incumbrances.—In case there shall be, at any time stipulated for the commencement of the work, any earth, rubbish, or other incumbrance on the line of the work, the same is to be removed at the expense of the contractor.

4. *Bridge-stones.*—When new bridge-stones are required they are to be furnished in conformity with the following description, to wit:

The new bridge-stones to be of blue stone equal in quality to the best North River blue stone, free from seams and imperfections. Each stone to be not less than 4 nor more than 8 feet long, except in cases where especially permitted, and 2 feet wide, and of a uniform thickness, which may vary from 5 to 8 inches, and dressed to a face on top not varying in evenness by more than one-fourth of an inch, and on the bottom bedded, with sides square and full, and ends cut to such bevel as shall be directed by the Commissioner of Public Works. The new stone to be in quality and workmanship equal to the pattern at the office of the Department of Public Works, and to be cut so as to lay to a joint not exceeding $\frac{1}{4}$-inch from top to bottom on the ends, and $\frac{1}{8}$-inch on the sides. Old bridge-stones shall be relaid in accordance with the specifications for laying new bridge-stones.

5. *Manhole-heads, etc.*—All the frames and heads for sewer man-holes, and for Croton water or gas stop-cocks, on the line of the work are to be reset or new ones set if required, on a level with the new pavement, by the Department of Public Works or the gas company. The sewer man-holes, if below the grade, will be built up to the proper height by said department.

6. *Old Curb and Gutter Stones.*—The old curbstones along the line of the work, the pavement in the intersections that may be retained, and the pavements adjoining, and also the gutters of the adjoining pavements, as far as may be necessary to obtain proper drainage and adjustment between the old and new pavement, shall be readjusted and brought to the grade and lines of the proposed pavement, as, and to the extent required, without extra charge therefor.

7. The stone-block pavement will be prepared in the following manner for a

Foundation.—The surface of the stone-block pavement must be thoroughly swept with stiff brooms until all dirt and fine particles have been removed from the surface, and from the joints to a depth of two inches.

The surface must then be brought to a uniform grade and cross-section, not to exceed a crown of 5 inches on a roadway 30 feet wide, by filling all depressions with a fine bituminous concrete or binder, to be composed of clean broken stone not exceeding $1\frac{1}{4}$ inches in their largest dimensions, thoroughly screened, and coal tar residuum, commonly known as No. 4 paving composition.

The stone must be heated by passing through revolving heaters, and thoroughly mixed by machinery with the paving composition in the proportion of one gallon of paving composition to one cubic foot of stone.

This binder must be hauled to the work and spread with hot iron rakes in all holes or inequalities and depressions below the true grade of the pavement, to such thickness that after being thoroughly compacted by tamping and hand-rolling the surface shall have a uniform grade and cross-section, and the thickness of the binder at any point shall be not less than three-fourths of an inch.

The upper surface shall be exactly parallel with the surface of the pavement to be laid.

Upon this foundation shall be laid the wearing surface, or paving proper, the basis of which or paving cement, will be pure asphaltum, unmixed with any of the products of coal tar.

The wearing surface will be composed of :

1. Refined asphaltum.
2. Heavy petroleum oil.
3. Fine sand, containing not more than one per centum of hydro-silicate of alumina.
4. Fine powder of carbonate of lime.

The asphaltum shall be specially refined and brought to a uniform standard of purity and gravity, of quality to be approved by the Commissioner of Public Works.

The heavy petroleum oil shall be freed from all impurities and brought to a specific gravity of from 18 to 22° Beaume, and fire test of 250° Fahr.

From these two hydro-carbons shall be manufactured an asphaltic cement which shall have a fire test of 250° Fahr., and, at a temperature of 60° Fahr., shall have a specific gravity of 1.19, said cement to be composed of 100 parts of pure asphalt, and from 15 to 20 parts of heavy petroleum oil.

The asphaltic cement being made in the manner above described, the pavement mixture will be formed of the following materials, and in the proportions stated :

Asphaltic cement...................................... from 12 to 15
Sand........... ... .from 83 to 70
Pulverized carbonate of lime........................from 5 to 15

The sand and asphaltic cement are to be heated separately to about 300° Fahr. The pulverized carbonate of lime, while cold, shall be mixed with the hot sand in the required proportions, and then mixed with the asphaltic cement at the required temperature, and in the proper proportion, in a suitable apparatus, which will effect a perfect mixture.

The pavement mixture prepared in the manner thus indicated, shall be laid on the foundations in two coats. The first coat, called cushion coat, shall contain from 2 to 4 per cent. more asphaltic cement than given above; it will be laid to such depth as will give a thickness of $\frac{1}{4}$ inch after being consolidated by a roller. The second coat, called surface coat, prepared as above specified, shall be laid on the cushion coat ; it shall be brought to the ground in carts, at a temperature of about 250° Fahr., and if the temperature of the air is less than 50 degrees, iron carts, with heating apparatus, shall be used in order to maintain the proper temperature of the mixture ; it shall then be carefully spread by means of hot iron rakes, in such a manner as to give a uniform and regular grade, and to such depth that after having received its ultimate compression, it will have a thickness of 2 inches. The surface shall then be compressed by hand-rollers, after which a small amount of hydraulic cement shall be swept over it, and it shall then be thoroughly compressed by a steam roller weighing not less than 250 pounds to the inch run ; the rolling to be continued for not less than five hours for every 1,000 yards of surface.

The powdered carbonate of lime shall be of such degree of fineness that 5 to 15 per centum of weight of the entire mixture for the pavement shall be of an impalpable powder of limestone, and the whole of it shall pass a No. 26 screen. The sand shall be of such size that none of it shall pass a No. 80 screen, and the whole of it shall pass a No. 10 screen.

In order to make the gutters, which are consolidated but little with traffic, entirely impervious to water, a width of 12 inches next the curb will be coated with hot pure asphalt and smoothed with hot smoothing irons in order to saturate the pavement to a certain depth with an excess of asphalt.

If rock asphalt be used, it shall be natural bituminous limestone rock (1) from the Sicilian mines at Ragusa, equal in quality and composition to that mined by the United Limmer and Verwohle Rock Asphalt Company, Limited, (2) from the Swiss mines at Val de Travers, equal in quality and composition to that mined by the Neuchatel Rock and Asphalt Company, Limited, or (3) from the French mines at Seyssel, equal in quality and composition to that mined by the Compagnie Générale des Asphaltes de France, and it shall be prepared and laid as follows :

(1) The lumps of rock shall be finely crushed and pulverized, the powder shall then be passed through a fine sieve. Nothing whatever shall be added to or taken from the powder obtained by grinding the bituminous rock. The powder shall contain 9 to 12 per cent. natural bitumen, 88 to 91 per cent. pure carbonate of lime, and must be free from quartz, sulphates, iron pyrites or aluminum. (2) This powder shall be heated in a suitable apparatus to 200° or 250° Fahr., brought to the ground at such temperature in carts made for the purpose, and then carefully spread on the foundation previously prepared, to such a depth that after having received its ultimate compression, it will have a thickness of two inches. (3) It shall be skillfully compressed by heated rammers and rolled until it shall have the required thickness of two inches. (4) The surface to be rendered perfectly even by heated smoothers and rolled with a steam roller weighing not less than 250 pounds to the inch run, the rolling to continue for not less than five hours for every 1,000 yards of surface.

8. *Laying the Crosswalks, etc.*—The crosswalks adjoining the new pavement are to be laid, or the present bridge-stones shall be relaid, as the

said Commissioner of Public Works may direct, in which last case they shall be dressed or redressed so as to form one-quarter of an inch joints from top to bottom when laid. All the new bridge-stones, and such of the present bridge-stones as may be retained, are to be well and firmly bedded on a foundation of sand or gravel, not less than six inches in thickness, and laid with joints not exceeding one-quarter of an inch in width from top to bottom on the ends. The courses to be so laid that the transverse joints will be broken by a lap of at least one foot. The bridge-stones and pavements adjoining, and also the gutters of the adjoining pavements, as far as in the opinion of the said Commissioner may be necessary to obtain proper drainage, will be taken up, brought to the grade of the new pavement and relaid, without extra charge therefor. The contractor shall lay one or two rows of paving blocks between the courses of bridge-stones, when directed so to do by the Water Purveyor.

9. *Old Materials.*—All old materials which it becomes necessary to remove, excepting the trap and granite blocks, the sewer manhole-heads, and the frames and heads to Croton water or gas stop-cocks, shall be considered as the property of the contractor, and the same shall be immediately removed by him from the line of the work.

10. *Clearing Up.*—All surplus materials, earth, sand, rubbish and stones, except such stone as shall be retained by order of the Water Purveyor, are to be removed from the line of the work block by block, as rapidly as the work progresses. At any time within one month after the completion of the pavement of each block, or of the entire work, if so required by the Commissioner of Public Works, all material except building material covering the stone pavement shall be swept into heaps and immediately removed from the line of the work; and unless this be done by the contractor within forty-eight hours after being notified so to do, to the satisfaction of said Commissioner, the same shall be removed by the said Commissioner of Public Works, and the amount of the expense thereof shall be deducted out of any moneys due or to grow due to the party of the second part under this agreement.

11. *Loss or Damage to be Sustained by Contractor.*—It is further agreed that all loss or damage arising out of the nature of the work to be done under this agreement, or from any unforseen obstructions or difficulties which may be encountered in the prosecution of the same, or from the action of the elements, or from incumbrances on the line of the work, shall be sustained by the said contractor.

In case any injury is done along the line of the work in consequence of any act or omission on the part of the contractor or his employees or agents in carrying out any of the provisions or requirements of this contract, the contractor shall make such repairs as are necessary in consequence thereof, at his own expense and to the satisfaction of the Commissioner of Public Works, and in case of failure on the part of the contractor to promptly make such repairs, they may be made by the Commissioner of Public Works, and the expense thereof shall be deducted out of any moneys to grow due, or retained from, the party of the second part under this contract.

12. *Work may be Suspended.*—The prosecution of the work shall be suspended for such periods as the Commissioner of Public Works may from time to time determine; no claim or demand will be made by the contractor for damages by reason of such suspensions in the work, but the period of such suspensions, to be determined in writing by the said Commissioner,

will be excluded in computing the time hereinafter limited for the completion of the work. During such suspensions all materials delivered upon, but not placed in the work, shall be neatly piled or removed so as not to obstruct public travel.

13. *"Contractor," etc., to Mean.*—Whenever the word "contractor," or the words "party of the second part," or a pronoun in place of either of them is used in this contract, the same shall be considered as referring to and meaning the party or parties, as the case may be, of the second part to this agreement.

13a. *Security to be Retained for Repairs.*—And it is further agreed, that if, at any time during the period of *fifteen years* from the date of the acceptance by said Commissioner of the whole work under this agreement, the said work, or any part or parts thereof shall, in the opinion of said Commissioner, require repairs or sanding, and the said Commissioner shall notify the said party of the second part to make the repairs or do the sanding so required, the said party of the second part shall immediately commence and complete the same to the satisfaction of said Commissioner ; and in case of failure or neglect on his part to do so within forty-eight hours from the date of the service of the aforesaid notice, then the said Commissioner of Public Works shall have the right to purchase such materials as he shall deem necessary, and to employ such person or persons as he may deem proper, and to undertake and complete the said repairs or sanding, and to pay the expense thereof, out of any sum of money due the contractor or retained by the said parties of the first part, as herein mentioned. And the parties of the first part hereby agree, upon the expiration of the said period of *fifteen years*, provided that the said work shall at that time be in good order, or as soon thereafter as the said work shall have been put in good order to the satisfaction of the said Commissioner, to pay to the said party of the second part the whole of the sum last aforesaid or such part thereof as may remain after the expenses of making the said repairs in the manner aforesaid shall have been paid therefrom. And it is hereby further agreed, between the parties hereto, that if the termination of the said period of *fifteen years* after the completion and acceptance of the work done under the agreement shall fall within the months of December, January, February or March, then and in that case said months of December, January, February or March shall not be included in the computation of said period of *fifteen years*, during which the work is to be kept in repair by the contractor, as aforesaid ; and also in that case the payment to be made under the provisions of this paragraph shall not be made before the 15th of April next thereafter, unless otherwise specially permitted by the Commissioner of Public Works.

The party of the second part further agrees that he will lay and restore the pavement over trenches made for laying water and gas pipes, sewers and for other purposes permitted by the Commissioner of Public Works at the contract price, and when once so laid and restored, maintain the same in the same state of repair as agreed to for the other parts of the pavement. He further agrees not to demand additional or further payment on account of injury or sinking of the pavement so laid and restored.

14. *Measurement.*—The said party of the second part further agrees that the return of the engineer appointed by the Commissioner of Public Works to survey the work, shall be the account by which the amount of work done shall be computed.

15. *Work to Commence.*—The said party of the second part hereby further agrees that he will commence the aforesaid work on such a day and at such point or points as the Commissioner of Public Works may designate, and progress therewith so as to fully complete the same, in accordance with this agreement, on or before the expiration of fifty days next thereafter ; and that in the computation of said days the time (aggregated in days and parts of days) during which the work required by this contract has been delayed in consequence of the condition of the weather, or by any act or omission on the part of the parties of the first part (all of which shall be determined by the said Commmissioner of Public Works, who shall certify to the same in writing), and also Sundays and holidays on which no work is done, and days on which the prosecution of the whole work is suspended by written order of the said Commissioner, shall be excluded.

But neither an extension of time, for any reason, beyond the date fixed herein for the completion of the work, nor the doing or acceptance of any part of the work called for by this contract, shall be deemed a waiver by the said Commissioner of the right to abrogate this contract for abandonment or delay in the manner provided for in paragraph (17) of this agreement.

Damages for Non-Completion.—And the said party of the second part hereby further agrees that the said parties of the first part shall be and they are hereby authorized to deduct and retain out of the moneys which may be due or become due to the said party of the second part under this agreement, as damages for the non-completion of the work aforesaid within the time hereinbefore stipulated for its completion, or within such further time as in accordance with the provisions of this agreement shall be fixed or allowed for such performance or completion, the sum of twenty dollars for each and every day the aggregate time of all the inspectors employed upon said work may exceed the time stipulated for its completion, or such stipulated time as the same may be increased as hereinbefore provided, which said sum of twenty dollars per day is hereby, in view of the difficulty of estimating such damages, agreed upon, fixed and determined by the parties hereto, as the liquidated damages that the parties of the first part will suffer by reason of such default, and not by way of penalty.

16. *Personal Attention.*—The said party of the second part hereby further agrees that he will give his personal attention constantly to the faithful prosecution of the said work ; that he will not sublet the aforesaid work, or any part thereof, without the previous written consent of the Commissioner of Public Works indorsed on this agreement, but will keep the same under his own control ; that he will not assign, by power of attorney or otherwise, any of the moneys payable under this agreement, unless by and with the like consent, to be signified in like manner ; that no right under this contract, nor to any moneys to become due hereunder, shall be asserted against the Mayor, Aldermen, and Commonalty of the City of New York, or any department, bureau, or officer thereof, by reason of any so-called assignment, in law or equity of this contract, or any part hereof ; that no person other than the party signing this agreement as the party of the second part has now any claim hereunder ; that no claim shall be made excepting under this specific clause, or under paragraph 18 of this agreement ; and that he will punctually pay the workmen who shall be employed on the aforesaid work in cash current, and not in what is denominated store

pay. If at any time any overseer or workman employed by the contractor shall be declared by the Water Purveyor to be unfaithful or incompetent, the contractor, on receiving written notice, shall forthwith dismiss such person, and will not again employ him on any part of the work.

17. *Contract May Be Declared Annulled for Violation, etc.*—The said party of the second part further agrees that if at any time the Commissioner of Public Works shall be of opinion, and shall so certify in writing, that the said work, or any part thereof, is unnecessarily delayed, or that the said contractor is willfully violating any of the conditions or covenants of this contract, or is executing the same in bad faith, or if the said work be not fully completed within the time named in this contract for its completion, he shall have the power to notify the aforesaid contractor to discontinue all work, or any part thereof, under this contract, by a written notice to be served upon the contractor, either personally or by leaving said notice at his residence or with his agent in charge of the work, and thereupon the said contractor shall discontinue said work, or such part thereof, and the Commissioner of Public Works shall thereupon have the power to place such and so many persons as he may deem advisable, by contract or otherwise, to work at and complete the work herein described, or such part thereof, and to use such materials as he may find upon the line of said work, and to procure other materials for the completion of the same, and to charge the expense of said labor and materials to the aforesaid contractor, and the expense so charged shall be deducted and paid by the party of the first part, out of such moneys as may be then due, or may at any time thereafter grow due, to the said contractor, under and by virtue of this agreement, or any part thereof ; and in case such expense is less than the sum which would have been payable under this contract if the same had been completed by said contractor, he shall forfeit all claim to the difference ; and in case such expense shall exceed the last said sum, he shall pay the amount of such excess to the parties of the first part.

18. *Claims for Labor, etc.*—And it is further agreed by and between the parties hereto, that if, at any time before or within thirty days after the whole work herein agreed to be performed has been completed and accepted by the parties of the first part, any person or persons claiming to have performed any labor or furnished any materials towards the performance or completion of this contract, shall file with the said Department of Public Works, or with the Bureau having charge of said work, and with the head of the Finance Department of the said City of New York, any such notice as is described in the Act of the Legislature of the State of New York, passed May 22, 1878, entitled "An Act to secure the payment of laborers, mechanics, merchants, traders and persons furnishing materials towards the performing of any public work in the cities of the State of New York," then, and in every such case, the said parties of the first part shall retain, anything herein contained to the contrary thereof notwithstanding, from the moneys under their control, and due or to grow due under this agreement, so much of such moneys as shall be sufficient to pay off, satisfy and discharge the amount in such notice alleged or claimed to be due to the person or persons filing such notice, together with the reasonable costs of any action or actions brought to enforce such claim or the lien created by the filing of such notice. The money so retained shall be retained by the said parties of the first part until the lien thereon created by the said act and the filing of the said notice shall be discharged, pursuant to the provisions of the said act.

And the said party of the second part hereby further agrees that he will furnish said Commissioner with satisfactory evidence that all persons who have done work or furnished materials under this agreement, and who may have given written notice to the said Commissioner, at any time within ten days after the completion of the work aforesaid, that any balance for such work or materials is still due and unpaid, have been fully paid or satisfactorily secured.

Amounts Claimed Retained.—And in case such evidence be not furnished as aforesaid, such amount as may be necessary to meet the claims of the persons aforesaid shall be retained from any moneys due the said party of the second part under this agreement until the liabilities aforesaid shall be fully discharged or secured, or such notice be withdrawn.

19. *Indemnification of City.*—And the said party of the second part further agrees that during the performance of said work he will place proper guards upon and around the same for the prevention of accidents, and at night will put up and keep suitable and sufficient lights, and that he will indemnify and save harmless the parties of the first part against and from all suits and actions, of every name and description, brought against them, and all costs and damages to which they may be put for or on account or by reason of any injury or alleged injury to the person or property of another, resulting from negligence or carelessness in the performance of the work, or in guarding the same, or from any improper materials used in its prosecution, or by or on account of any act or omission of the said party of the second part or his agents ; and the said party of the second part hereby further agrees that the whole or so much of the moneys due to him under and by virtue of this agreement, as shall or may be considered necessary by the Commissioner of Public Works, shall and may be retained by the said parties of the first part until all such suits or claims for damages as aforesaid shall have been settled, and evidence to that effect furnished to the satisfaction of the said Commissioner.

20. And the said party of the second part hereby agrees to indemnify and save harmless the parties of the first part against and from all suits and actions of every nature and description arising out of the claim or claims of any person or persons claiming to be patentees of any process connected with the work herein agreed to be performed, or of any material or materials used upon said work. And the said party of the second part hereby further agrees to execute, with two sufficient sureties, the bond, in the sum of $10,000 attached to this agreement, for the indemnification of the parties of the first part against and from all such suits and actions as aforesaid.

21. *Prices for Work.*—And the said party of the second part hereby further agrees to receive the following prices as full compensation for furnishing all the materials and performing all the labor which may be required in the prosecution of the whole of the work to be done under this agreement and in all respects performing and completing the same, to wit:

For completed asphalt pavement, per square yard, the sum of.......
....

For the new bridge stone, per square foot, the sum of..............
....

It being expressly understood that the measurement shall be taken after the laying and setting of the pavement, and that the aforesaid prices cover the furnishing of all the different materials and all the labor, the maintaining of said pavement in good order, and sprinkling with clean sharp sand, such portions of said pavement, and as often as the said Commissioner shall direct, and for the period of fifteen years, and the performance of all the work mentioned in this specification and agreement.

Examinations.—After the completion of the work, should the Water Purveyor require it for his more perfect satisfaction, the contractor shall make such openings and to such extent through such part or parts of the said work as the Water Purveyor shall direct, and he shall make the same good again to the satisfaction of the Water Purveyor. Should the work be found faulty in any respect, the whole of the expense incurred thereby shall be defrayed by the contractor, but if otherwise by the parties of the first part to this agreement.

22. *Payments when Made.*—And the said party of the second part further agrees that he shall not be entitled to demand or receive payment for any portion of the aforesaid work or materials until the same shall be fully completed in the manner set forth in this agreement, and such completion shall be duly certified by the Engineer, Inspecter and Water Purveyor in charge of the work, and until each and every of the stipulations hereinbefore mentioned are complied with, and the work completed to satisfaction of the Commissioner of Public Works, and accepted by him ; whereupon the parties of the first part, under chapter 346 of the Laws of 1889, will pay, and hereby bind themselves and their successors to pay, to the said party of the second part, in cash, on or before the expiration of thirty days from the time of the completion of the work and the acceptance of the same by the Commissioner of Public Works, 70 per cent. of the whole of the moneys accruing to the said party of the second part under this agreement, and the balance of the moneys that may be due to said party of the second part, under this agreement, as follows : Three per cent. of the whole amount of money accruing to said party of second part on the expiration of the sixth year, and a like further sum of three per cent. at the expiration of each succeeding year thereafter until the whole or as much as may remain due of said contract price shall be paid, should the party of the first part perform the work stipulated under section (13A.) of this agreement. But in case the amount payable under this contract shall be five thousand dollars or over, payments will be made to the said party of the second part, in conformity with and subject to the terms and conditions of an ordinance of the Mayor, Aldermen and Commonalty of the City of New York, passed December 30, 1854, and amended March 8, 1861, entitled, " An ordinance to authorize the issue of Bonds upon Contracts payable by Assessments in pursuance of the Act of the Legislature, passed April 16, 1852," by monthly installments of seventy per cent. on the amount of work performed, and also on the quantity of materials furnished and delivered, should the Commissioner of Public Works deem it advisable so to do, in which case, however, the quantity returned shall be such that the amount paid will be fairly due and in accordance with the provisions and stipulations of this agreement ; *provided*, the amount of work done on each installment shall not be less than fifteen hundred dollars ; *and provided*, that the parties of the first part may at all times reserve and retain out of said installments, or any of them, all such sum or sums as by the terms hereof, or of any act of the Legislature of the State of New York, or of any ordinance or resolution of the Common Council of the City of New York, passed prior to the date hereof, they are or may be authorized to reserve or retain ; *and provided*,

that nothing herein contained be construed to affect the right hereby reserved of the said Commissioner to reject any return or certificate of the Engineer or Inspector having charge of the work, should such return or certificate be, in the opinion of the Commissioner of Public Works, not in accordance with the facts of the case, or the requirements of this agreement, or be otherwise improperly given, and to reject the whole or any portion of the aforesaid work, should the same or any part thereof not be in accordance with the requirements of this contract; *and provided also*, that where the contractor, although the lowest bidder in the gross calculation, is to receive unusual or extraordinary prices for the different items, or any of them, of the work, when considered separately, nothing herein contained shall be construed to affect the right of the Commissioner hereby retained to determine the amount that may be due, from time to time, not necessarily by the rates agreed upon in this contract, but by causing an estimate to be made of the value of the work done, taking as a basis of the calculation the whole amount of money that will have become due, according to the terms of this contract, when the whole work shall be completed.

23. It is further expressly understood and agreed by and between the parties hereto, that the action of the engineer or surveyor by which the said contractor is to be bound and concluded according to the terms of this contract, shall be that evidenced by his final certificate, all prior certificates upon which seventy per cent. payments may be made being merely estimates, and subject to the corrections of such final certificates, which may be made without notice to the contractor thereof, or of the measurements upon which the same is based.

24. And it is hereby expressly agreed and understood by and between the parties hereto, that the said parties of the first part, their successors and assigns, shall not, nor shall any department or officer of the City of New York, be precluded or estopped by any return or certificate made or given by any engineer, inspector or other officer, agent or appointee of said Department of Public Works, or said parties of the first part, under or in pursuance of anything in this agreement contained, from at any time showing the true and correct amount and character of the work which shall have been done and materials which shall have been furnished by the said party of the second part or any other person or persons under this agreement.

In witness whereof, the... .... Commissioner of Public Works has hereunto set his hand and seal on behalf of the said parties of the first part, and the said paty of the second part has also hereunto set his hand and seal ; and said.......... Commissioner and party hereto of the second part have executed this agreement in triplicate, one part of which is to remain with the said Commissioner, one other to be filed with the Comptroller of the City of New York, and the third to be delivered to the said party hereto of the second part, the day and date herein first above written.

Signed and sealed in presence of

.... Commissioner of Public Works.

............Contractor.

ASPHALT PAVEMENT IN NEW YORK.*

The specifications for laying asphalt pavement on Park Avenue, in New York City, over existing stone pavements, which we printed in full October 5, are attracting some attention, and the *Tribune*, which has criticised Mr. Gilroy, Commissioner of Public Works, calls them a refreshing illustration of the way in which public work ought to be done. A point of interest noticed was the provision for retaining 30 per cent. of the contract price subject to the contractor's keeping the pavement in repair fifteen years.

The *Tribune* commends this, even though it may naturally involve an increase in the price paid by the city for the work.

The Department, though, has valid precedents for this provision, as will be seen by reference to the articles on Pavements and Street Railroads in these columns.

The usual requirement of the City of London, as will be seen in our issue of February 4, 1888, was that the contractors maintain the pavement in good repair two years without cost to the city, and after that, fifteen years more at a price named at the time of bidding; the pavement to be in good condition at the end of seventeen years and then to weigh not less than a given amount per square yard. In Paris the practice has been rather to compel the contractors to guarantee the condition of the pavement for long terms of years outright, as appears in our issues of April 7 and October 13, 1888. This is entirely reasonable, and any community can afford to pay responsible parties an extra price as an insurance premium.

For any work experimental in its nature, or not fully established in character by actual test, the method proposed is unquestionably proper.

It is also proper to keep in view the importance of exacting from the contractors what they promise, and especially is this true in a city where politics sometimes interferes with administrative work.

THE ASPHALT PAVEMENTS OF BERLIN.†

The annual reports published by the Municipality of Berlin afford some interesting reading with regard to the cost, etc., of the asphalt paving in that city since its introduction. The *first experiments* with Val de Travers asphalt were made in 1873. These proving satisfactory, a commencement on a small scale was made in 1877, when 2,556 square meters were laid down. From that year the *extension* of asphalt pavement grew rapidly. At the end of 1878, 23,586 square meters of the public streets of Berlin were paved with asphalt; at the close of 1879, 63,258 square meters; 1880, 106,223

---

* Ed. xx, 282. † xvii, 346.

square meters; 1881, 125,034 square meters. By April 1, 1883, 187,672 square meters of asphalt pavement had been laid down; at the same date in 1884, 253,586 square meters; in 1885, 322,042 square meters; in 1886, 359,409 square meters; in 1887, about 412,000 square meters. At present the superficies of asphalt paving in Berlin is about 470,000 square meters, and further 76,000 square meters have been ordered to be executed. The above figures show that asphalt pavement, notwithstanding its reported disadvantages, is in great favor at Berlin, at any rate. As to *expenses*, according to the *Builder*, a square meter of asphalt paving costs there, on an average, 17 marks 50 pfennigs (17*s.* 6*d.*), while the cost of maintenance for twenty years, by contract, is 7 marks 50 pfennigs (7*s.* 6*d.*). The *total cost* per square meter of asphalt pavement for twenty years is, therefore, 25 marks (£1 5*s.*). This is only 75 pfennigs (9*d.*) more per square meter than the cost of granite pavement. From the report of the Municipality of Berlin, it appears that the complaints with regard to the slippery condition of asphalt pavement are gradually disappearing. On the one hand, great improvements, by the light of experience, have been introduced in the treatment of the asphalt; on the other hand, both "drivers and horses are getting more and more used to the new pavement." The *cleansing* of the pavement also leaves little to be desired, although only forty-five boys are engaged in the work of cleaning nearly 500,000 square meters.

COST OF ASPHALT PAVEMENTS IN LIVERPOOL.*

Mr. Dunscombe, City Engineer of Liverpool, sends the following table as to the cost in Liverpool of pavements of compressed natural asphalt:

Thickness of Asphalt.	Price for construction and maintenance of carriage-way for three years per superficial yard per annum.	Price for maintenance of carriage-way from year to year by contractor, per superficial yard per annum.	Price for maintenance of carriage-way by contractor for a further period of thirteen years per superficial yard.	Price for repairs by contractor for carriage-way, per superficial yard repaired.
1	2	3	4	5
⅞ inch.	....	....	....	....
1 inch.	5s. 6d.	3d.	3s. 3d.	9s.
1¼ inch.	7s.	4d.	4s. 4d.	11s. 3d.
1½ inch.	8s. 6d.	4d.	4s. 4d.	13s. 6d.
1¾ inch.	10s. 3d.	9d.	9s. 9d.	16s.
2 inch.	11s. 3d.	6d.	6s. 6d.	18s.
2¼ inch.	12s. 3d.	6d.	6s. 6d.	20s.

*xv, 485.

The reason for the high prices in the fifth column is the small amount of repairs required and the loss of time consequent on this.

By the terms of the contracts with those putting down pavements, all repairs are to be made on twenty-four hours' notice from the City Engineer, and such repairs do not pay the contractor, often, for the work done. Column 4 of the table is obtained by multiplying the quantities in the previous column by thirteen.

COST OF MAINTAINING ASPHALT PAVEMENTS IN LONDON.*

Return showing the annual cost of maintaining the carriage-way pavements in some of the principal thoroughfares of the city of London. Average for fifteen years, from City Surveyor William Haywood's report, January, 1882 :

Name of Thoroughfare.	Description of Pavement.	Annual cost of maintenance per yard superficial.	
		s.	*d.*
Bishopsgate Street Within..............	Val de Travers Asphalt.	1	3
Cheapside and Poultry	"	1	6
Fenchurch Street, between Gracechurch Street and Railway Place............	"	1	6
Finsbury Pavement and Moorgate.......	"	0	9
Gracechurch Street.....................	"	1	0
Gresham Street.......................	"	1	3
King William Street, narrow portion.....	"	1	3
London Wall...	"	0	9
Moorgate Street, between Coleman Street Buildings and London Wall..........	"	0	9
Moorgate Street, between Lothbury and Telegraph Streets..................	"	1	0
New Broad Street and Old Broad Street.	"	0	9
Paternoster Row.......................	"	1	0
Queen Street, between Cheapside and Pancras Lane......................	"	0	9
Queen Street, between Pancras Lane and Queen Victoria Street................	"	1	3
Threadneedle Street....	"	1	3
Queen Victoria Street, from Mansion House to Cannon Street................	"	0	6
Aldgate........	Limmer Asphalt.	0	9
Cornhill.............................	"	0	9
Lombard Street.......................	"	0	9
Mark Lane...........................	"	1	0
Mincing Lane.........................	"	0	9
Moorgate Street, from Telegraph Street to Coleman Street Buildings..........	"	0	9
Newgate Street.......................	"	0	9
Fenchurch Street (Eastern end).........	Société Francaise des Asphalt.	0	9
King Street, Cheapside.................	"	0	9
Princes Street (part of)............. ..	"	1	3
Philpot Lane.........................	"	0	6
Milton Street........................	"	0	6

Average of 16 Val de Travers, 1s,   Average of 7 Limmer, 9½d.
Average of 5 Société Francaise, 11d.

* xxi, 98.

IMPROPER LAYING AND MAINTENANCE OF PAVEMENTS.*

BOSTON, April 23, 1887.

SIR :—I desire to call attention to the asphalt pavement on Columbus Avenue.   This is the principal piece of asphalt in Boston, I believe, and was laid only about three years ago.   But it shows what result may be expected from improper laying and caring for road-bed surfaces.   There is a double line of horse car tracks down the middle of this street, thus dividing the surface in two strips of asphalt separated by a strip of stone-block pavement. Now, instead of protecting each side of the track by block pavement to the full width of gauge of large trucks, the block extends only about 20 inches on one side (the north-west) and from 4 to 8 inches on the south-east (that is, one header and one stretcher).   The consequence is that the asphalt is all worn and rotted down next to the track on the south-east side for a width of 14 to 16 inches, and on the north-west side about the width of a wheel-tire, the marks of the heavy truck wheels measuring seven feet out from the edge of the inside rail of track on each side.   Now, if the pavement of stone blocks had been carefully laid out to the full width of seven feet from inside rail, and laid on a solid concrete foundation, so as to support the outer wheel of the heavy truck while the other wheel was running in the groove of the inside, the asphalt would have worn as evenly next to the pavement as farther away from it.   Just at present these worn ruts, some 2 to 4 inches deep on the southeast side of the track, are being *repaired.*   That is, the asphalt is cut out in sections, and the hole filled with *cement concrete* to the level of the old asphalt.   No repairs have been made on the general surface, and it presents lots of cradle holes, which should have been cut out and relaid as soon as they appeared ; then, in many places the gutter is all worn down and is a mass of wet, pasty substance, frequently extending out in patches six or eight feet from sidewalk, thus destroying the quality of the surface.   It is this kind of carelessness in regard to asphalt (or any other pavement) which brings it in disrepute, and really increases the cost of an already costly pavement, as its life is shortened by one-half or two-thirds by lack of judicious care in keeping it in repair.

The avenue is sufficiently wide to allow of all of one side of the tracks being laid, in sections, at one job ; then the other side for the whole length of street, and thus the new surface would, when put in use, receive equal wear, which does not happen when the surface is laid in sections on opposite sides of street.   If restrictions are put on to the width of stone pavement each side of track by horse rail-

---

* xv, 661.

road company or by the city, any reputable firm ought to refuse to lay their pavement in such unpromising and unfavorable conditions, and should see to it that the wheels of heavy trucks running in the groove of one of the rails should not extend on to the asphalt surface, which has less resistance than the rail surface, and will therefore be quickly destroyed by this constant wear in the same rut.

Y. D. S.

RENEWALS OF SHEET ASPHALT PAVEMENTS.*

At a meeting of the Civil Engineer's Club of Cleveland, April 10, 1888, Captain D. Torrey, of New York, read a paper in favor of sheet or asphalt pavements compared with stone. The following discussion took place, as reported in the *Journal of the Association of Engineering Societies:*

Durability — Repairs. — Mr. Morse said that sheet pavements were fine pavements if they could be made to last, but that the cost of keeping them in repair was very great.

Mr. Richardson asked if information could not be obtained from London, Paris, Washington and other cities with regard to the durability of these pavements and cost of repair.

Captain Torrey stated that on some of the streets of London that ran from ten to six times the tonnage per foot of width of Superior Street in Cleveland, the sheet pavement had been found economical. Leadenhall Street, in London, has never been renewed, but repairs have been made. The oldest pavement of this kind in America has been in use twelve years. The asphalt, apparently, is not worn at all.

Mr. Richardson said that he had been informed that in Paris small depressions were at once filled up. They were not allowed to grow large.

Mr. Baker asked whether repairs were made in these London pavements by the contractors or by the municipal authorities.

Captain Torrey said that the repairs were sometimes made by the authorities and that sometimes the contractors offered to keep the pavements in repair for a term of years. In Washington repairs are made twice a year, spring and fall. The city of Washington makes a contract for repair twice a year for a five years' guarantee. It costs thirty-three and a-third cents per foot — three dollars per yard.

Mr. Morse asked how much it would cost for a guarantee of ten years.

Captain Torrey said that after five years the contractors would agree to repair for ten cents a yard for a long term of years. That

---

includes renewals. The pavement never wears out. A pavement that has been down twelve years is said to be good for twelve years yet. A record taken from Fifteenth Street, in front of the Treasury Building, showed the daily average of vehicle tonnage to be 66 tons per square foot. The tonnage on Pennsylvania Avenue must be between 40 and 50 tons per day. The tables will show that there is little traffic on the streets of Cleveland except with light-weight vehicles.

Mr. Whitelaw asked if new covering could be put on old asphalt pavements.

Captain Torrey said that he had never seen this done, but that the asphalt can be easily separated or peeled off from the concrete. If any one could discover a way of warming it up without burning so as to separate, it might be used again.

Mr. Whitelaw said that he supposed that when the asphalt was entirely worn out, the surface might be entirely renewed.

Mr. Morse said that it had been stated that the pavement between Erie and Perry Street was much worn. It was not so much worn, but it was uneven in consequence of being taken up to lay water-pipes.

Mr. Richardson stated that the St. Paul's Church people protested against the asphalt pavement being taken up and stone laid.

A Voice : There is 75 per cent. of coal tar in that pavement in front of St. Paul's Church.

Mr. Morse said that in some of the light traffic streets a stone pavement would last fifty years. He did not know any sheet pavement that would last half that time.

Mr. Whitelaw said that River Street was the first street laid with Medina stone. It was laid in 1857, and he thought the stone would be good to take up and lay again.

Mr. N. B. Wood said that in this discussion no account had been taken of the loss of life of horses and destruction of vehicles. In considering the question of economy of pavement that would last fifty years, it should be asked how many horses has it killed and how many vehicles has it worn out ?

Mr. Morse said that it would be difficult to answer that question correctly.

A Voice : Would Captain Torrey give as nearly as he can the amount of tonnage carried per square foot per day by the oldest Trinidad asphalt pavement and compare that with the tonnage of the viaduct ?

Captain Torrey said that a street paved with asphalt in Philadelphia carries 142 tons per square foot per day ; that is, about three and a half times the tonnage of Superior Street and about four times that of the viaduct. The pavement of Superior Street viaduct will have to be taken up and reset before it has carried the amount of the tonnage of the street in Philadelphia.

Mr. Morse said that the stone pavement on the viaduct was laid in the fall. It was expected that it would settle, and it has settled.

A Voice: I disagree with Mr. Morse with regard to the inequalities of the pavement over the viaduct being due to settlement. They are caused by the horses' hoofs chipping off the edges of the blocks.

Mr. Morse said that the gentleman was a little mistaken. Blocks of stone in the railroad track were worn turtle-back, but in other places the pavement was worn smooth.

Mr. Baker said that the convenience of persons who were taxed on the abutting property should be considered.

A member stated that he had recently interviewed a number of people in Buffalo and found that the persons who most wanted the Trinidad asphalt pavement were those who had to pay for it. They desired to have it on account of its beneficial effect on their property.

Mr. Wood said that it was claimed for the Nicholson pavement that it increased the value of property, but people thought differently when it had to be renewed.

Captain Torrey said that a great deal of wood pavement was in use in London and Paris. In Paris a large amount of work has been done, and a man has made a contract to take care of the pavement for twenty years. Sometimes as high as a dollar per square yard is paid. In London wooden pavements are put down and renewed as soon as they begin to be a little uneven. To keep asphalt pavement perfectly clean and in good repair for ever would cost about 40 cents per foot of property. On an average street in Cleveland it should cost a man who lives on a 40-foot lot about $16 per year.

Mr. Morse said that if all the streets in the city were paved and were all paid for out of the general fund it would be well, but the pavement was a tax on the adjoining property.

Mr. Strong said that he would guarantee the pavement on Wilson Avenue for fifty years. Any depressions there have been caused by the taking up of sewers, etc.

Captain Torrey said that there was no such trouble with asphalt. It could be taken up and relaid so that there would be no mark on the street. Like a patch on plaster, it can be put on so that the place cannot be discovered.

ASPHALT PAVEMENTS INJURED BY ESCAPING GAS.*

In a recent issue of the *Centralblatt der Bauverwaltung* attention is directed to the fact, observed in some of the streets of Frankfort-on-the-Main, Germany, that the asphalt pavement in the immediate neighborhood of large gas mains is rapidly destroyed by escaping gas, deep cracks being formed. This has been found to be particularly marked at places where the underlying layer of beton was imperfect, due to interruption of the work over night while laying. If this is true, it furnishes an additional reason for preventing that escape of gas from the mains in New York City which has already given so much trouble by explosions in subways and sewers.

ASPHALT PAVEMENT INJURED BY ESCAPING GAS.†

In the issue of *The Engineering and Building Record* of October 5, 1889, a brief reference was made to the injury done to asphalt pavements in some of the streets of Frankfort-on-the-Main, Germany, by gas escaping from street mains. The *Centralblatt der Bauverwaltung* of recent date gives further particulars bearing on the subject.

In removing the asphalt covering at one of the more seriously damaged places it was found that the underlying bed of beton had cracked in the direction of one of the mains of the Frankfort Gas Company. Breaking through the layer of beton, it was further discovered that the crack extended almost vertically clear through the bed, and contained a sticky substance resembling tar. If further evidence had been wanted that the damage was due to gas, it would have been furnished by the exceedingly offensive smell of gas which was encountered. Continuing the excavation and laying bare the gas mains, a break in the pipe was found After having made the necessary repairs, and before again putting down the beton and pavement proper, iron ventilating pipes were provided for which extended to the street surface and afforded a convenient means of escape for accumulations of gas. The results attending this change proved eminently satisfactory, the new asphalt covering having thus far remained in excellent condition. The crack in the layer of beton is explained by the fact that the asphalt pavement and the beton bed originally terminated at the place where the crack was found, and in extending it subsequently, the union of the old and new layers of beton was imperfect.

It is probable that the whole matter would thus have been considered as satisfactorily disposed of had not the possibility of damage to asphalt paving from illuminating gas been seriously questioned

---

* xx. 258. † xxi, 34.

by several authorities. To set the matter at rest two specimens of the asphalt covering used, one damaged as found and the other in good condition, were sent to the Chemical Pharmaceutical Institute, at Marburg, for examination. In the report which was submitted it was pointed out that past experience did not sustain the view that illuminating gas was responsible for the trouble. As a matter of fact asphalt-coated pipes were used at Paris for gas mains. An experimental investigation was therefore instituted. Pieces of the undamaged asphalt the size of nuts were placed in a large glass tube, and the latter was connected with a gas-pipe allowing a current of gas to pass through it. After having been thus subjected to the action of the gas for eight days the asphalt specimens were taken out and found to have become soft, readily crumbling to pieces under the pressure of the hand. Similar specimens exposed to the gas regained their original condition after several weeks' exposure to the air. Under the circumstances there seemed to be little question that the damage to the Frankfort pavement was due to gas. The explanation offered is that a portion of the carburetted hydrogen of the illuminating gas is absorbed by parts of the asphalt, destroying cohesion. This view is sustained by the observation that the asphalt regains its original condition after exposure for a time to the atmosphere.

It is interesting to note in conclusion that similar destructive action of illuminating gas on asphalt pavement was observed several years ago at Leipsic. The Chief Engineer of the Department of Canals and Streets, at Copenhagen, moreover, recalls the fact that, while pursuing a course of studies at Paris fifteen years ago, attention was drawn by one of the lecturers to the great importance of protecting asphalt street pavments from attacks of illuminating gas.

SLIPPERINESS OF PAVEMENTS.*

The Horse Accident Preventive Society, of London, published some statistics to show the slipperiness of asphalt pavements with a view to condemning their use and recommending wood as preferable. In reply to this the London *Times* publishes a letter from William A. Vivian, the Secretary of the French Asphalt Company, in which the following interesting points are made. He says :

The deputation appears to have based its arguments chiefly on some statistics as to the relative slipperiness of asphalt, wood, and granite pavements, according to which asphalt is made to appear the most cruel and dangerous to horses. But it was not pointed out that these statistics were compiled some fifteen years ago, at a time when asphalt was comparatively a new thing and its proper treatment very insufficiently understood. It was not then recognized that asphalt requires to be constantly and thoroughly cleansed in order to do justice to itself.

---

* xxi, 116.

In the years which have passed between 1874 and 1889 great improvements have been made in this direction, so much so that the statistics taken in the former year are now the very reverse of the truth. Granite sets constitute without a doubt the most slippery form of pavement, and at the same time the noisiest and the heaviest. It is, however, unfortunately true that much yet remains to be done to our asphalt roadways before they can be considered as really representative of asphalt—that is, such asphalt as is to be seen in Paris and Berlin, where it is flushed with water each night, and in consequence is kept sweet, clean and safe.

London has to thank the ill-advised parsimony and indifference of its vestries for the condition of its roadways. The fault lies with the dirt and not with the asphalt. As an instance of the half-and-half way in which things are done, it may be mentioned that street orderly boys leave their work at five o'clock in the afternoon, while the traffic and consequent dissemination of refuse continue practically unabated until a much later hour.

Colonel Colville did not hesitate to boldly denounce asphalt as the very worst species of pavement in existence. His opinion is chiefly remarkable as being quite the reverse of that held by those who are best competent to pass judgment on the matter. Berlin is not a city which would be content, not merely to put up with, but to continually extend the worst paving in existence. Yet nearly 100,000 square yards of new asphalt are laid annually in Berlin, and the authorities, supported by the public, do not mean to rest till the whole of their streets are asphalted. Not long since a petition was presented, signed by more than 1,500 principal horse owners, asking for an extension of the asphalt, on the ground that it was better suited to their animals than any other known description of paving. Recent statistics, very carefully compiled, show that whereas in 1885 4,403 horses fell on 398,000 square yards of asphalt pavement, the number was reduced in 1887 to 2,456, while the area had increased to 485,000 square yards. The horses and their drivers had learned to understand the asphalt, and the authorities had learned how to keep it in a properly cleansed condition. The improvement has gone on since 1887, and it is not unreasonable to expect that street accidents in Berlin will soon be reduced to a minimum. What is possible in Berlin and other large towns should surely be possible in London.

It may be noted, in conclusion, that the deputationists who so warmly advocated the use of wood pavement while condemning asphalt, are not reported to have made any reference to the relative hygienic qualities of the two materials. The smell arising from the roadway of Piccadilly on a hot summer afternoon should of itself suffice to prevent the further use of this most unhealthy substance in public thoroughfares where the traffic is at all considerable. Asphalt, with its smooth, impermeable surface, may at least claim to be the most wholesome covering for our streets as yet introduced.

ASPHALT AND COAL TAR.*

Coal-Tar Pitch.—The confusion of mind respecting these is first mentioned, and then the defects of the latter are pointed out. These consist in the fact, that if the boiling is continued long enough to remove the volatile matters, the material becomes brittle

---

* From a Report to the Common Council of Topeka, Kan. xv, 375.

and soon crumbles after being laid ; while if the volatile portions are *not* removed, the pavement is too soft in hot weather. "Coal-tar softens at 115° Fahr., and is very brittle at the freezing point ; while true bitumen is tough at 20° Fahr., and will not soften at 170° Fahr." General Gillmore and Captain Greene reported against the use of coal-tar for pavement, on the ground of the gradual oxidation of the cementing substance, and the disintegration that is sure to follow. Other authorities say that when this occurs, the only remedy is to patch them, and this is never satisfactory.

Other cities give testimony much like the above. General M. C. Meigs writes to a similar committee from Philadelphia : "The average annual cost of coal-tar pavements in Washington is about six times as much as that of good asphalt pavements."

Asphalt Pavement in Various Cities.—The Engineer Commissioner of Washington, in 1885, reports asphalt as the standard pavement for seven years past.

Buffalo laid over 16 miles between 1882 and 1886, and is laying many miles more, and it has given the highest satisfaction.

At St. Joseph, Mo., they are now substituting it for the limestone macadam, which has failed. The specification there used is the same as that of Washington. After grading the street, a layer of concrete six inches thick is laid down, and when fully set a "cushion coat" of asphalt-mastic ½-inch thick is spread. This is followed by another layer 2½ inches thick, containing 60 to 70 per cent. of fine sand, with a small percentage of limestone dust.

Streets with grades of 4 to 5 per cent. have been laid, but it is not considered good practice on account of slipping of horses in wet or icy weather.

For this reason in Kansas City, where the grades are mostly high, the standard pavements are stone blocks of granite or Colorado sandstone and cedar blocks on a concrete base.

For the same reason in Cincinnati the Board of Public Works recommend granite blocks on all streets of over 2½ per cent. grade, and asphalt-mastic on streets of less grade.

In Omaha, where the climate in winter is especially severe, this pavement has borne the test for four years thoroughly well. There are now over seven miles of it in use there, on business and semi-business streets, and it is giving entire satisfaction.

ASPHALT PAVEMENTS AFFECTING VALUE OF PROPERTY.*

Some ten or twelve years ago an effort was made to pave Fifth Avenue with the bituminous pavement that wore so well in front of the Worth Monument, until it was replaced in the fall of 1879 by

---

* Ed. xiv, 463.

one of an inferior description; but the property-owners on the avenue, with great unanimity, petitioned the Legislature against the projected improvement, asserting that it contained a job, and charged complicity between the projectors and the Health Department that endorsed the plan. Their opposition was effective, and since that time they have enjoyed the advantages of the worst pavement in the city, and probably the worst in any city of a million inhabitants in the world.

Finally, when after trials and tribulations the citizens of that thoroughfare have got a noisy and dirty pavement, which in due time will be broken up for the laying of steam, gas and electric-wire pipes, they may well look to the example of Cincinnati and see how blind and short-sighted they were ten years ago, when they so persistently and successfully opposed the laying of the asphalt pavement referred to. If not comforting for them to know, it may be valuable to other citizens to learn that Cincinnati, which is now spending about four million dollars in putting down decent pavements, about three months ago completed the pavement of Race Street with asphalt. A gentleman residing on the street informs us that immediately after the completion of this pavement the wheel traffic from the city was apparently concentrated there; that there is no noise, no dust, and every night the street is thoroughly cleaned. Moreover, property has been sold at an advance of 33⅓ per cent. over any figures heretofore obtainable, and with the prospect of still higher prices.

This enhancement of the value of property along the line of a well paved street is only a repetition of the experience of every city that has laid down and maintained decent pavements.

In thus endorsing an asphalt pavement we do not wish to be understood as recommending it for every street and under all circumstances; and though we would like to have seen it on Fifth Avenue, yet we admit that there are one or two spots on that avenue where the grade would demand a different pavement. But we have no hesitation in predicting that if any of the cross-town streets in the upper part of the city should secure a properly laid asphalt pavement from river to river, property would be enhanced in value, as was the case in Cincinnati.

ASPHALT PAVEMENTS IN NEW YORK AT COST OF PROPERTY OWNERS.*

We have frequently urged the trial of asphalt pavements properly put down on some of the residential streets of New York, pointing out the advantages of this noiseless and cleanly pavement as recorded in the experience of other cities. It was with considerable satisfaction, therefore, that we noticed in a recent ride on the west

---

* Ed. xvii, 334.

side in this city, where so much building has been in progress, that on West End Avenue and Seventy-second Street the enterprising property owners had contracted with a responsible company who had laid this pavement. On inquiry we learned that some thirty different owners have co-operated in this matter, Messrs. W. E. D. Stokes and W. J. Merrill, who own a number of buildings in the locality, having been the active promoters of the movement.

The excellent appearance of the street will doubtless enhance the value of their houses more than it has cost the owners. We hope the fact may speedily be demonstrated so that the example they have set may be followed elsewhere.

ASPHALT PAVEMENT FOR NEW YORK'S BOULEVARD.*

We had occasion, last May, to note with approval that certain property owners on the west side of the upper part of New York City had laid some asphalt pavement at their own expense, with a view to improving the value of their neighboring real estate, and expressed the hope that the result would justify their expectations. It would seem to have done so, for they now wish to have the city do something more in the same line, and to that end W. E. D. Stokes, who was one of the leaders in the first movement, has prepared a bill, which has been introduced in the Legislature by Mr. Connolly, to provide for laying an asphalt pavement on the Boulevard. Francis M. Jencks, Charles T. Barney, Jacob Lawson, David B. Ogden, the members of the West End Protective Association, and other west side property owners are interested with Mr. Stokes in promoting this bill, which has been so drawn that any asphalt paving company can contract for work under it ; and the charge that the whole thing is a job for the benefit of some particular paving interest seems, therefore, to be utterly without foundation. It is much more likely that the opposition to this measure proceeds from those who are interested in some description of stone pavement. The Boulevard having been already macadamized, can be prepared at very little expense to receive the asphalt, and we trust nothing will interfere to prevent the passage of the bill and the speedy commencement of the work.

ASPHALT BLOCKS.†

Asphalt Blocks, in size 4x5x12 inches, composed of genuine asphalt and broken stone compacted by hydraulic pressure, have in some places failed rapidly. In other places they are approved for light traffic.

The patentees do not consider them suited for anything else, and also consider it essential that they be made at or near the place used.

They are therefore reported against.

---

*Ed. xix, 118. †From a Report to Common Council of Topeka, Kan., xv, 375.

CHAPTER IV.

BRICK PAVEMENTS.

BRICK STREET PAVEMENTS.*

Clay.—As towns grow into cities, and their street traffic increases, pavements become a necessity.

In parts of Illinois, where timber is scarce and stone and gravel almost wanting, hard-burnt bricks have been used for paving, and the cheapness of the material has commended its use to the people.

In Bloomington, Ill., fifteen years ago, a courtway was paved with bricks as an experiment, and this courtway, exceptional for a first experiment, is still in a fair condition, after scarcely any repairs.

Bricks fit for paving are so different from common bricks, that we have thought some facts about the former might be interesting.

Clay, which will make good paving bricks, should be able, without fusing, to withstand a sufficient degree of heat, and for a long enough time to render the bricks hard and impervious to water. Yet, while the value of a clay for paving brick depends largely upon its refractory nature, fluxing agents, in proportions yet undetermined, are necessary to render the bricks homogeneous and tenacious.

Fire clays are generally hydrated silicates of alumina. Potash, soda, lime, magnesia, iron and other elements are sometimes present in small quantities, and render the clay less refractory the greater the quantity of them present. When 6 to 10 per cent. is found the clay will generally melt. When the clay is silicious, 3 to 4 per cent. makes it fusible. When it is aluminous, 6 to 7 per cent. of oxide of iron does not render it non-refractory, because aluminites are mostly of a refractory nature.

Lime alone is completely infusible ; but in very small quantities in clay, it makes a brick fusible at high temperatures. It is especially to be avoided in clays for paving-bricks, because the heat of burning reduces the carbonate to caustic lime, which in the pavements will absorb moisture and cause disintegration of the brick.

Magnesia in small quantities is said by Egleston to make clay fusible.

The alkalies, from 1 to 3 per cent., make clay fusible.

---

*xxi, 23. A paper read by Charles E. Green, before the Tenth Annual Convention of the Michigan Engineering Society, and published in the Michigan Engineers' Annual for 1889.

The change of color in burning is due to iron, or to a small percentage of bituminous material. Iron, beyond 6 to 10 per cent., if the clay is silicious, usually renders it fusible.

Silica and alumina constitute the refractory parts, and the impurities the fluxing part.

Silica is infusible alone, except at the highest temperatures of the oxyhydrogen blow-pipe. It possesses no binding property, and a clay having an excess of it is rough and lacking in plasticity and tenacity, thus rendering the bricks weak.

Alumina is highly refractory and is binding. When combined with iron or other bases alone, it makes infusible aluminites; but with silica present, it fuses more or less readily.

The plasticity of clay is due to the fineness of the particles, to the presence of alumina, and to the water of combination, which may be driven off at a red heat, the clay losing its power of combining with water.

The refractory nature of a clay is then due to the presence of alumina and silica in excess, and to the absence of potash, soda, lime, magnesia and iron.

The characteristics of all fine clays may be said to be, that they do not effervesce with acids: that they make a paste with water, which paste can be drawn out without breaking, and is very plastic; that when dry they are solid, and break into scales when struck; have a soapy feeling, are easily scratched or polished, can be cut into long ribbons with a knife, and appear somewhat like horn.

The commercial test is the only sure way of ascertaining the value of a clay.

Some clays which we will mention have been used more or less successfully in making a paving material.

"*Mingo*" *Fire-Clay, Haydenville, Ohio.*—The clay from which the Hayden patent Paving Block (of which a sample is submitted) is made, has nearly the compactness and density of fine grained sandstone, and yet is semi-plastic. It can be characterized neither as a hard nor as a plastic clay. Nitro-glycerine is used in mining it; yet, while apparently very hard, it is readily reduced to powder in the dry pan under heavy rollers. The clay when ground is excellent for laying fire-bricks, mixing easily, working smoothly, and resisting great heat.

The paving block in burning is subjected for ten days to 2,000° Fahr. This paving block is square in plan, 5¾ inches on a side, and 4¾ inches deep, hollow below to facilitate burning and save material, and having a flat upper surface broken by indentations intended to provide a firmer foot-hold for horses. These indentations, however, seem to be of little permanent value, as under heavy traffic they are soon obliterated. The blocks are salt-glazed. The outer top edges are slightly beveled, to admit readily to filling of sand, pitch or other material in the joints after laying. One thousand blocks will lay 26 square yards, at 50 cents per hundred, or $1.92 per square yard. The sidewalks are 1¼ inches thick and top 2 inches. The endurance of the block is therefore limited to the wear of the top. A block weighs about 9½ pounds, giving about 360 pounds to the square yard.

An analysis of this clay, by Professor E. M. Reed, is subjoined, and a comparison is made with the well-known Stourbridge clay of England.

ANALYSIS OF CLAY.

	Mingo Fire-Clay, Haydenville, Ohio. Per cent.	Stourbridge Fire-Clay. Per cent.
Silica.............................	72.24	73.82
Alumina...........................	16.87	15.88
Iron..............................	0.16	2.95
Lime	0.50	trace.
Magnesia..	trace.	trace.
Alkali............................	1.09	0.90
Water.............................	5.44	6.45
Total...................	100.00	100.00

Paving Brick Clay, Virden, Ill.—This clay is found in a bed 25 feet thick, lying just beneath a stratum of slate, and 275 feet below the surface of the ground. It is compact, with a distinct cylindrical cleavage, of a light steel-gray color, and emits, when first mined, a characteristic odor resembling that of slate. Although it is blasted out, it has only the hardness of soft slate, and can be reduced to an extremely fine powder free from grit. It has a specific gravity of 2.5595. A piece kept at a white heat in a hard coal fire for twelve hours was not disintegrated, and concentrated nitric acid failed to produce effervescence.

It is reduced to powder by being first passed through a Blake crusher, then finished in a dry pan crusher. The process is imperfect, as particles of uncrushed clay weighing as much as 50 milligrams have been taken from a green brick. These particles had not apparently been affected by the water used for tempering, and they remain intact through the process of burning, being easily discerned in a burned brick, and rendering it heterogeneous. When tempered with water, the clay becomes plastic, tenacious and firm, being readily formed into different shapes. A sample is here submitted.

Clays from Bloomington, Ill.—These may properly be called "surface clays," and are of three varieties.

" A " is a yellow, aluminous clay, very fine and plastic, found three feet below the surface, in a bed 12 feet in thickness. It is subjected to no process of grinding or reduction, being simply tempered in a vertical pug-mill directly from the bank. It burns to a grayish-red color.

" B " is a hard-pan clay just beneath the " A " layer, and from one to one and a-half feet thick. Great difficulty is experienced in reducing this clay to an impalpable powder. When dug and exposed for one winter, to "weather," or disintegrate, very good results are claimed for it. But when this clay is reduced at once, uncrushed particles occur which cause unequal shrinkage in the burned brick, leaving small cavities where, in the pavement, water may accumulate and cause disintegration by freezing. Hard-pan clay generally burns from pink to cream in color.

" C " is a blue clay directly beneath the hard-pan clay, and has a depth of from one to six feet. It is plastic or semi-plastic, fine, firm and very tenacious, and usually burns to a chalk white. In many respects it is similar to the light blue fire clay used for fire and paving brick at Empire, O.

From the experience of the best manufacturers of paving brick, we are led to conclude that fineness of grain in clay is most essential, and where that property is found in a plastic clay, provided the clay is sufficiently refractory, care only is needed in tempering and burning to produce good results. When fine-grained clay has a compact and hard structure, no pains should be spared to reduce it, in a dry state, to an impalpable powder.

Where the clay is neither dry nor fine, but of desired composition, it may be calcined and then finely ground. This may appear a trivial matter, yet, if two samples of the same clay are ground, the one finely and the other coarsely, the latter will require much more heat before vitrifying to a homogeneous mass; while the more finely it is ground, the stronger and harder the bricks will become, the better they will stand abrasion, and the less liable will they be to break.

Manufacture of Brick.—The great diversity in the clays makes the process of manufacture a difficult problem, for the experience gained in one place may be of little or no use in another.

In several towns, notably Ottawa and Virden, Ill., manufacturers use the plant commonly employed for making tile, modified for bricks. At Virden, Ill., where the capacity is 1,600 per hour, the process is in substance as follows :

The clay hoisted from the mine is delivered from tram-carts to a Blake crusher, and then finished in a dry-pan crusher. It is then delivered to a tempering machine, and forced from the latter by a screw through the desired mold, whence it issues in one continuous bar, 8¼ inches by 4¼ inches in a section. It runs over a series of rollers, and is cut up by wire spaced two inches apart, in a hinged frame worked by hand. The bricks are removed on tram-cars to a dry-house or shed, and left until sufficiently dry. They are then placed in down-draft cylindrical kilns having hemispherical tops, and burned from 70 to 90 hours with bituminous coal, on grates, at intervals, around the outside base. A 80 horse-power engine suffices for this plant. The heat is sufficient to melt tin, and to cause pieces of copper to run together.

"A" and "B," of dimensions 8¼x4¼x2 inches, are burned in the same kiln, but differ in color and quality. "A" is dark, almost black, rings more and appears stronger than "B," which is a light red. They are quoted at $8 per thousand free on boards cars.

"C" is a brick from the same clay, and made in the same way, but intended for a finished brick for face work on buildings.

At Bloomington, Ill., the "A" clay is simply tempered in a vertical pug-mill, molded into bricks by hand, dried in a yard, racked in rectangular stacks, and burned for from 96 to 120 hours with wood and bituminous coal placed in the usual arches. It is a grayish red in color, 7x3½x2¼ inches, quoted at $8.50 per thousand at the yards, or furnished in sufficient numbers to lay *two* courses, one flat and the other on edge, for $1.12½ per square yard. Bloomington "B" and "C," quoted at the same figures, have about the same dimensions. All have a sharp, clear ring when struck.

Bricks are made at Ottawa, Ill., from fire-clay which is mined, and from common surface clay. They are burned some 50 hours with bituminous coal in kilns similar to those used at Virden. When burned the interior is dark gray in color, very hard, and cannot be cut with ordinary tools. They are of two dimensions, 4x5x12 inches, and quoted at $1 per square yard, which is about $35 per thousand, and 4x4x12 inches, at 75 cents per square yard, or about $28 per thousand.

At Empire, O., a light blue fire-clay is taken from a drift in the side of a mountain, ground in a pan, molded by hand, and burned in up-draft open-top kilns, with natural gas. The bricks are hard, ring like steel, and are quoted at $10 per thousand. Much the same process is employed at New Brighton, Pa., where the bricks made from fire-clay are claimed to be as hard as granite, to emit a sharp, steely ring when struck, and to resist all effects except natural wear.

In all the above cases no addition of sand or other material is made. In Wheeling, W. Va., a brick, in form a truncated pyramid, is made from clay and iron slag.

As the condition and life of a pavement depend upon uniformity of the bricks, they should all be burnt to a uniform hardness, which cannot be obtained in ordinary kilns. If soft and salmon bricks are found at the top of a kiln, while those next the arches are over-burned, warped and melted, the cost of the rest, suitable for use, is much enhanced by the waste.

Crushing Strength.—All specimens were crushed by the direct application of weights. The specimens were all $\frac{3}{4}$x$\frac{3}{4}$x$1\frac{1}{2}$ inches, placed on end between pieces of soft pine about $\frac{1}{4}$-inch in thickness. The length chosen for the samples permits the shearing action to have its full effect. All were ground to true size on emery wheel and grindstone, and carefully calipered.

In Virden "A," the short axes were parallel to the $8\frac{1}{4}$ inch and $4\frac{1}{4}$ inch dimensions of the brick from which it was taken. It was compact, having no holes or spongy places, took on sharp corners and smooth surfaces, was excessively hard, and would glaze at a white heat on the emery wheel. With 5,266 pounds, it crushed by a scale of about $\frac{1}{8}$ inch by $\frac{1}{4}$ inch by 1 inch, first flying off from one vertical edge, accompanied by a sharp report, immediately followed by splitting into nearly prismatic fragments, parallel to the action of the force.

A second specimen from the same brick, with short axes parallel to the $4\frac{1}{4}$-inch and 2-inch dimensions of brick, crushed at 5,625 pounds.

Virden "B" did not take on as true dimensions as "A," though free from holes. Crushed at 5,160 pounds. Virden "C," like the preceding, crushed at 4,985 pounds.

It was difficult to bring Bloomington "A," as well as the next specimen, to the required dimensions, because of a tendency to scale when heated under the action of the emery wheel. Several holes occurred in them, as may be seen in the bricks. They also glazed somewhat under the emery wheel. Crushed at 5,693 pounds.

The second specimen crushed in smaller prismatic pieces, at 7,268 pounds.

Bloomington "B" was excessively hard to reduce to size, would also glaze, but not scale off, took on smooth surfaces and sharp edges, and had spots which contained a yellow powder. Crushed at 5,565 pounds.

Bloomington "C" was much like the above. Crushed at 5,280 pounds.

Good Haverstraw stock bricks of a fair average, test at East
River Bridge, by Mr. Abbot, with no packing between brick and
machine, gave:

Crushing Strength.

Whole brick on end ....
- max., 3,060 lbs. on sq. in.
- min., 1,600 " "
- average, 2,065 " "

Half brick on flat side ....
- max., 4,153 " "
- min., 2,669 " "
- average, 3,371 " "

Half brick on small side ....
- max., 6,400 " "
- min., 2,900 " "
- average, 4,612 " "

Experiments by Prof. W. A. Pike, Minneapolis, Minn., on
building bricks. Half bricks between pieces of pasteboard.

Failed with.

St. Louis, Mo., brick ....
- flatwise, 6,417 lbs. on sq. in.
- edgewise, 4,080 " "

Hastings red brick..............
- hard, 2,017 " "
- medium, 2,012 " "
- soft, 1,748 " "

In the following table are given the specific gravity (mean of
5 determinations), the absorption for three, five, and ten days, after
the specimens had been heated to a red heat, cooled in a desiccator
and weighed, and finally the crushing strength per square inch of
section.

NAME.	Specific Gravity.	Per Cent. Absorption.			Crushing Strength, lbs. on sq. inch.
		3 days.	5 days.	10 days	
Bloomington "A"	2.1852	3.79	3.92	4.06	12,920 max. 11,520 mean. 10,120 min.
" " B "......	2.3045	3.18	3.19	3.37	9,893
" " C "......	2.2269	1.87	2.20	2.70	9,386
Virden " A "	2.3166	3.10	3.12	3.82	10,000 max. 9,680 mean. 9,361 min.
" " B "	2.4150	7.67	7.96	8.71	9,173
" " C "...........	2.3601	7.11	7.48	8.70	8,862
Hayden Paving Block ..	2.2861	3.25	3.38	3.64	.....

It may not be amiss to compare the crushing strength of the
above with that of some common or building bricks.

Work in Different Cities.—Paving bricks are sometimes laid on a
foundation of broken stone covered with cement mortar, and sometimes on sand. It will hardly be expedient to go into details now.

At St. Louis, ten years ago, strips of pavement 22 inches wide
were laid down as a test, and a two-wheeled cart with tires 2⅛ inches
wide, and loaded to two tons, or 800 pounds per inch of width, was
rolled back and forth by machinery. The heaviest traffic at that
time in St. Louis was 75 tons per day per foot of width, and the

average for business streets was 35 tons. Estimating the effect of horses' shoes at one-third of this amount, 50 tons per foot were taken as a standard. The samples were weighed before and after testing, and were subjected to an amount of travel by the above cart equivalent to eight and one-half years on the street.

The total abrasion of the fire-brick pavement was 9 per cent., or a depth of ¾ inches. but about one-half of the bricks were broken. Asphaltum blocks under the same test wore 14 per cent., and but one was broken. Granite block were scarcely worn appreciably, while limestone had lost 10 per cent. The action of the elements would have affected the limestone still more, as would also probably have been the case with the bricks.

As a result, granite blocks were chosen for streets having heavy traffic. The tests seem to warrant the claim that brick pavements will answer well for moderate and light traffic.

Soft bricks intermixed may give trouble; but the first cold season will usually determine whether the bricks are hard enough, as, if not, the moisture absorbed by them will freeze, and crumble the bricks by expansion.

At Nashville, Tenn., are used creosoted or bitumenized bricks, prepared by saturating common kiln bricks of medium hardness, under a high degree of heat, with liquid pitch, obtained by distilling coal-tar. These bricks have been laid about four years, on a good foundation of rolled macadam, bedded on edge in 1½ inches of sand, and rammed tight. They have been subjected to a considerable amount of traffic, and show very little, if any, wear. The cost was $1.80 per square yard for brick and laying on finished foundation.

At Yougstown, O., fire-brick on cement concrete base and sand with pitch in joints, cost for all $1.73 per square yard. Their best bricks came from Empire, O., and New Brighton, Pa.

At Allegheny City, Pa., fire-clay bricks from the West Virginia Fire Brick Company have been laid in two layers on rammed and rolled gravel and sand, at a cost of $1.95 per square yard, exclusive of grading. The pavement, in use three years, is stated to be very smooth, easily cleaned, and of a handsome appearance.

At Bloomington, Ill., some two and one-half miles of brick paved streets have been laid with bricks of home manufacture, "A," "B" and "C." The sample herewith was under traffic there twelve years. The corner and edge of one end, and also the large scale on the upper edge of the same side, were broken off with a pick in removing the specimen from the pavement. It does not appear to have lost appreciably in width, although the majority did not present so favorable an appearance, having generally lost the upper edges and corners. Such pavements, however, are not noticeably rough in driving over them, and afford good foot-hold for horses. The brick pavements in Bloomington, generally speaking, are in good condition, and are being constantly extended. They have stone blocks, Nicholson, and macadam pavements also, and prefer brick, on the score of cost, durability, lack of noise, home production, ease of cleaning and repairs. The entire pavement, two layers of brick, including *all materials*, grading and stone curbs, cost $2 per square yard, with laborers at $1.60 per day, teams $3.25, and sand 80 cents to 90 cents per cubic yard, delivered.

Jacksonville, Ill., has laid over three miles of pavement with Virden brick. Some down for five years is now in good condition. One piece of one-half mile, two layers, cost, including everything, engineer's services and all, $1.92 per square yard. Sand was filled into the joints in place of pitch.

Cincinnati had a pavement on rolled sand. A failure.

Peoria, Ill., paved two blocks in 1886, and five blocks in 1887, with Ottawa paving brick, and Chicago used the same brick three or four years ago on Sixty-third Street car tracks.

The intersection of Washington and Dearborn Streets, Chicago, was paved with Hayden paving blocks in 1886. In March, 1888, it was found to have succumbed to the very heavy traffic, and crushed.

Galesburg, Decatur, Champaign, and Springfield, Ill., have laid more or less brick pavement within five years, the major part of which has given satisfaction.

New Orleans, and Wheeling, W. Va., Steubenville, O., Camden, N. J., and other places have laid experimental sections.

The disadvantages of brick pavements arise from lack of uniformity in the material, and the liability of incorporating in the pavement brick of too soft and porous structure, which crumble the first winter under the action of frost.

When laid on a solid foundation, the advantages are cheapness, smoothness and cleanliness, freedom from noise, and sanitary excellence as non-absorbent.

Taking all things into consideration, it is superior to macadam, wood, and, for a maximum traffic of three tons, superior to granite.

BRICK PAVEMENTS IN NASHVILLE, TENN.*

CINCINNATI, O., January 8, 1890.

SIR: In the article on brick pavements in your issue of the 4th inst.,† I notice a statement in reference to brick pavements in Nashville, which is not quite accurate.

---

* xxi, 104.

† The article referred to was a paper read by Charles E. Green before the tenth annual convention of the Michigan Engineering Society.

The first sample of "creosoted brick" pavement laid in Nashville was at the intersection of Union and College Streets. This was put down by the inventor or patentee of that pavement as a sample, using carefully selected brick, and doing the work in the best possible manner. Notwithstanding this fact the pavement did not wear satisfactorily, and was taken up in the summer of 1888 when it was almost worn out, and replaced with granite, after being in use about three years.

In 1887 a part of Union Street, beginning at the intersection named above, was paved with creosoted brick by the same person or contractor who paved the intersection of College and Union Streets, but the work was evidently not done with the same quality of materials or care as the first sample put down, and this pavement is practically worn out, and will be replaced with something more permanent in the coming summer. This general information was obtained by me during a recent visit to Nashville, but I prefer, before you publish the facts, that you should write to Mr. J. A. Jowett, the City Engineer of Nashville, and obtain further information as to dates and details, which I am not able to give.

I may add in this connection that a part of Seventh Street, between Market and Georgia Avenue, in Chattanooga, Tenn., was paved during last season with this creosoted brick. It has not of course been long enough in use to demonstrate whether or not it will prove durable and satisfactory. The bricks used were selected ordinary red brick, boiled in coal tar and laid upon a foundation of hydraulic concrete.

S. WHINERY,
General Manager of the Warren-Scharf Asphalt Paving Company.

FIRE-BRICK PAVEMENTS FOR CARRIAGEWAYS.*

Paving Brick.—Brick pavements have been used for a long time in Holland, where the "clinkers" of ordinary bricks are laid in the wheelways; but those, though hard enough, are too brittle for pavements carrying heavy traffic. In this country fire-brick have been used for some time, giving a regular, clean and nearly noiseless pavement, with possibly the best foot-hold for horses of any pavement known.

Used in Wheeling, W. Va.—Wheeling, W. Va., probably has more of this pavement than any other place, and the *Ohio Valley Manufacturer*, of that city, indorses the pavement very strongly. The first of these pavements was laid in New Cumberland, W. Va., some six years ago, and in November, 1883, a part of Chapline Street, in Wheeling, was laid with fire-brick, where, on account of the bad condition of the streets on each side of it, there was a concentration of traffic. The *Manufacturer* says :

This pavement has not received a dollar's improvement since it was laid, over five years ago, and is apparently as good now as when it was laid. Every year since the area has been extended, and now fire-brick pavement covers several miles of our streets. To say that it gives satisfaction very mildly expresses it. We believe it receives the approval and indorsement of every citizen of Wheeling, without a single exception.

* Ed. xix, 183.

Construction.—In laying, a bed of compact earth is covered with four inches of good gravel and that with two or three inches of sand, on which the brick are laid transversely on edge, and then thoroughly rolled with a heavy roller. A light coating of sand is then given and the whole treated with a top dressing of coal-tar and pitch heated to 300° Fahr. This dressing with coal-tar and pitch should be repeated annually as it prolongs the life of the pavement. The bricks should not be too large for thorough burning, 8½x4x2½ inches is found to be the preferable size, and a fire-clay is recommended containing more iron than is admissible in bricks for furnace linings.

Cost.—Bricks of the above-mentioned size are furnished at Wheeling for $10 per thousand, which will lay about 17 square yards, and the total cost varies with the labor of preparing the road-bed, etc., from $1.25 to 95 cents per square yard ; the last mentioned price is without the pitch dressing, which costs about 12½ cents per square yard.

Brick and Granite Blocks.—The *Manufacturer* adds :

The most serious obstacle that has prevented the adoption of this pavement by many of the largest cities, notably Buffalo, Cleveland, Columbus and Philadelphia, are the cliques and rings formed by the owners of stone quarries who have Belgian and granite blocks for sale, and who see in the fire-brick an enemy that will eventually retire them from this branch of trade.

This is doubtless partially true, but the most serious obstacle might well be said to be the supineness of our more intelligent citizens, who seldom take any interest in the public improvements unless it may be to deprecate any expenditure for them. It was a combination of these two influences which has resulted in the noisy pavement now afflicting Fifth Avenue in New York City.

BRICK AND GRAVEL PAVEMENTS.[*]

Suitable Brick.—According to reports large areas of brick pavement have been laid during the past year in Bloomington, Decatur, Jacksonville and Galesburg, Ill., and experimental sections have been laid in Camden, N. J., and New Orleans.

Mr. George F. Wightman, in a paper before the Illinois Society of Engineers and Surveyors, says : "Brick pavements have been laid which show remarkable qualities of endurance. There are but few localities in which clay has been found that will stand, *without fusing*, the amount of fire necessary to make a brick hard enough for street pavement. Brick in every way suitable have been made at Haydenville, O., Bloomington and Ottawa, Ill. Some of these pavements, although laid ten years ago, present now not only a uniform, but a very smooth surface."

---

[*] xx, 658.

Construction.—He prescribes the method of laying to be : to excavate ten inches and roll hard, then fill with four inches of sand, on which a course of bricks are laid lengthwise on the flat and so as to break joints, this layer to be tamped, and joints swept full of sand. Then cover with one inch of sand, and place a second course of brick, laying them crosswise and on edge. Sweep the joints full of dry sand and cover one inch thick with sand. This upper course to be burned to a flint hardness.

In the discussion which followed, Mr. Bell exhibited samples of bricks which, after fourteen years' wear at Bloomington, had lost but half an inch in width. He says the brick pavements are the first to dry after a rain. That in Nashville, Tenn., a creosoted brick is in use, so as to make it non-absorbent. The Bloomington pavements are laid with 7-inch crown in 36-foot width between curbs.

Mr. Mead stated that in Wheeling a brick was made of clay and iron slag in the form of a truncated pyramid.

Gravel Pavement.—Mr. Wightman gave the preference to a rolled gravel pavement, over either macadam or brick. He would make a lower layer, 6 inches deep at centre and 4 inches at sides, of stone or coarse gravel, from curb to curb, well rolled ; then lay on this gutters 3 feet wide of cobble-stone, well rammed, and with spaces filled with fine gravel. The roadbed between to be finished in gravel of ½ to 1½ inches diameter, and enough clay, loam, or sand to make it bind, and thoroughly compacted with a five to six ton roller.

FIRE-BRICK PAVEMENTS FOR ROADWAYS.[*]

Not Suitable for Heavy Traffic.—A report has been made public to the effect that a specimen piece of fire-brick pavement put down in 1885 on Pearl street, Cincinnati, has failed utterly. The description given states merely that the foundation was sand, well rolled, but says nothing as to the thickness of the brick-work. It has never been claimed by the advocates of brick pavements that it was suitable for very heavy traffic ; and it is possible the mistake has been made in this case of subjecting it to a traffic for which it is not fitted. A *thin* pavement on a sand foundation would be in such case pretty sure to fail.

Paving Brick.—It may also be that the *bricks* were not of the right sort, as they need to be hard-burned and *vitrified* throughout. Such bricks are reported to have stood a crushing strain of over 8,000 pounds per square inch, approaching in this respect the lower grades of granite. It would be manifestly unfair to condemn the use of brick altogether in consequence of an indefinite report such

---

[*] Ed. xvi, 318.

as this is. We are inclined to believe that, under right conditions, there is a wide field to be usefully occupied by well-made brick pavements.

BRICK STREET PAVEMENT.*

Material and Cost.—The following is a portion of a letter addressed by W. Voorhies, Jr., to the Editor of the *Lexington* (Ky.) *Daily Press:*

The city of Decatur, Ill, commenced paving the streets with brick four years ago, and is still using the same material, having paved about two miles each year. So we have had only four years' use in which to test its durability. As to appearance and smoothness there is no question, for it gives universal satisfaction except to the abutting property owners, who pay the entire expense, the city only paying at the intersection of streets. The first laid here was on the public square, which is probably used as much as any street in the city. I do not think the wear on this will amount to 10 per cent. This work was laid with Bloomington, Ill., brick, made of the yellow clay underlying the black soil. They cost about $10 per 1,000 delivered here. We are now using home-made brick, which are equally as good, steam-pressed and burned in ovens (same ovens used for burning tile). They are burned very hard; cost, about $10.

The entire cost of this brick paving is $1.40 per square yard, which includes the necessary grading, which consists of removing about one foot of the surface dirt from the road-bed. This is filled with coarse gravel, the center rounded about a foot higher than the edges. On this is laid the first course of brick, flat side down; then a layer of sand one inch thick; then the top, or finishing course of brick, laid edgeways, and this again covered with sand to fill up the cracks and wedge the brick solid, completing the work.

Durability.—So far they have withstood the heaviest loads without any serious damage. Barnum's circus wagons, some of them weighing 20,000 pounds, passed over them after a heavy rain without making any impression except where gas pipes had been recently laid and the replaced earth not yet thoroughly settled. Bloomington, Ill., has had brick streets in use for thirteen years, and I am assured that they are in good condition now, without having been repaired.

FIRE-BRICK PAVEMENTS.†

Durability and Cost.—A committee, after visiting Steubenville and Wheeling to examine the pavements there, reported that it is smooth, durable, cheap, and free from noise. On the entire length of one street examined in Wheeling they did not find a broken or crushed brick. Part of the paving had been down two years and another part three years, and one was apparently as perfect as the other. The bricks are made from a softer clay, and are burned harder than ordinary brick. A pressure of eighty tons was not sufficient to crush one.

---

*xviii, 146. † xv, 375. From Cleveland *Leader.*

Mr. J. F. Holloway, civil engineer, read a paper strongly indorsing the use of this material. A pavement in Allegheny City
examined by him was very smooth, and he could not hear the rumble of carriage-wheels until they were within fifty feet of him. The
street was easily cleaned, and made a handsome appearance. The
pavement can be laid for from $1.20 to $1.60 per square yard. Not
one of a number of 2-inch cubes failed under sixteen tons pressure ;
or, in other words, they bore safely a strain of 8,000 pounds per
square inch, which is considerably more than the crushing strength
of the best pressed brick.

The bricks are 9x5x2 inches or 3 inches, and are made in West
Virginia. The upper layer is laid on edge with the spaces filled
with pitch or tar and covered with sand.

BRICK PAVEMENT—CONSTRUCTION AND DURABILITY.*

The first pavement considered was *brick*. As laid at several
places, the graded street is covered with several inches of sand, on
which is placed two layers of brick, the first on the flat, the second
on edge, making about six inches thickness of brick. This they
report against, principally on the ground of lack of uniformity in
the material. They found no place where the pavement was satisfactory except on streets having very light traffic. Bloomington,
Ill., favored it where not more than 100 teams per day passed over
it. Decatur, Ill., and Cincinnati reported against.

A recent report upon the wear of fire-brick pavement in Steubenville, O., shows the wear to have been less than $\frac{1}{16}$-inch annually,
but the amount of traffic is not stated. An experiment reported as
to absorption shows it to be inappreciable at the end of twelve hours.
Fire-brick for this purpose are now manufactured in West Virginia
and in Clearfield County, Pennsylvania.†

TESTS OF PAVING BRICK.‡

Noticing a newspaper clipping referring to some tests of paving
brick made by City Civil Engineer Walter P. Rice, of Cleveland, O.,
we wrote for verification of the item and received the following copy
of the report, which is dated May 6. The foot-note being added by
Mr. Rice for publication in *The Engineering and Building Record*:

CLEVELAND, O., May 6, 1889.

In the specifications for paving Bolton Avenue with fire-clay brick the
following clause was inserted :

"*Testing Bricks.*—The city reserves the right, after the bids are received
and before the awarding of the contract to any bidder, to make such tests
of the sample bricks deposited in the engineer's office as said engineer shall
recommend or deem necessary to determine the durability and fitness of

---

* xv, 375. From a Report to Common Council of Topeka, Kan.   † xv, 658.
‡ xix, 343.

the material proposed to be furnished by such bidder, as shown by his samples, and reserves the right to hold the bids, not to exceed two weeks, for such purpose, and any bricks not fulfilling the requirements of the specifications or that may be deemed unfit for paving purposes, as shown by such test, will be rejected and the bid considered as not in conformity with said specifications. The cost of the test made upon all samples submitted, not to exceed $50, to be paid by the party or parties receiving the contract."

In conformity with above clause tests were made on : (1) absorption ; (2) attrition ; (3) impact ; (4) freezing and impact. In all probability the samples furnished are a little better than the general run, and the number of test pieces in each case being small. The results are, therefore, relative, not absolute, but may be safely taken as a basis of judgment. I have no hesitation as a result in recommending the Porter's Union brick, manufactured in West Virginia, for this street. As a result of the different tests, I would accord the following rank to the different specimens submitted, Medina stone being put through the same tests as a standard for comparison Medina sandstone, 1; Porter's Union, fire-clay brick, 1; Porter's Ætna, fire-clay brick, 2 ; Massillon Star, fire-clay brick, 3 ; United States, fire-clay brick, 4 ; Williams' Canton, fire-clay brick, 5 ; McReynold's Patent, brick, 6 ; Rielley's, Pa., fire-clay brick, 7 ; Claflen Paving Company's red brick, 8 ; Claflen Paving Company's Malvern fire-clay brick, 9. Ordinary building brick stands about the same as Claflen Paving Company's sample of red brick. Respectfully submitted,

WALTER P. RICE,
City Civil Engineer.

NOTE—In comparing Porter's Union with Medina sandstone I would not have it inferred from above that the same rank would hold good in actual practice. WALTER P. RICE.

SPECIFICATIONS FOR BRICK PAVEMENT.*

From the specifications for paving Wick Avenue in Youngstown, O., with vitrified fire-clay brick, we abstract as follows :

The pavement is to consist of vitrified fire-clay brick laid on a foundation of concrete composed of furnace slag, or cinder and sand, with joints filled with hot paving cement or coal-tar. The bricks are to be hard-burned and vitrified entirely through, free from cracks, true and smooth, and are to stand a crushing load of 36,000 pounds on pieces two inches cube. The street is to be excavated 11½ inches below finished surface of pavement, and all soft places made good. The filling in sewer trenches, etc., to be probed with a rod, and where soft to be removed and carefully refilled and rammed. Where ground is wet, agricultural tile-drain to be laid at two feet below finished surface, and carried to nearest catch-basin. The concrete foundation to be six inches thick. The slag or cinder is from Bessemer furnaces, and contains considerable lime, making it almost equal to hydraulic cement. The fresh material is brought while still hot, dumped into a box, and thoroughly mixed with water to make the concrete, and then spread and rammed in the usual way.

On the bed thus made two inches of sand are spread to a uniform thickness by means of guide strips, which are then removed, and the bricks

* xv, 658.

laid in lines crossing the roadway, and so as to break joints by at least three inches. The bricklayers are to stand on the finished work while laying.

The bricks are then to be rammed with a wooden rammer weighing twenty pounds, after which the joints are to be filled with the paving cement heated to at least 300° Fahr. This cement to consist of coal-tar pitch and creosote oil. The surface to be covered with sand. All broken bricks to be removed. The contractor to lay at least 300 square yards per day. Then follow the usual requirements as to care of the work, clearing away the rubbish, etc.

BRICK PAVEMENTS IN OHIO CITIES.*

City Engineer S. W. Parshall, with the Street Committee of Akron, O., have made a report on the brick pavements of several Ohio cities. They recommend them for their city, giving, among other reasons, that being the centre of the earthen pipe and brick-making industries the paving brick can be made at home, increasing the employment of local labor. Among the cities visited were Canton, Cleveland and Columbus. In Canton they found two kinds of brick pavement in the process of construction, one of a hard West Virginia and the other of Williams' Canton road brick. Both of these pavements are laid on gravel foundations, rolled with a steam roller, and the joints of brick filled with asphaltic paving cement. Being only a short time in use, no opinion is offered as to their durability. A short piece of brick pavement in Canton, put down two or three years ago with Williams' Canton fire brick, shows signs of disintegrating and crumbling, and could be readily cut with a knife, because, as was said, of the softness of the brick made at that time.

In Cleveland but four short streets had been paved during the past three years with brick, and these with cement joints. Experience there impressed the Committee with the imperative necessity of properly preparing and rolling the foundation and cementing the joints of both stone and brick pavements. The brick mostly used were Porter's union brand and West Virginia fire-clay brick.

"All pavements in Columbus," the Committee says, "are constructed with great care, as it appears to us. After excavation has been made to a proper depth, broken limestone is scattered over it in such a way that after it is rolled with a 10 or 15-ton steam-roller it will be six inches thick and of the same cross-section as the street when finished. After this is done a course of sand two or two and a half inches thick is put on to make a bed for brick or stone pavements. When the paving blocks are laid they are rammed down and the joints are filled with paving cement. This makes the entire pavement one homogeneous mass, firm, impervious to water, in a

* xxi, 50.

sanitary view of the best, doing away with filth and disease-breeding deposits that collect in the joints of pavements not cemented.

In Columbus we did not see a poor pavement that was laid by the city. The granite block sheet asphalt were laid as usual with asphalt on old cobble-stone pavement as a base. The brick pavements were in the best condition, all being pleasant to ride on, attractive to look at, and were in keeping with the thrift of the city, in which all her citizens seemed to take pride."

Opinion as to durability is withheld, as the pavements have not yet been down three years.

BRICK PAVEMENTS AT FREMONT, O.*

Mr. H. Coleman, President of the United States Fire Clay Company, of Pittsburg, Pa., in a communication to a newspaper gives some figures of cost of brick pavement at Fremont, O., which may be of interest to contractors and city officials. He says :

Contractors at Fremont, O., who have just completed laying a large quantity of pavement at this place this season, report the cost per square yard of our fire brick pavement laid complete to be $1.58.

In laying this pavement they first excavated 12 inches of earth, then used for ballast 8 inches of crushed stone ; this was well rolled with a 6 to 8-ton roller, and on top of this stone was used 4 inches of sand. The brick are then laid on the sand transversely with the street, every other course being started with a half brick, thus breaking joints throughout the entire pavement. The sand, however, before laying the brick, is made a little crowning in the centre, thus giving an opportunity for surface water to run into the gutters. The pavement is then rolled with a heavy roller. The crushed stone cost on the street at Fremont, O., $1 per cubic yard or 2 cents per square yard, 8 inches deep ; the sand cost $1 per cubic yard delivered on the street, or 11 cents per square yard 4 inches deep.

The contractors above referred to say that 6 or 8 inches of good gravel or sand without the crushed stone would make a first-class pavement.

BRICK PAVEMENTS IN CHARLESTON, S. C.†

Mr. T. A. Hugenin, Superintendent of Streets, Charleston, S. C., writes us in response to our reference to a brick pavement in Cincinnati, on page 318, issue of August 20, that brick prepared in accordance with a method patented by him have been "laid in Charleston for three years, and subjected to the heaviest traffic" of that city "with entire satisfaction, showing less wear than the granite block immediately adjacent."

PRIORITY OF USE OF BRICK FOR PAVEMENT,

CHARLESTON, W. VA., July 18, 1888.‡

SIR : I notice an inquiry about "brick street pavements" in your issue of June 30. This city was the first in America, or in the world, to adopt brick paving for street roadways, and now has several miles of it in use.

---

* xxi, 67.   † xvi, 374.   ‡ xviii, 103.

One street, laid in 1873, after fifteen years' constant and hard service, and practically without repairs, is almost as good to-day as when first laid. The city is extending it every year to other streets as fast as they are able to pay for it. Several other cities have now introduced it.

J. P. HALE.

186 DEVONSHIRE STREET, BOSTON, August 1, 1888.*

SIR: In your issue of July 28, Mr. J. P. Hale claims for the city of Charleston, W. Va., the honor of being the *"first city in the world"* to use brick pavement for streets.

His statement may be correct, but at The Hague, in Holland, brick pavement is used very extensively for roadways, and most things in Holland claim some antiquity at least.

An examination of the bricks proved very few indeed to have been fractured by the traffic upon them. A later experience in trying to break one in pieces in order to bring home a sample, explained their perfect condition. I have never seen an American brick approaching them in *toughness*, and am sure that any of our common bricks would be crushed to fragments with a fraction of the hard usage which they seem to stand with impunity.

They are very small ; more or less chocolate in color, and in the streets are laid on edge.

It opens a problem of compounding or selecting a clay with a special view to toughness which may open the road to fortune to some of our native brick-makers.

W. G. PRESTON, Architect.

[Our correspondent is probably correct as to the priority of use of bricks for roadways in Holland. Had Mr. Hale confined his claim to cities in this country we think he would have been correct.

Reports on London (Eng.) pavements dating back to 1839 (see pp. 150 and 195, Vol. XVII.), only once mention the use of brick pavements—viz., in 1877.

As to toughness, we are led to believe, from descriptions published, that the bricks used for this purpose here have been of excellent quality, but have not had opportunity for personal examination.

Those who have seen the small Holland bricks found in old buildings in New York, can bear witness that those who made them certainly understood their business; but we have no doubt equally good can be and are made here when a sufficient price is paid.]

EARTHEN TILE PAVEMENT.†

A pavement of earthen tile blocks, about 8 inches square and 4 inches thick, impregnated with bituminous products and laid with hot tar joints on 6 inches of concrete, has been put down experimentally in Berlin. It is not expected to be subjected to the American process of tearing up every few months to lay down a new line of pipes in the street, and to be relaid by a contractor who is not responsible to anybody.

---

* xviii, 129. † ix, 544.

PAVEMENTS FROM BLAST-FURNACE SLAG.*

In the Hanover district of Germany the blast furnaces make a specialty of a road-metaling material called slag-stone. For its preparation the slag as tapped from the furnace is run into a bottomless cast-iron mold, which is tapering to the top and stands on an iron truck. After the slag has cooled enough to have an outer shell sufficiently strong to support the melted mass inside the mold is lifted off and placed on another truck ready for use, while the loaded truck is run off to the dump-heap, where the truncated cone is pierced and the slag run on to the ground and covered with cinders and ashes and allowed to cool very slowly. This makes the slag much tougher than when allowed to cool in the open air, and it is said to wear very well when broken properly and placed on roads.

The ordinary brittle slag makes a very good foundation for a road, particularly on clay and wet soils, as by rolling, the top pieces form a powder that fills the interstices between the lower fragments so thoroughly that neither clay nor mud can work up through the layer, and on this the more durable wearing layers can be placed. It was found impossible to form any roads on the soft clay surface of the Centennial Fair Grounds at Philadelphia until their beds had been prepared by a layer of well-rolled furnace-slag, after which they stood very heavy teaming without underdraining ; the bonding of the fragments of slag with the thorough filling of the interstices preventing any mud from working through the first or lower layer, thus keeping the road from breaking up.

STEEL PAVING BLOCKS.†

A substitute for granite blocks for paving purposes is a steel paving block claimed to have superior durability and whose cost is said to be somewhat less than the stone. It is thus described :

The block is made of steel strips, some $2\frac{1}{2}$ inches wide by 1 thick, with a rolled channel on the side exposed to traffic, and containing notches about half a foot apart. The weight of these strips is 11 pounds to the yard ; they are laid across the street, a distance about 5 inches between centres, and as their length is sufficient only to extend to the middle of the street the proper slope from the centre to the gutters is easily secured. To insure their not slipping sidewise they are bolted together and fastened to wooden sills The support for the new pavement is composed of a firmly constructed bed of gravel, while between the steel strips a compound of pitch and cement is poured, filling the interstices to a level with the tops of the strips, and rendering the surface comparatively smooth.

* xvi, 547.   † xix, 212.

* Our Milwaukee correspondent sends a newspaper clipping which says : " The North Chicago Rolling Mill Company has completed an order for fifty tons of steel rails, of a peculiar pattern, to be used in paving a block in Chicago. The rails were rolled in the Merchant Mill of the Bay View Works. They are 16 feet 10 inches in length and have a grooved surface on top, to render them rough, so that horses will not slip upon them. The rails will be placed a few inches apart, and the space between will be filled with a patent composition, which is said to be very hard and durable. The pavement is to be used where there is very heavy teaming. If the experiment proves successful it is probable that many of the Chicago business streets will be paved in a like manner."

EXPERIMENTING WITH A COMPOSITE PAVEMENT IN PARIS.†

A correspondent in a letter to *Le Genie Civil*, calls attention to a composite pavement laid by an American syndicate on the Rue de Rivoli in Paris, which he states in eight days after it was open to traffic began to break up. It seems that this pavement was laid on the Rue de Rivoli, where the traffic amounts to 33,000 vehicles per twenty-four hours, and the contractors who had agreed to keep it in repair for two years failed to take into account this unusual amount of traffic. Hence they are relaying the pavement with greatest care to meet these extreme conditions. Another year's trial will doubtless demonstrate whether this kind of pavement will be suitable for such heavy traffic.

PROPORTIONS FOR CEMENT PAVEMENTS.‡

The *Moniteur de la Ceramique* gives the results of some experiments by M. Bohme on the durability and resistance to wear of cement pavements.

In a series of experiments, with twenty-eight different mixtures of cement with a perfectly uniform sand, he eliminated the variations usually due to irregularities in quality, and second results indicating that the durability of the pavement does not depend materially upon the tensile or crushing strength of the cement as usually determined.

In this series of experiments the proportion of sand to cement giving the best results varied from 1 to 1 to 2 to 1, 1½ to 1 proving best in the majority of cases and being recommended for use with cements whose properties have not been specially investigated.

The resistance to wear of pure cement is in general equal to that of a mortar composed of 2 parts cement and 7 parts sand, but this proportion varied considerably, ranging in the experiments from 1 to 2 to 1 to 5.

---

* xvi, 685. † xv, 521. ‡ xviii, 175.

ASHES AND CINDERS.*

Professor I. O. Baker, in his address before the Illinois Society of Engineers and Surveyors, mentioned an experimental section of iron pavement in Chicago; also the experiment in Toledo, of covering unpaved streets with six inches of cinder and ashes.

This covering becomes so compact that in winter heavy loads do not cut through it, and in summer it is less dusty than an ordinary earthen roadway.

RUBBER PAVEMENT.†

For paving streets India-rubber threatens to enter into competition with asphalt. A recent number of *Kuhlow's* gives a fuller account of this new pavement, which was briefly noted in our issue of January 12. It is the invention of Herr Busse, of Linden, who has introduced it in Hanover. He used it first in the summer of 1887 for paving the Gœthe Bridge, which has a surface of about 1,000 square meters, or 10,764 square feet. The new pavement, it is stated, proved so satisfactory that 1,500 square meters (16,146 square feet) of ordinary carriageway in the city were paved with it last summer. The Berlin corporation, being favorably impressed with the new pavement, has had a large area paved with India-rubber as an experiment, and the magistracy of Hamburg is likewise trying the pavement. It is asserted that the new pavement combines the elasticity of India-rubber with the resistance of granite. It is said to be perfectly noiseless, and unaffected either by heat or cold. It is not so slippery as asphalt, and is more durable than the latter. As a covering for bridges, it ought to prove excellent, as it reduces vibration; but a question may be asked as to its cost. The expense must be heavier than that of any known pavement.

---

* xv, 658.† xix, 157.

CHAPTER V.

CURBS, SIDEWALKS AND TRAMWAYS.

A NOVEL FORM OF STREET CURB.[*]

As a matter of interest we reproduce from *La Semaine des Constructeurs* of September 18, a description of a novel form of street curb, recently introduced at Verdun in France. This application of cast iron is the invention of Mr. Nicot, a contractor for asphalt paving, and it is now in use on about two miles of sidewalks. The curb is cast in lengths, as shown in Fig. 2, of an L-shaped section,

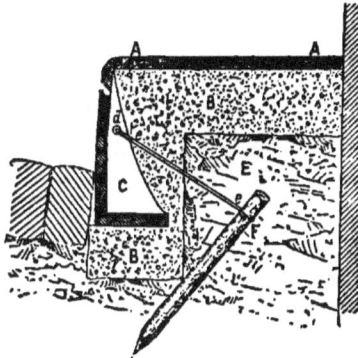

Fig. 1.—

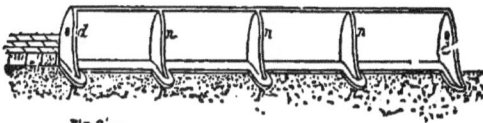

Fig. 2.—

with ribs or knees, the bearing of the latter being enlarged by an expansion of the bottom flange. The segments rest on a bed of concrete, and the concrete is also filled in upon the bottom flange, as shown in the sections, so that the curb is held firmly in place. The surface of the sidewalk is finished in asphalt to the level of the top of the curb.

[*] xv, 658.

In addition, a peg of wood or iron (Fig. 1) is driven in the earth behind, opposite the end ribs, and secured to the ribs by a galvanized iron wire.   This serves to keep the curb to place while the process of filling is progressing.   The reviewer criticises this as smacking too much of the gardener.   The suggestion is made that a "fish-tail" rod screwed into an enlargement of the ribs at the joints in the curb, and buried in a small mass of concrete, as shown in Fig. 3, would be more in accordance with sound ideas of construction.

Fig. 3.

The reviewer considers the curb as one well worthy of investigation and trial, and we would add that with such improvements as experience may suggest it may be worthy of introduction in localities where good stone for curbs is not obtainable.

AN ARTIFICIAL STONE FOR STREET CURBS, ETC., USED AT GALESBURG, ILL.*

From an engineer traveling in the West we learn that in some of the Western States no stone can be obtained near by suitable for street curbs, rendering it often necessary to transport material a long distance for this purpose, or pave the street without curbs.   When recently in Galesburg, Ill., he noticed a method there in use which may be of interest to some of our readers, and which he thus describes :

Ohio stone is used to some extent for curbing, but has a rough appearance.   The best material used is known as Asbestine Building Stone, and is prepared of German Portland cement, sand and broken stone.

The dimensions of each piece of this curb are uniform, having a length of 4 feet, and in section 6 inches at bottom and 5¼ inches at top, with the two upper edges rounded.   The ends are fitted to make a close joint, with mortise and tenon molded in the material.

The material forming this curb is rammed into wooden molds, the molds being made in sections, permitting them to be easily removed when the material has set ; the usual time allowed for this purpose is nine days, the curb being fit for use in a few days after.

*xx, 231.

The inner face of the mold is varnished to prevent the material from adhering to the wood. No broken stone is visible on either face of this curb, it therefore has the appearance, when set, of fine axed natural stone.

Narrow strips of wood, one inch in thickness, are laid on the bottom of the trench on which the curbing is set to keep the grade line uniform ; wood is used for this purpose as there is no suitable stone in the neighborhood.

This curb is furnished and set for fifty cents per lineal foot. It is manufactured by W. H. Walburn, of Galesburg, Ill.

SIDEWALKS.

FOOTPATHS IN ENGLAND.*

We interrupt our description of the pavements of Liverpool to notice a paper entitled " Footpaths " (sidewalks), by H. Percy Boulnois, M. Inst. C. E., Borough Engineer of Portsmouth, which is of interest in connection with this series.

Government Requirements.—The paper first remarks on the proportion between the widths of footways and carriageways adopted by the Local Government Board. The requirement is that every new street shall have a carriageway at least twenty-four feet wide, with a footpath each side " of a width not less than one-sixth of the width of such street," (meaning, as is explained, one-sixth of the width of the carriageway). The width of roadway should be, if possible, some multiple of eight feet, " since this is the allowance of vehicles passing each other at a rapid rate."

The gutters at each side to be not less than three inches deep, or more than seven inches. Curbs to be set on " clean sharp ballast," and well rammed.

Materials.—The materials used for footpaths may be classified as: (1) Natural stones; (2) Natural asphalts; (3) Artificial asphalts and concretes ; (4) Brick ; (5) Gravel and stone chippings.

Requirements.—The requirements of a good covering for sidewalk are that it shall not be slippery and shall not scale or flake. It must be durable, not easily abraded, strong, of uniform quality, dry rapidly after rain, and dust should not readily adhere to it. Its absorbent powers should be tested. Its quality will also be in a measure indicated by its microscopic appearance and specific gravity. Its wearing qualities may be tested by rubbing and drilling, but better by actual test under a known traffic.

Natural Stones.—Of natural stones granite is durable, but wears very slippery, is difficult to work, and is expensive.

Yorkshire flags are much more largely used. They are readily cut, do not wear smooth, and are never slippery from wear. They

---

* xiv, 393.

are more liable to breakage, absorb considerable moisture, and tend to laminate when exposed to frost. (The same is true of many of our American flagstones.) The wear under heavy traffic in the city of London was about one-sixteenth of an inch per year, and this is given as the rate in the Strand for every 9,000,000 foot-passengers.

Slate wears very slippery ; most of the freestones are too soft ; and some of the limestones are very brittle and very slippery.

Natural Asphalts.—Of natural asphalts, that is recommended which contains about 7 per cent. of bitumen, and, where possible, that it should not be laid except by the compressed method. The "mastic" (which contains more bitumen) is considered only "suitable for broad paths with light traffic." The former should not be laid of less than 1-inch thickness, and the latter of $\frac{1}{2}$-inch; it being essential that a foundation of cement concrete three inches thick be first provided, "as asphalt is like a mineral leather or elastic skin, and has no strength in itself, but acts solely as a cover to the concrete," which actually carries the traffic. "It is almost impervious to moisture, and is the best pavement that can be used from a sanitary point of view." In Birmingham it requires relaying every five or six years, and compressed asphalt paths have lasted ten years in some of the busiest thoroughfares in London. In Leicester, mastic paths (uncompressed) have lasted fifteen years. Under considerable traffic it is estimated by the author that compressed asphalt one inch thick will last twelve years.

Artificial Asphalts and Concretes.—Of artificial pitch mastics the composition of one of the first used is given as 50 parts waste products of tar-oil, 20 of caustic lime, 200 of pitch, 30 of sawdust, and 700 of iron slag, grit, or chalk ; the materials being simply mixed, spread in layers, and rolled. Another laid in 1840 was Stockholm tar three parts, chalk two parts, and sand one part, boiled together and spread hot.

One of the best modern mixtures is of gravel or stone chippings screened to the several sizes, passing through sieves of $1\frac{1}{4}$-inch, $\frac{3}{4}$-inch, $\frac{1}{2}$-inch, and $\frac{1}{4}$-inch mesh, and heated on iron plates ; when dried and thoroughly heated, the following mixture was made and well-boiled : 12 gallons of tar, 56 pounds of pitch, and two gallons of creosote, this quantity, while still hot, being added to about a ton of the hot screened materials. A thin layer of the largest material is laid first and rolled by a $\frac{1}{2}$-ton roller, and the others similarly in succession, the finest on top. The chippings make the best pavement. The foundation must be dry, and the pavement is best laid "in the spring or winter, if dry, as a hot sun draws the composition away from the stone on to the surface of the path." The surface should be dusted over with fine grit or stone-dust, "and this facing, accompanied by a thin 'painting' with tar, should be repeated at least every other year."

The life of these pavements is given variously at from five to twenty years.   The disadvantages named are its dark color, its wearing gritty and bumpy, and its "rather difficult repair."   In very hot weather it becomes sticky or soft.   " Tar pavement must only be reckoned as a substitute for ordinary graveled footpaths.   It must not be compared with paved or asphalted paths."   It is largely used for suburban footpaths.

Monolithic pavements of Portland cement concrete are gaining in favor, "and may be looked upon as the pavement of the future," and these when properly laid, with joints at about six feet intervals, are found to be very durable.   Shingle or gravel does not work so well as crushed granite, since the former is apt to become dislodged or else protrude and make a rough surface.   A bottom layer, $2\frac{1}{2}$ inches thick, of one part Portland cement to six parts shingle or stone, with a $\frac{1}{2}$-inch layer on top of one part cement to two parts finely broken and well-washed granite, are recommended.   Great cleanliness and care in mixing are needed, with thorough ramming of the concrete and troweling of the surface.   Traffic should be kept off until it is thoroughly set by a covering with boards for a fortnight, and afterward by a covering of wet sand.   This description of pavement is rather slippery and should have a light cross-fall. It may be molded in blocks and laid in this form if desirable.   After the upper surface is worn these can be turned and will present a fresh wearing-surface.

Brick.—Brick pavements are undesirable on account of the multiplicity of joints and their consequent insanitary character, their harshness to the foot, and the necessity for a concrete foundation to keep them level.   Special vitrified bricks have a life under heavy traffic of thirty years and upward.   Gravel and similar materials are only desirable for cheapness.

In conclusion the author states that "asphalt has suffered in repute because inferior materials have been used under that name. Tar pavement is an excellent substitute for gravel paths, and is a good pavement for light traffic.   Concrete, monolithic, and flagged pavements are every day gaining in favor and have much to recommend them."

Comments on Mr. Boulnois' Paper.—We wish to remark, by way of comment on the proportion between width of roadway and sidewalk given in the paper, that, while it might be an advisable one in streets almost entirely devoted to wheel traffic, it should not be adopted in other localities.   In towns of moderate dimensions where *shade*-trees are planted the most pleasing results will be obtained with the walks nearly or quite one-half the width of the roadway, and even in busy streets it will be found that about one-third, exclusive of stoops or other projections, is not too great a width for the sidewalks.

With the necessity in this latitude for providing for heavy snows the depth of gutter mentioned is too small. The general rule holds good that the better the surface of a roadway the less crown is required to carry water to the gutters. The experience of the present writer on streets with a graveled surface where the longitudinal fall obtainable was from 0.1 of a foot to 0.2 of a foot per 100 feet was that a crown of 18 inches on a roadway 40 feet wide, reducing to 1 foot on one 20 feet wide, on newly made streets, was not objectionable in use and became rapidly less from wear. On the contrary, crowns of about 6 inches are quite sufficient on well-paved streets which are kept clean and in good surface.

It is a very common fault to give too steep a cross-slope to flag and other footpaths. The sole object of this slope is to carry water to the gutter. Since the surface is required by municipal ordinances to be kept clean from snow, there is sure to be at times a thin coating of ice formed upon it, and under such circumstances a cross-slope of ½-inch per foot becomes dangerous to passengers. It will be found in practice that ¼-inch per foot will give the most satisfactory results. In comparing practice in our cities with that abroad it should be borne in mind that much greater care is usually exercised there to obtain accuracy of alignment and smoothness and perfection of surface ; also, that much greater attention is paid to keeping the surface in repair. We may well take a lesson from them in this important matter.

ASPHALT AND CONCRETE FOOT PAVEMENTS.*

The object of this paper is to draw the criticism of the members of the association upon the experiments and experience of the writer and others, on asphalt and concrete as materials for foot-pavements, and, if possible, to induce others to carry forward experiments with a view to perfecting the use of these pavements.

The writer desires to place in the forefront of the paper the fact that the credit of the Hornsey experiments herein referred to is due to Mr. T. de Courcey Meade, A. M. I. C. E., who most readily placed them at the disposal of the writer for this paper; and also his obligations to the French Asphalt Company, the Val de Travers Asphalt Company, the Imperial Stone Company, and others, for the

---

*A paper by George R. Strachan, A. M. Inst. C. E., of Chelsea, London, S. W., read at the annual meeting of the Association of Municipal and Sanitary Engineers at Leicester, July, 1887. xvi, 266.

information given. Every pavement described has been personally examined by the writer, and the exact locality of each is stated, so that any one may examine them for his own information.

Asphalt, its Nature and Occurrence.—Asphalt, properly so called, is a natural compound of carbonate of lime and bitumen, and is found principally in volcanic areas. Men of erudition have asserted that it was the "pitch" used to make the Ark watertight, and that it was the "slime" used as a mortar in the construction of the Tower of Babel and the city of Babylon. If such ancient uses of this substance are facts, its virtues were strangely lost sight of in the intervening centuries, for it is not till 1700, A. D., that its use became common. It was then used for the purpose of extracting "balm" from its beds, which was used for medical purposes and was credited with superior healing powers. The origin of the asphalt beds has given rise to much speculation. A Swiss geologist has made an effort to explain their formation in a striking manner. Starting from the observation that all organic matter exudes bitumen in decomposing, he suggests that the beds are the remains of huge banks of oysters, the shells of which furnished the carbonate of lime, and the oysters themselves furnished the bitumen. As the asphalt beds are in some cases 27 feet thick, and their areas are measured by square miles, it is evident that oysters were plentiful in those days.

The works of the French Asphalt Company are described, as that company has executed all the asphalt works in Chelsea; but the writer wishes to say that the Val de Travers Company, Claridge's Asphalt Company, and others, do equally good work.

The mines of the French Asphalt Company, from which their English supply of asphalt is obtained, are situated at St. Ambroix, in the south of France. The asphalt is in seams, which lie nearly horizontal, and which have their faults, bends, etc., like coal seams. The bed and roof of the seams are of pure carbonate of lime rock, presumably the same as that of which the asphalt is largely composed. The seams vary from three to five feet in thickness, and are worked by drifts from the outcrop on the hillside. The rock is mined by blasting and hand labor, and comes from the drifts in pieces measuring one cubic foot and downwards. The asphalt then has a very dark chocolate color, and appears to be a tough homogeneous substance, with striations of white matter running through it parallel to its natural bed, which are probably narrow seams of carbonate of lime. When it is exposed to a hot sun the surface will glisten with small fatty beads of bitumen, but at ordinary temperatures it is dry. After exposure to the air the surface turns a dull white color, owing to the evaporation of the bitumen; but this change is only skin deep. The rock is conveyed to Marseilles just as it leaves the mine, and is shipped to England. Formerly the beds of asphalt yielded a supply of natural bitumen, but twenty years ago the supply ceased, owing. it is believed, to some widespread cause, as several mines were affected in the same way at that time.

Preparation.—At the depot of the company at Stratford, the rock as delivered by ship is passed through a crusher which acts like a Blake's crusher, save that the solid jaw is replaced by a series of knives, and by which it is reduced to pieces not exceeding three inches in length. The pieces are poured into a Carr's disintegrator, which has spindles revolving 800 times a minute in opposite directions, and which reduces the asphalt to powder. The powder is screened through a rotatory cylindrical sieve (144 to the square inch) and is then stored in sacks. The asphalt varies in the proportion of bitumen it contains. The richer parts are ground and stored separately from the other, and are afterwards mixed in suitable proportions for the particular use to which they are to be applied.

Composition.—The following are analyses of asphalt rock of average richness at this stage:

	No. 1 Sample.	No. 2 Sample.	Average.
Bitumen...........................	10.70	10.60	10.65
Carbonate of lime................	88.05	88.15	88.10
Silica	0.55	0.40	0.48
Alumina..........................	0.10	0.15	0.12
Peroxide of iron	0.20	0.10	0.15
Moisture.........................	0.40	0.60	0.50
	100.00	100.00	100.00

Making the Footpath.—When the asphalt is about to be used, the powder is poured into revolving roasters, and roasted for three hours at a temperature of 280° Fahr., during which operation the moisture is driven off. As the asphalt chars at 320° Fahr., care has to be exercised as to the proper temperature. . It is loaded direct from the roasters into carts lined with sheet-iron, covered with hemp cloths, and thus protected it retains its heat till it is taken to the site where it has to be laid. It is carried from the carts in baskets, spread over the foundation by means of a rake and rammed solid by a series of blows from heavy heated rammers. The surface is ironed by a heated iron, which draws bitumen to the top, and in a few hours it is ready for traffic. This form of asphalt is known as compressed asphalt, and is the form always used for carriageways and frequently for footways.

Mastic Asphalt.—The other form of asphalt is known as mastic asphalt, and is a manufactured compound made up of natural asphalt, artificial bitumen, and grit. The asphalt is reduced to a powder as described. The artificial bitumen is used because of the scanty supply of natural bitumen. Its principal component is Trinidad pitch, to which is added from five to seven per cent. of shale oil. The mixture is boiled for twenty-four hours; the top liquid is ladled out, and is the artificial bitumen. It is a soft, viscous, black substance which softens under the sun's rays. Its quality is tested by taking a piece between the fingers and drawing it out to a string; if it does not snap until drawn out very fine it is of good quality. The grit is obtained from Bridgeport, is wholly composed of flint, very clean, and the pieces do not exceed ¼-inch in size.

Its Preparation and Use.—The mastic asphalt is prepared as follows: From 5 to 7 per cent. of artificial bitumen, from 20 to 30 per cent. of grit, and the balance in powdered asphalt are placed in a covered caldron and heated for four or five hours. The mixture liquefies at 280° Fahr. to 300° Fahr. If it is to be used near the works (within ten miles), it is run into locomobiles (boilers on wheels), with a fire under them, and drawn to the site. When it is used, it should be hot enough to vaporize a drop of water. It is carried in pails and spread over the foundation by means of a float. Silver sand is then spread sparingly over the surface and rubbed in by floats. In six hours the footway is ready for traffic. One ton of asphalt covers 20 square yards when laid one inch thick.

When mastic asphalt is to be laid at a distance from the works instead of running it from the caldrons into the locomobiles it is run into molds, and molded into flat cylindrical pieces weighing about 56 pounds each. These are taken to the site, placed in a caldron, from 3 to 4 per cent. of additional bitumen added to make up for the loss by evaporation, and heat applied to reduce it to a liquid condition. The laying is then performed in the same manner as before described.

This description may be taken as applicable to the method adopted by the Val de Travers Company, with a few variations in the proportions used.

Analyses of Rock Asphalts.—The following analyses of asphalts are of interest, as they are those of rocks of average richness·

	Val de Travers Co.	French Asphalt Co.
Bitumen.....................	9.75	10.65
Carbonate of lime, etc...............	89.75	88.85
Moisture........................	.50	.50
	100.00	100.00

The following analyses, in a different form, were placed at the writer's disposal by Mr. Meade.

	VAL DE TRAVERS.		FRENCH ASPHALT CO.		
	Rock.	From Cheapside.	From Hornsey Lane.	From Crescent Road.	Rock.
Silica............................	0.6	0.5	0.3	0.4	0.4
Volatile organic matters (tar, oils, etc.)...	5.8	5.8	6.5	6.0	8.2
Non-volatile organic matters...	13.0	9.8	13.6	16.9	16.8
Lime, etc............................	80.6	83.9	79.6	76.7	74.6
	100.0	100.0	100.0	100.0	100.0

This detailed description and the numerous analyses of good asphalts have been given, so that spurious asphalts may be avoided.

Traffic—Durability.—In Chelsea there are 16¼ miles of footways paved with mastic asphalt, having an area of 68,290 square yards. On the Queen's Park Estate there are 41,500 square yards, which have been laid five years, and which are now in good condition, not having cost one penny for repairs. In King's Road, at Walpole Street, a length has been laid for seven years. The foot traffic over it is 7,500 persons in eighteen hours. At the end of the first five years it was cut open, and the wear was found to be such as had reduced the thickness to a spare ¾ of an inch, the original thickness being 1 inch full. On the east side of New Bond Street a length of mastic asphalt was laid thirteen years ago between Oxford Street and Conduit Street, the thickness being ¾ of an inch. The asphalt is now wearing through on to the concrete in the line of traffic at the forecourt line. The cost for repairs has been so trifling that it may be neglected. In this case the concrete foundation is as sound as before, and all that is necessary to restore the footway is to relay the asphalt at about two-thirds of the original cost, when the pavement will be good for another thirteen years. As the traffic here is very severe and the footway narrow, it is reliable evidence of the durability of asphalt.

Foundation.—The foundation of the asphalt footway is made with 3 inches of Portland cement concrete (6 to 1) of very good quality. The surface is smoothed with the shovel, and four days are allowed for drying. The concrete has been laid hitherto without any joints. The mastic asphalt is floated over the surface, and the path is then completed. Mastic asphalt does not show any cracks on the surface. The concrete foundation, when the asphalt is removed, shows the regular tree-like cracks all along its length, branching from the kerb to the back line, but the elasticity of the mastic asphalt is sufficient to resist the tearing action of the concrete as it contracts.

Expansion and Contraction of Concrete.—A study of the asphalt question resolves itself principally into a study of the movements of concrete when laid in long lengths, narrow widths, and small thicknesses. The writer inclines to the opinion that concrete has in itself a small power of contraction, apart from any considerations of temperature. The experiments of Dyckerhoff, which show that neat cement (slow-setting) had an average expansive power over twelve months of .0734 per cent., and quick-setting cements of .2019 per cent., and that concrete (3 to 1, sand) had an expansive power of .0264 per cent. (slow-setting) and .0320 per cent. (quick-setting), seem to show the contrary to be the case. The writer laid down a length of concrete (6 to 1, ballast), 52 feet long, 12 inches wide and 3 inches thick, under a shed which had an open front, but so that the sun did not touch the concrete. The strip was laid on sand so as to give it freedom of movement. Another strip 26 feet long, of the same width and thickness (3 to 1, pebbles), and a third of the same dimensions (3 to 1, sand), were also laid under the same conditions. The only movement discernible, at the end of one month, was a slight contraction in length in all the samples. The uniform experience of concretes under asphalt is that cracks occur, which would tend to show that contraction and not expansion is the rule. At the same time, the writer has experience that concretes do expand, but this he attributes to the action of temperature. It is no uncommon thing to see the surface of an asphalt path raised crosswise in an irregular line, as though a small tree root was under it. In every case where the asphalt has been uncovered at these points by the writer, he has found the concrete crushed and the concrete on the falling level thrusting itself under the concrete on the rising level. This effect is most marked on hot days. In January last the writer laid some thousands of feet of asphalt path in St. Luke's Gardens, Chelsea. The sun is on it all day, and during the hot weather at the beginning of June, the number and size of these raised lines was astonishing. Shortly after midday they were most pronounced, and towards night they were less prominent. As a further evidence of the expansion of concrete under the sun's rays the streets in the city can be named. The footway and carriageway are in asphalt on concrete. The expansion of the concrete in the carriageway presses the kerb at the bottom ; the expansion of the concrete in the footway presses the kerb at the top on the opposite side, and the two have tilted up the kerb in a marked manner. The writer has on a hot day taken up asphalt on a footway and has found the heat much greater under the asphalt than on the surface. In order to avoid the expansion showing itself in footways the concrete should be laid in sections, and the joints between them filled with some compressible substance.

Compressed Asphalt. Cracks.—Compressed asphalt has about one-third longer life than mastic asphalt under the same conditions. The cost is the same, but the use of compressed asphalt for this purpose has not been universally followed by reason of the cracks that appear on its surface. The cracks do not tend to spread under traffic, nor does the asphalt wear more at these parts than at others. They are unsightly, however. It is found that these cracks are exactly of the shape and in the position of the cracks in the concrete foundation. Compressed asphalt has no elasticity in itself, and when subjected to the contracting force of the concrete it is torn through. It is an admirable tell-tale of the movements of the concrete. Much ingenuity has been displayed in endeavors to avoid the cracks. The first step was to localize them. This was done by laying the concrete in 12-feet bays and in alternate bays, and filling up the screed space with fine concrete. The contraction then showed its effects at these places, with a result that a series of regular straight cracks appeared instead of the irregular tree-like cracks when the concrete was laid in one piece. These effects can be seen at many places in London without specifying any particular place. Having localized the cracks, an experiment was made at Hornsey to avoid them. A strip of bituminous felt six inches in width and ¾-inch thick was placed on the concrete over the whole length of the screed mark. This felt has much elasticity, and the object of the experiment was to ascertain whether it would take up the contracting movement of the concrete and absorb it. The length is laid at Crouch Hall Road, between Coolhurst Road and Clifton Road. The result has been that instead of one crack at each screed mark there are two, one on each side of the narrow strip of felt. It is evident that the concrete in contracting compresses the asphalt longitudinally, and that the cracks appear at the points where the opposing motions meet ; and as the strip of felt represented a narrow area which was free from these forces a crack appeared on each side where the forces took effect.·

In Archway Road, Hornsey, another experiment was made by covering the screed mark with a strip of mastic asphalt nine inches in width and ½-inch thick, just as in the last case with felt. For three months no cracks appeared ; then a few slowly and at irregular intervals showed themselves ; but during the severe winter of 1886-7 every screed mark showed its crack. These cracks were irregular in line, but they are confined in each case to the area covered by

the mastic asphalt.   These footways are laid on a 3-inch foundation
of concrete.   In Marlborough Road, Chelsea, an experiment was
made on different lines.   A foundation of concrete six inches thick
was laid, and the compressed asphalt laid on it.   For four months
no cracks appeared, but after that time they occurred at frequent
intervals, though they are fewer than usually appear on a 3-inch
foundation.   When the asphalt and concrete were removed at the
cracks it was found that the crack extended through the whole
thickness of the concrete.   This experiment was based on the
observation that cracks do not appear in compressed asphalt carriageways, and as the principal difference between the foundations
in the footways and carriageways is the thickness of the concrete,
it was assumed that it was the cause.   The observation is, however,
an incomplete one.   In streets of light traffic the cracks do appear
in the asphalt, as in Little Blenheim Street, Chelsea, and elsewhere.
In streets of heavy traffic the cracks in the concrete tear the asphalt
as they slowly form, but the traffic welds the asphalt together again
before they show on the surface.   In footways of heavy traffic there
are fewer cracks than in those of light traffic, for a similar reason.
At Muswell Hill, Hornsey, the experiment of covering the whole
area of the footway between Onslow Rise and Grosvenor Gardens
with bituminous felt was tried.   The felt was in 3-feet widths, and
was laid longitudinally with butt joints.   At the circular kerb the
pieces were necessarily somewhat patched.   The result has been
that cracks have appeared at every joint of the felt with marvelous fidelity, owing to the movement of the concrete.   In White-
head's Grove, Chelsea, between Marlborough Road and Keppel
Street, on the north side, a length was laid in 1885 on a 3-inch concrete foundation, which was covered with mastic asphalt ¼-inch in
thickness.   On this ¾-inch of compressed asphalt is laid.   The
mastic asphalt was laid to absorb by its elasticity the movement of
the concrete without transmitting it to the compressed asphalt.   It
has survived two winters of great severity, and has lived twenty-one
months without any cracks appearing.   From this it would appear
that the principle of a material between the concrete and the compressed asphalt which will absorb the effects of the movements of
the concrete is a correct one, and the writer invites the members of
the association to experiment on cheapening the method.   The
present result is an increase of life of 33 per cent. at an increase of
cost of 12 per cent.

Durability. — As evidence of the durability of compressed
asphalt in footways, those in Cheapside may be mentioned.   They
were laid in 1876 at a thickness of 1 inch, and are now wearing
through.   On the south side of the Strand east of Wellington

Street 1 inch of compressed asphalt was laid in 1881, and has had only the most trifling repairs. The streets in the city with the heaviest foot traffics in the world are paved with compressed asphalt on the footways.

Advantages of Asphalt.—The advantages of asphalt foot pavements are durability, a smooth surface unbroken with joints, a good foot-hold, even and regular wear, their impervious character, and the readiness and neatness with which they are repaired. They have a sombre appearance, and show water on their surface longer than stone pavements. They wear to the last thickness without breaking up, and give a useful wear for the whole of their thickness. When a stone or other pavement has worn 1 inch it may not be half worn through, but its usual life is over. Where there are cellars under footways, asphalt as a material for footways is unrivaled. In ordinary traffics, such as those named in King's Road (7,000 to 8,000 persons per day), an asphalt pavement can be laid 1 inch thick with the certainty that for at least ten years it will need no repairs whatever. This pavement has also the advantage that the foundation is always preserved and good for use again when the wearing surface of asphalt has to be renewed.

Cost.—The following abstract of the present contract schedule in Chelsea will give the prices for these pavements.

	Per sq. yd.		
	£	s.	d.
Compressed or mastic asphalt 1 inch in thickness on 3 inches of concrete.	0	6	3
Ditto, ditto, ¾ inch thick.	0	5	6
Compressed asphalt ¾ inch thick on ¼ inch of mastic asphalt, laid on 3 inches of concrete.	0	7	0
Compressed or mastic asphalt 1 inch thick on existing concrete foundation (relay).	0	3	6
Ditto, ditto, ¾ inch thick.	0	3	0

These prices carry a guarantee of free maintenance for ten years. The Vestry prepares the foundation for the concrete in new work at a cost of 2*d.* per square yard. The specification provides that the asphalt will be cut open at distances not exceeding 50 feet apart, and the thickness measured. Five out of every six of these measurements must be at least the specified thickness, and the average of every six must be at least the specified thickness. The specification is strictly adhered to. The cost of the foot pavement in New Bond Street, already referred to, over twenty-six years, would be as follows at per square yard :

	£	s.	d.
Preparing foundation	0	0	2
Laying concrete foundation and ¾ inch mastic asphalt.	0	5	6
At end of 13 years relaying mastic asphalt (life 13 years).	0	3	0
Repairs.	0	0	0
Total	0	8	8

or a cost per year, not including interest, of 4*d*. per square yard. It should be mentioned that the small cost of renewing asphalt foot-ways is due to the fact that the asphalt taken up is as good as new asphalt after it has been cleared and prepared, and is re-used for footways.

Concrete Footpaths—In dealing with the question of concrete as a material for foot pavements, the writer cannot claim such an experience of it as he has had with asphalt. In 1880, however, the Vestry of Chelsea had laid by the respective makers a series of pavements in King's Road against the Royal Military Asylum wall, and in 1885 a report was made on them by the writer. So many applications for copies have been made that it is now out of print, and as applications continue to be made, the results of the experiments are herein set forth for the information of the Association.

Construction.—The writer holds that a knowledge of the manufacture of these pavements is useful information, and therefore he incorporates a description of the manufacture of imperial stone pavements. The depot of the company is at East Greenwich, on the banks of the Thames. The aggregate used by the company is broken granite. Flint has been used, but the pavement in wear became very slippery. Kentish rag has also been used, but its wear was too rapid. The granite is broken, so that it passes a $\frac{1}{16}$ sieve. After screening it is carefully washed, as the dust acts as a coat round the piece, and prevents the cement laying hold of the granite. The washing machine is a slanting archimedean screw, working in a trough, with openings in the thread of the worm. The water is run in at the high end, and the screened granite pieces put in at the lower. The screw churns the pieces over and over each other, and carries them up by its motion. The clean water meets the cleanest granite, and thus the pieces are not soiled by the dirty water they make. About 8 per cent. by weight is washed out as waste. The importance of a clean aggregate is seen when it is stated that briquettes made from washed pieces have a tensile strain of 15 to 20 per cent. higher than those made from unwashed pieces, when tested under similar conditions. The cement used is the best Portland, and is required to stand a tensile strain of 350 pounds per square inch after seven days' immersion in water. As a matter of fact, the cement used runs to an average of 425 pounds on the square inch. Hitherto a residue not exceeding 10 per cent. on a 50-mesh was allowed, but the value of increased fineness is recognized, and preparations are being made for a cement that will not leave a residue of 10 per cent. on a 76-mesh. The weight runs from 116 to 120 pounds per striked bushel. Before use, the cement is laid out to cool for fourteen days, and is turned frequently in that time. Great care is taken to keep the direct rays of the sun off the cement. The company exposed part of a sample of cement to the sun, and exposed the other part to the air. It was found that the part exposed to the sun showed loss of strength equal to 50 per cent.

The soluble silica used for the induration of the stone is a clear viscous substance made from pure flint and caustic soda, which are digested by heat under pressure in Papin's digester or an analogous machine. Its strength is technically known as 140° Twaddle, which shows 1,700 on a hydrometer. The silica is diluted with water until it shows 1,250 or 1,300 on the hydrometer, and is then a clear copper-colored liquid.

The stone is made of three parts by measure of washed granite and one part of cement. They are thoroughly incorporated in a dry state in a horizontal cylinder by machinery, and when this is secured, water is sparingly added, and the mixing continued. At each mixing there is made sufficient concrete for a 3′x2′x2″ slab only. When it is ready for putting in the mold, the concrete does not appear to the eye to be sufficiently wet. The molds are metal-lined, true in shape, with clearly defined arrises. Before the concrete is placed into them they are oiled all over, and then placed on a trembler. This is a machine which gives a rapid vertical jolting motion to the mold. When the machine is started, the concrete is placed into the mold by small shovelfuls at a time, and two men with trowels spread it over the mold. When the mold is filled they pat the concrete with the trowels, the water rises to the surface and an even smooth face is secured. The mold and concrete are then removed to a rest for two days. The whole operation of mixing the concrete and making the slab in the mold is completed in six minutes. Machine-made stones are of necessity homogeneous, as veneering is impossible during the process. The slabs when taken from the molds at the end of the two days are air-dried for seven or nine days, and then immersed in a silica bath for seven or nine days more. They are then stacked in the open for some months before use. The value of the silica bath is in hastening the hardening of the stone. At the end of a month the stone will stand from 30 to 40 per cent. more tensile strain if silicated than if air-dried only. It is doubted by Mr. Faija whether silicating increases the ultimate strength of concrete.

The stone so prepared stands a tensile strain of 650 lbs. on the square inch when three months old. Strains of 1,000 lbs. have been obtained, but they are not the average. The value and excellence of this concrete is shown by the fact that though the neat cement has 75 per cent. of aggregate added to it, yet in three months the mixture bears nearly double the tensile strain of the neat cement after seven days.   The writer was shown the original of a report by Kirkaldy, dated September 6, 1882, in which one sample of stone took 8,075 lbs. per square inch to crush it, and another reached the marvelous strength of requiring 9,492 lbs. per square inch to crush it.

Comparison of Different Pavements.—In the King's Road experimental pavements the makers laid their pavements with the knowledge that they were competing with their rivals.   They were laid in 1880.   Asphalt, York stone, Ferrumite stone, Victoria stone, and Imperial stone were laid side by side and subjected to the same traffics.   The writer includes the York stone results, though not strictly coming within the subject of the paper.

In June, 1884, the Ferrumite stone was removed, as its slipperiness had become a source of danger.   Its area was taken by Wilkinson's granite concrete pavement and by Shap-stone pavement.   After five years' wear of the original stones, the following results were obtained :

The York stone occupied an area of 87 square yards ; the original thickness was 3 inches ; the number of stones 123, of which, after five years' wear, 10 had broken edges, 16 had broken corners, 21 had their surfaces peeling off, and 6 were worn so as to be dangerous.   The wear was not measurable by reason of the uneven thickness of the stones, but it was unmistakable.   The foot-hold was good in all weathers.   The actual cost was 8s. 7d. per square yard laid.

The Victoria stone occupied an area of 81$\frac{1}{4}$ square yards ; the original thickness was 2 inches ; the number of stones 168, of which 35 had broken edges and corners, 2 had their surfaces peeling off, and 3 were visibly cracked across.   The wear was not quite $\frac{1}{4}$-inch at the greatest traffic line.   The joints of the stone were pleasing, and the color cheerful.   The foot-hold is not so sure as that of York stone in dry weather, and in a drizzling rain it approaches slipperiness.   The cost was 6s. 4d. per square yard laid.

The Imperial stone occupied an area of 90$\frac{1}{4}$ square yards ; the original thickness was a full 2$\frac{1}{4}$ inches ; the number of stones 187, of which 18 had their corners and edges broken, and 1 had its surface partly peeled off.   The wear was not quite $\frac{1}{4}$-inch at the greatest traffic line.   The regularity of the joints is pleasing to the eye, and the color is light and cheerful.   The foot-hold is more sure than that of the Victoria stone sample, but it becomes somewhat slippery in drizzling rain.   The cost was 6s. per square yard laid..

The experience gained with the Shap-stone laid in 1884 is more limited, but the stone does not appear to be better than the concrete stones described.   The length laid *in situ* by Messrs. Wilkinson is subject to the same remark, but the aggregate is already wearing up. It is laid in 8-feet bays. A repair made in it is a great dissight to it. The effect of the traffic in wear is visible.

Mr. Walker, of Leeds, has laid down (1886) in King's Road a short length, in bays about 4 feet square. It has a very smooth surface, and appears to be a very good pavement. He lays a foundation of ballast concrete $2\frac{1}{4}$ inches thick to within $\frac{1}{4}$-inch of the finished surface, and cuts roughly through it so as to form the bays. He then floats the surface with fine rich concrete, made of one part of cement to one or one and a quarter parts of crushed granite or slag which pass through a $\frac{1}{10}$ sieve. This is then cut through with a knife into bays of small areas. The foot-hold is fairly secure and the wear very satisfactory. No cracks appear when the bays are made less than 10 feet square. At 322 Oxford Street a piece has been down three years, and no wear is visible. He claims that his fine rich concrete on the surface attains more nearly to the texture of York stone than any other pavement. The joints, however, have a tendency to spread.

Concrete pavements laid in slabs have two serious disadvantages. The writer's experience refers to Victoria stone, but there is no reason to suppose that slabs of other make are free from them. The first is the annoyance caused by the hard metallic sound of the footfall on them, which is especially noticeable at night. The other is their brittle nature and the presence of hidden cracks after wear. When taken up to relay, it is often found that a stone which was apparently sound, breaks before it can be relaid.

Concrete pavements *in situ* have the serious disadvantage that they cannot readily and cheaply be repaired. A patch in them shows for the whole life of the pavement, unless a whole bay is removed.

The writer is of opinion that concrete pavements *in situ* are preferable to concrete pavements laid as slabs. There is, however, no reason why this pavement should not be laid by surveyors themselves, without the aid of a contractor. Much has to be learned before concrete as a material for foot-pavements is perfected, but that can most readily be learned by each engineer doing his own work, and exchanging experiences. The prices paid for concrete pavements are very great. One stone is advertised for 7*d.* per square foot, which equals £4 14*s.* 6*d.* a cubic yard. This is an extravagant price to pay for concrete, even if it be of the very best kind, and silicated, too. Concrete for foot-pavements has its own field, and within well-marked limits its use is advisable, but until its capabilities are more largely tried and its usefulness increased, the full advantage will not be obtained from it. The price stated is large enough to cover the cost of experimenting, and the writer trusts that the members will carry out such a series of tests and trials as will complete the knowledge of concrete as a material for foot-pavements.

Discussion. A Remedy for Cracks.—In the discussion following the reading of Mr. Strachan's paper on "Asphalt and Concrete Foot Pavements," we learn from the *Builder's* report that Mr. Vawser, of Manchester, mentioned the advantage which was experienced at Macclesfield by laying concrete footpaths in about 4-foot squares, and putting between each a piece of plank. It was found that the elasticity of the planking was sufficient to prevent any cracking.

Mr. Fowler, of Manchester, agreed as to the great value of concrete paving, which, seven years ago, he used for the Newcastle Cattle Market. There he placed a lath round the boundary of each section, and there was not a single crack.

Mr. Ellice Clarke, of London, said his experience of an asphalt footpath $\frac{3}{4}$-inch thick was that at the end of three years it required very considerable repairs. The unsightliness of cracks in asphalt was their principal objection ; they did not detract at all from its wearing well. The practice mentioned by Mr. Vawser had been discontinued by all who had had much experience of asphalt. Now it is usual to lay two slabs of concrete on alternate days, or two or three days.

Mr. Boulnois, of Portsmouth, said that some of the earlier samples of ferrumite stone had too much iron in them, and being thus too hard, had a very slippery surface. The secret of the wear of asphalt lay in a nutshell. Unless it got sufficient traffic upon it it would never consolidate properly.

Mr. Lemon, of Southampton, said that twenty years ago he was very enthusiastic about asphalt, and his Corporation laid a great deal of it down. He had come to the conclusion now that there must be a very large traffic over it, or at least sufficient to keep it together, otherwise it was bound to fail. Some laid down $\frac{3}{4}$-inch thick soon wore out, but some laid 2 inches thick in a roadway where there was great traffic had been down for fifteen years, and had never cost sixpence for repairs. With respect to concrete, he thought concrete made and laid *in situ* was the footpath of the future.

TRAMWAYS.

LIVERPOOL TRAMWAYS.*

Tramways Built by the City.—We propose in this and subsequent articles to give quite full details of the important work under this head that has been done in Liverpool. The contrast of methods in the construction of tramways, as compared with those usually pur-

*xiv, 369.

sued in this country, is violent. As will be seen, instead of these important franchises being given away, as they are in our large cities, the city of Liverpool builds the roads and leases them at the handsome figure of about eight per cent. on their entire cost.

Their Extent and Cost.—The change in the method began in 1880, when the city, under powers granted by Parliament, purchased the then existing lines. Since that date the lines have been extended, until now the total length of single lines is 45¾ miles. In connection with the construction of the lines 198,452 square yards

FIG. 1.—LAYING THE ROAD.

of stone block paving within the statutory width has been set on a foundation of Portland cement concrete, and the total cost, including engineering and other incidentals, has been $1,394,736, or $30,486 per mile of single track. The whole street has also been paved in tramway streets, requiring in all about three times the amount of paving mentioned, and the monthly progress has been about one mile of tramway and 13,400 yards of "impervious" paving, all executed by corporation workmen under the supervision of the City Engineer, Mr. Clement Dunscombe, M. I. C. E., and staff.

Construction.—In doing the work all old stone blocks were taken up and redressed, occupying forty-two stone-dressers. The fragments and waste were used in concrete, and in the cost given the value of all old material is included.

Prior to 1879 the reconstructions and extensions were made on the system of Mr. Deacon. Since that time they have been made on a modification of this known now as the "Lyver" system. The cost of maintenance is spoken of as very small.

Figure 1 is a general view showing the work of laying the road in progress.

Figure 2 shows a longitudinal elevation of the rail.

In constructing the work the street is fully excavated to a depth determined upon as it progresses. A bed of Portland cement concrete is then formed to within $7\frac{1}{2}$ and $6\frac{1}{2}$ inches respectively of the finished roadway. The concrete is made of one part by measure of cement, six parts of gravel, and eight parts of broken stone. The

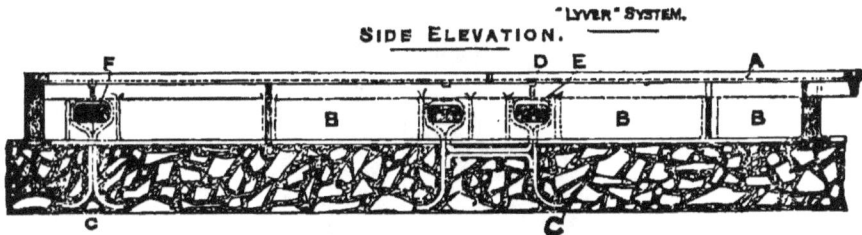

FIG. 2.

cement is first mixed dry with four times its bulk of gravel and then wet and mixed with the remainder, only water enough being added to make the mass just adhere together when pressed in the hand.

On the bed prepared for the concrete, molded blocks of Portland cement concrete, eight inches square at the base, are laid with their upper faces on a level with the under side of the sleepers B (see Fig. 3, which is cross-section of one rail). The sleepers are then laid on the blocks, the rails A being placed on them, and the wrought-iron jaws C secured to the rails by bronze bolts D and wrought-iron nuts E. A small space is maintained by temporary washers between the upper surfaces of the jaws and the rails. As soon as the rails are leveled up and in proper position the concreting is proceeded with up to the level of the bottom of the sleepers, as shown. After the concrete is set the bolts are unscrewed, the temporary washers removed, and the rails A and sleepers B firmly screwed down to the jaws. The recesses or hand-holes are then completely filled with plaster-pitch which keeps the nut E from turning round.

The jaws are placed three feet between centres, except at the ends of rails, where the space is nine inches. Points and crossings are of annealed crucible steel, secured to special cast-iron sleepers, a layer of roofing-felt being laid between them and the sleepers.

Paving the Roadway.—Between the rails, and for 18 inches on either side, the roadway is paved with syenite blocks. The blocks are cut $3\frac{1}{4}$x5 to 7 inches, and in two depths—viz., $6\frac{1}{4}$ and $7\frac{1}{4}$ inches—also $3\frac{1}{4}$x$3\frac{1}{4}$x$6\frac{1}{4}$ inches deep. They are squared throughout, accurately gauged, with a maximum allowed variation of one-quarter of an inch, and laid in straight and properly bonded courses on an

ENLARGED SECTION.

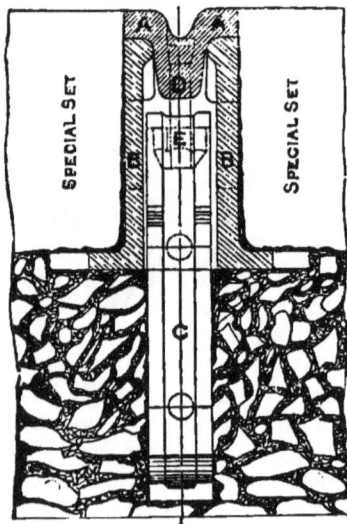

FIG. 3.

even bed of fine gravel not exceeding half an inch thick. After paving, the joints are filled with clean dry shingle passing through a $\frac{3}{4}$-inch sieve and retained by a $\frac{3}{8}$-inch sieve. The sets are then rammed and new shingle applied until the joints are full, after which they are grouted with hot pitch and creosote oil of the best quality and covered with half an inch of sharp gravel.

Along each side of the rail there is laid a course of alternate long and short blocks, shaped in plan as shown in Fig. 4. These are finely cut on the side next the rails so as to make close contact ; they are also cut to touch each other for one and a half inches from the rail.

The weights per mile of single track are given as follows :

Bessemer steel rails, 40 lbs. per lineal yard.... .. 62.8 gross tons.
Cast-iron sleepers, 80 lbs. per lineal yard.........125.6 .- ◄ ''
Wrought-iron jaws, single, 4.28 lbs. each.........115.5 cwt.
Wrought-iron jaws, double, 10.5 lbs. each......... 35.5 "
Bronze bolts, 4 ozs. each................ .... 9.83 "
Wrought-iron nuts, 5 ozs. each.... .... 10.5 "

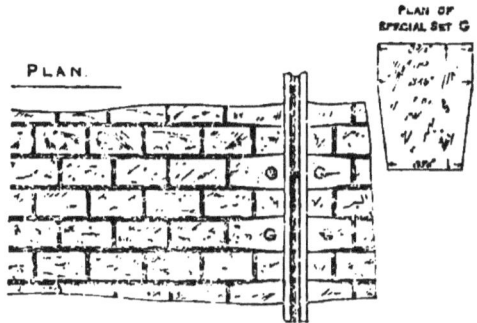

FIG. 4.

Testing Materials.—The Portland cement was all tested. A 50-gauge wire sieve (2,500 meshes per square inch) must not return more than 10 per cent.

When tested neat, after twenty-four hours' immersion in water, it must stand 1,000 pounds on $2\frac{1}{4}$ square inches. Slow-setting cement must take an impression from a needle having an area of $\frac{1}{100}$ of an inch, and loaded with $2\frac{1}{4}$ pounds, at any time within three hours after molding, and quick-setting cement must give an impression at any time within half an hour.

The Bessemer rails were tested by the drop test. With groove upward, rail supported at three feet between supports, weight 2,240 pounds, dropped ten feet, the average deflection was 8.51 inches and none were broken. Tensile tests were made on pieces 18 inches long cut from rails. Allowed limit twenty-eight to thirty-two gross tons per square inch, with elongation of 20 per cent. on $6\frac{1}{4}$ inches. The average strength was 31.62 tons, and elongation $22\frac{1}{2}$ per cent. Hot bending tests by heating to cherry red and cooling in water at 80°, and bending afterward double, to a curve with inner radius of three times the thickness of specimen, gave no fractures. Short lengths were bent cold under steam hammer to a radius of six inches.

Deflection tests, with supports three feet apart, showed maximum elasticity of 10 tons and a minimum of 5 tons.

Cast-iron was tested on supports three feet apart, bars one inch square to support a centre load of 800 pounds. The maximum was 1,800, minimum 640, and average 960 pounds.

Two per cent. of the bolts were tested by a load of three tons, the nut being turned while the load was on, about 4 per cent. failing under test from inferior workmanship.

Sample nuts were tested by heating to a cherry red and flattening on an anvil. The result of reducing the thickness from one inch to one-eighth of an inch was to show no splitting at the edges.

CHAPTER VI.

STREET OPENING.—MAINTENANCE.

LIVERPOOL STREET EXCAVATION CONTRACT.*

Excavation of Trenches for Electric Supply-Tubes and Temporary Reinstatement of Pavements.—Clement Dunscombe, M. Inst. C. E., City Engineer, July 11, 1888.

FORM OF TENDER.

To the Chairman of the Health Committee, Liverpool.

SIR : We hereby agree to execute and perform such works as may from time to time be ordered in writing by the City Engineer in connection with the excavation of and reinstatement of pavements in Liverpool, consequent upon the operations of the Liverpool Electric-Supply Company, under their license, at the rate of the several prices we have affixed to the items in the annexed schedule, duly signed by us.

And further, we are prepared to enter into a contract for the due fulfillment of the work in accordance with the annexed specification.

......propose Mr... ..of.....and Mr..... of......as proper persons who are willing to be bound as sureties for the due performance of this contract in accordance with the terms of this specification.

We are, Sir,

Your obedient servants,

Signature......Address......Date......

SCHEDULE OF PRICES REFERRED TO IN THE ANNEXED TENDER.

Prices inclusive of all charges whatsoever per lineal yard of trench two feet wide, and given separately for each of the following conditions and depths, not exceeding 3 feet, 3 feet 6 inches, 4 feet, 4 feet 6 inches or 5 feet.

Footways.—To take up the pavements and curbs in footways where required, excavate trench for pipes, fill in and ram the ground over pipes, and temporarily relay the footway permanent and where disturbed.

Carriageways.—To take up the channels or impervious pavements, excavate the trench through the cement or bituminous concrete foundation as required for laying the pipes, fill in and ram over pipes, and temporarily block in the pavements.

The same with impervious pavement in hand-pitched rock foundation.

The same with gravel jointed set pavement on hand-pitched foundation.

The same with macadam on hand-pitched foundation.

The same with boulder pavement on hand-pitched foundation.

Extra excavation for connections or other necessary works measured at the time of execution of any of the foregoing depths, per yard, cube.

Add extra for work done at night, as per Clause 8.

Signature .....Address. ....Date......

---

* xviii, 293.

SPECIFICATION.

1. *Interpretation.*—In this specification the word "engineer" shall be held to mean the City Engineer of Liverpool for the time being, and the word "assistant" shall be held to mean the person whom the engineer may appoint from time to time to superintend the works.

2. *Extent of Contract.*—This contract comprises the execution of such works as the City Engineer may from time to time order in writing in connection with excavation of trenches for laying electric supply tubes under the public highways in the city of Liverpool, for which the Liverpool Electric Supply Company possess a license, and the reinstatement of such trenches and pavements as specified and any other works as may be required in connection therewith.

3. *Excavation of Trenches and Reinstatement of Pavements.*—The contractors shall, upon receiving notice in writing from the City Engineer, forthwith proceed to take up the existing pavements, concrete or other foundation where required for laying the electric supply tubes authorized by the Liverpool Electric Lighting License, 1888, and they shall carefully place on one side these materials until the pipes are laid. They shall then excavate to the required depths, and after the electric supply tubes are placed in position, fill in and well ram down, in layers not exceeding six inches deep, the trenches with the excavated material, watering the same where necessary in order to thoroughly consolidate it, and when the filled-in material is thoroughly consolidated the pavement over the trenches and also any flagging disturbed shall be temporarily reinstated. All surplus excavation and rubbish shall be removed by the contractors, on completion, and the surface left clean and in such a condition as shall be safe and free from danger to both vehicular and pedestrian traffic.

4. *Tools, etc.*—The contractors shall provide all materials, labor, tools, tackle, implements, etc., for the proper execution of the works executed under this contract, and shall pay or provide for all carriage and cartage of materials. The contractors shall exercise every care and precaution in taking up the existing pavements so as to cause as little damage as possible to the materials so removed, and any new materials that may be required shall be the best of their several kinds, and the same shall be applied in the most workmanlike and substantial manner possible and to the entire satisfaction of the engineer.

5. *Contractors to Pay Fees and Make Good Damage, etc.*—The contractors shall pay all fees and compensations, and make good at their own expense all damages of every kind which may occur by reason of the execution of the works for the performance of which the contractors are bound.

6. *Fencing, Hoarding, Watching, Lighting, etc.*—The contractors shall provide, make and maintain all necessary fencings, hoardings, struttings, shorings and bridgeways, temporary or otherwise, as may be necessary for and in consequence of any of the works. They shall also properly light and watch all the works in accordance with the requirements of the Public Health Act 1875, and to the satisfaction of the City Engineer and the police of the district.

7. All works shall be carried on in such a manner as to cause the least inconvenience to abutting occupiers, and in such manner lengthwise and widthwise as the engineer may direct, and which will least impede the busi-

ness of the neighborhood and the public traffic, and so as not to obstruct or endanger pedestrians, animals or vehicles.

8. The contractors shall, on receiving notice from the City Engineer, execute such sections of the work as may be considered desirable to be so executed during such hours of the night and early morning as may be prescribed by the City Engineer in such notice, and the same shall be paid for in accordance with the schedule attached to the tender without any further amount being chargeable beyond the sums therein stated.

9. *Screens.*—If the contractors shall neglect in any case to provide a suitable screen or screens wherever stones are being chipped for the purpose of protecting pedestrians they shall pay to the corporation as and for liquidated damages the sum of £1 per day for every day on which such default shall have been made, and the Corporation may deduct the same from any moneys in their hands due or to become due to the contractors.

10. *Notice to Gas Company and Water Committee.*—The contractors shall give due and sufficient notice to the gas company and water committee in order that the proper persons having charge of the mains, services, etc., may be enabled to attend and see that the said pipes are secured, relaid and reinstated in a proper and satisfactory manner, the contractors nevertheless to execute or pay for executing all such alterations and reinstatements and to be held chargeable and responsible for the proper protection and restoration of the same at their own expense.

11. *Facilities to Gas Company and Water Committee.*—The contractors, if so ordered by the engineer, shall allow the gas company and water committee by their workmen or agents to enter upon the site of the works for the purpose of relaying the gas and water-mains and services without any extra charges being payable by the Corporation.

12. *Measurement of Work.*—The contractors, their agent or foreman, shall attend weekly at such time and place as shall be named by the engineer for the purpose of measuring and ascertaining the quantity of work performed, and in default thereof the engineer shall be at liberty forthwith to measure and ascertain the quantity himself, and his decision as to the quantity shall be final, binding and conclusive upon all parties.

13. *Subletting.*—The contractor shall not assign or underlet or make over this contract, or any benefit or interest thereunder, to any other person without the consent in writing of the engineer.

14. *Occupation of Site of Works.*—The Corporation may, for any purpose, or at any time during the continuance of this contract, occupy any part of the site of any of the works simultaneously with the contractor for the execution of any works whatsoever without any compensation being payable to the contractor.

15. *No Claim to be Made for Delay.*—The contractors shall not be entitled to compensation should any portion of any of the works be delayed in its execution from any cause whatsoever.

16. *Extra Works.*—No allowance shall be made to contractors for any alteration in or addition to the work specified unless they can produce a written order of the engineer for the same.

17. *Disputes.*—If any dispute or differences arise during the progress of this contract or afterwards respecting the true intent of this specification the same shall be referred to the engineer, whose decision shall be final and binding upon all parties concerned.

18. *Instructions.*—The engineer shall have full power to issue such further instructions from time to time as he may deem necessary for the guidance of the contractors, and the contractors shall be bound by them.

19. *Expedition in Carrying on Works, Failing which Corporation may Employ other Contractors.*—The contractors shall commence and carry on the works specified herein with due diligence and as much expedition as the engineer may require, and in case the contractors shall fail to do so, or shall neglect to provide proper and sufficient materials, or to supply a sufficient number of workmen to execute the works which they shall be ordered to execute with due diligence or the dispatch required, then the engineer shall have full power without vitiating the contract, and he is hereby authorized to take the works, wholly or in part, out of the hands of the contractors and to engage any other person or workmen, and procure all requisite materials and implements for the filling in of the trenches opened up by the contractors whether the mains are laid or not, and the reinstatement of the pavements disturbed by their operations and due execution and completion of any works requisite in connection therewith, and the costs and charges incurred in so doing shall be ascertained by the engineer and paid for or allowed to the Corporation by the contractors, and it shall be competent to the Corporation to deduct the amount of such costs and charges out of any moneys due or to become due from them to the contractors under their contract, or in case there is no money due or to become due to the contractors, then the Corporation may recover such costs and charges by action at law or otherwise.

20. *Openings for Examination.*—Should the engineer or his assistant require it for his more perfect satisfaction the contractors shall, at any period during the continuance of this contract, make such openings and to such extent, through any part of the said works, as the engineer or his assistant may direct, and which the contractors shall, make good again to his satisfaction at their own expense.

21. *Superintendence, Skilled Workmen.*—The contractors shall employ only such foremen and workmen in or about the execution of any of the works under this contract as are careful and skilled in their various trades and callings ; and the engineer shall have full power to object to or dismiss any person who shall be found incompetent or who shall act in an improper manner. The contractors shall also have a competent representative upon the works.

22. *The Contractors to Suspend any Works when Ordered.*—The contractors shall suspend the execution of any work when ordered so to do in writing or otherwise by the engineer, and whatever expense may be occasioned by the suspension of the works shall be paid by the contractors and without charge to Corporation ; and the engineer shall have full power to direct any alteration in the manner of carrying on or finishing any such work, matters, or things, without thereby in any way affecting, vitiating or impairing the tenor or force of this contract.

23. *Precautions Against Accidents or Injury, Corporation and Officers to be Indemnified Against Action, etc., Corporation may Compromise Actions, etc.*—The contractors shall take every necessary, proper, timely and useful precautions against accident or injury to the works, or any of them, or to any other property or to any person, by the action or pressure of water, and whether the same shall arise from or be occasioned by floods, springs, rains, streams, accumulations, disruptions,

leakage, frost, or other natural or artificial causes whatsoever, and shall forthwith repair, make good and defray any loss, damage, cost, charge or expense, by, or in consequence of any accident, or by, or in consequence of the operations of the contractors occasioned to the Corporation or to the said works, or any of them or to any person or persons injuriously affected thereby and shall indemnify, save harmless and keep indemnified the Corporation and their officers from and against the same, and from and against all actions, suits, claims, penalties, liabilities, costs, expenses, and demands whatsoever by reason or on account thereof, and also from and against any claims for compensation under the provisions of the Employers' Liability Act, 1880, and the Corporation may deduct the expense thereby incurred, or to which the Corporation and its officers may thereby be put or be liable, or which may be incident thereto, from the amount of any money which may be or become due or owing to the contractors, or may recover the same by action at law or otherwise from the contractors ; and the Corporation may, if they shall see fit, compromise any such action, suit, or other proceeding, or any claim in respect of any such damage as aforesaid on such terms as they shall think fit ; and the contractors shall thereupon forthwith pay to them the sum, or sums paid by the Corporation on the occasion thereof, and shall in every case pay to them such sum or sums as shall fully indemnify them according to the present stipulation.

24. *Responsibility for Accidents, etc.*—The care of the entire line of works in each case shall until their permanent reinstatement by the Corporation remain with the contractors, who will be held responsible for all accidents from whatever cause arising, and for the making good of all damages and defects to the said works from bad or insufficient materials, bad workmanship, or any cause whatever.

25. *Payment.*—That subject to the conditions in this specification contained, payment shall be made to the contractors monthly upon the certificate of the engineer, and within three weeks from the date thereof at the rate of 75 per cent. on the value of the respective works executed as assessed by the City Engineer, and the remaining 25 per cent. within two months after the permanent reinstatement by the Corporation of the respective pavements disturbed, on condition that the terms of the contract shall have been fulfilled and upon the certificate of the engineer of the amount to which the contractors are entitled, and that the work has been executed to his satisfaction.

26. *Default of Contractors to Comply with Specification.*—The payments hereinbefore mentioned are subject to any deductions for expenses or costs to which the Corporation may be put in consequence of default on the part of the contractors to fulfill the terms of this specification.

27. *Tenders not Necessarily Accepted.*—The Corporation do not bind themselves to accept any tender, or to pay any expenses incurred by parties tendering, and reserve to themselves the right not to order any of the works herein contracted for.

OPENING AND REINSTATING PAVEMENTS IN LIVERPOOL.*

We would call the particular attention of our readers to what Mr. Dunscombe, City Engineer of Liverpool, says below on the sub-

---

* xv, 485.

ject of reinstating pavement over trenches and opening up of pavements :

"In cases where trenches are required for gas and water purposes, notice is given by the company to me; the trench is then opened by the company's workmen, and upon completion of the work the trench is filled in by them and the paving is also temporarily blocked in at the company's risk. The pavement is then reinstated in a permanent manner by Corporation workmen, and the cost of the same and charges incidental thereto charged to the company. The Corporation execute this work as contractors only, and the company are liable for, and are actually charged, the cost of any repairs necessary during a period of six months in order to maintain the surface in a perfect condition.

"By written permission of the Corporation only, openings in the pavements are allowed to be made for the purpose of examination of private drains, and no other interference with the pavements is allowed, either in the carriageways or footways, except under Parliamentary powers obtained for the execution of the work, in which case ample protective clauses in the Act of Parliament are obtained by the Corporation, dealing with the conditions upon which any undertaking shall be allowed to in any way interfere with the streets or pavements. These clauses are comprehensive and fully protect all interests of the Corporation against any undue interference with the streets, and provide for their reinstatement, in a proper manner, at the sole cost of the promoters, in cases where such interference is permitted.

"Interference with the pavements in Liverpool is of rare occurrence, inasmuch as both the gas company and the Water Committee of the Liverpool Corporation thoroughly examine and reinstate where requisite their mains and services concurrently with the repaving of a street, of the execution of which due notice is given to them by my department. In no case would streets be allowed to be opened up so as to prevent the free passage of traffic along it, except in cases of the most extreme urgency, and even in these cases the local police authorities would regulate it, and the promoters of any undertaking would not be allowed to interfere at all with the streets and roads except under the most stringent regulations. The street traffic is a matter which receives the first attention at the hands of the Municipality, as any interruption of it would be attended with serious inconvenience and loss."

HOW TO PRESERVE OUR PAVEMENTS.*

In his address to the new Board of Aldermen of this city, President Beekman made the following reference to the continual destruction of our pavements :

"The condition of the pavements of the streets and avenues in various portions of the city has long been a well-founded subject of complaint.

"Under authority from the Legislature various private corporations, in furtherance of their business objects, are constantly removing the pavements and excavating in the public streets.

"The result of this is that the pavement, insufficiently restored, becomes depressed, holes, ruts and channels are formed, and an irregular sur-

*Ed. xv, 179.

face is occasioned over which it is painful to travel, and which collects instead of shedding water and waste material.

" While, no doubt, the convenience and comfort of the public are greatly
increased by the advantages which these companies afford in their particular line of business, it is quite certain that the privileges they enjoy are of
exceptional value, and it can be no hardship to require of them all the speed·
care, and skill that can be exercised in the rapid and satisfactory completion
of their work.

" While this Board possesses no power to prevent the use of the streets
for such purposes, it may at least pass reasonable rules and regulations, as
it has in some measure in the past, defining the manner in which such excavations may be made and the street surfaces properly and permanently
restored. Having determined upon the most approved methods of doing
such work an ordinance should be passed strictly enforcing compliance with
such methods and requiring the greatest expedition in performing the work
—too often protracted beyond a reasonable time for its proper execution.'.

What he said may be news to an alderman, but it is no news to
any engineer or any one having the slightest familiarity with municipal engineering. It is our deliberate opinion that no matter what
ordinances may be passed, what regulations adopted, or however
good the inspection may be, the pavements of this or any other city
will never be in decent condition until all removals and restorations of pavements are done *solely by the city's own employees and
agents*, for which work any corporation or individual who desires the
streets to be disturbed should be compelled to pay the cost ; and
that cost should be sufficient to secure thoroughly sound work and
also the cost of a reasonable maintenance. We are under the impression that this is the practice in Liverpool, if not in other foreign
cities, and by this means, and this means only, will people avoid the
opening of a street except in most imperative cases. Indeed, there
are many occasions on which a street is now torn up, when, if there
was a sufficient tax placed upon such tearing up, the parties interested would find some way of tunneling or accomplishing the purpose without a disturbance of the surface.

ON OPENING UP PAVEMENTS.*

It is reported that a movement is on foot to secure legislative
authority to repave a large number of streets in this city, the work
to be under the control and direction of General Newton, Commissioner of Public Works. We hope it may be successful. *The Sanitary Engineer and Construction Record* has for some time been printing articles describing notable examples of pavements in other cities,
and a person only needs to read these, even if he has not an opportunity of visiting other cities, to become convinced of the wretched
character of the pavements of this great city. If authority is granted

---

* Ed. xv, 370.

General Newton's department to construct these pavements, and there are no needless conditions inserted in the act to tie the Commissioner's hands, we believe New York City will secure good pavements and at a moderate price. But these good pavements cannot be maintained unless additional powers are granted to the Department of Public Works, which will prevent in future any disturbance of them, except by the servants of this department and under its direction, such disturbances to be paid for by the companies, whoever they may be, in whose behalf the pavements are taken up. This city badly needs pavements, but if it is to retain any worthy of the name, after they are laid, some check must be placed upon the gas, steam and electric-light companies, who would otherwise keep them in a constant state of eruption.

THE REPAVING REQUIREMENTS OF NEW YORK CITY.*

We give herewith a brief report of the preliminary meeting of the Committee appointed by the New York Chamber of Commerce to confer with the authorities of that city to ascertain in what direction they can best co-operate to secure the repavement of its business portion. The disabilities under which the authorities now labor will doubtless be plainly brought out by this public action of the Chamber of Commerce. It is a hopeful sign that this influential body proposes to interest itself in a movement to put an end to the disgraceful condition of the streets of New York, and it is highly important that there should be concerted, intelligent action on the part of all good citizens, except the directors of surface railroads and corporations, who are constantly disturbing the streets without any regard to the convenience of the community and the pockets of taxpayers, and who hitherto have never failed to look out for their immediate interests, as has been frequently pointed out in these columns. Four things will have to be decided upon, and until they are done it seems useless to spend any money for new pavements. First.—The present form of street-rail should be abolished, and not another length of it should be permitted to be laid. Second.—Paving of every portion of a street should be done by the city authorities, or under their direction and control. Third.—No openings should be made in the pavements except under permits granted by the Commissioner of Public Works, which permits should state definitely the exact location and limits both of time and space within which the opening can be made. Fourth.—All restorations should be made by the Department of Public Works or under its immediate direction, the cost of which should be paid by the corporation or individual for whom the pavement was originally disturbed. This latter requirement is uni-

*ED. xvii, 31.

versal in every city in Europe that pretends to have a pavement worthy of the name. Without it responsibility cannot be placed on any one, nor is it possible to secure by present methods anything but the most worthless and inefficient work. As has been pointed out in the series of articles on "Pavements and Street Railroads," now appearing in *The Engineering and Building Record*, the custom abroad is to have corporations, like gas or water companies, who have occasion to frequently disturb the streets, leave a regular deposit with the department. Against this deposit charges are made for each piece of paving done at a fixed price per yard, according to the kind of pavement. Where special openings are made special deposits are required. The department in charge of streets, when it issues a permit for an opening, requires the space to be inclosed by means of iron pins and a rope, from which lanterns are suspended at night, and if it is a street which has a concrete foundation under the pavement, time is taken to allow the concrete to properly set before the pavement is relaid. When the necessary authority is given to the proper officials to institute and maintain the reforms here outlined, pavements suited to a civilized community may be secured and retained. To that end it is to be hoped the influence and co-operation of the Chamber of Commerce may be effectual.

The committee appointed by the New York Chamber of Commerce to confer with Mayor Hewitt and General Newton as to what steps could be taken in order to best co-operate with the authorities in securing new pavements for the business portion of this city, met by appointment in the Mayor's office on December 9. The committee consisted of Messrs. Francis B. Thurber, Secretary George Wilson, Silas B. Dutcher, James H. Seymour, Thomas Rutter, and Henry C. Meyer. General Newton, Commissioner of Public Works, presented a statement prepared by one of his staff, who had made a survey of the down-town streets, which indicated that 536,527 square yards of street surface required repaving, at an estimated cost of about two millions of dollars. Mayor Hewitt expressed himself as in favor of authorizing a loan of about five millions, to be expended by the Commissioner of Public Works, subject to the approval of the Board of Estimate and Apportionment, for repaving the streets of this city, provided authority were given the Department of Public Works to compel railroad and other corporations to obey its order with regard to the opening of streets and the obstruction of them by needless street rails put down by railroad companies to hold a charter, but on which they never ran cars. The Mayor cited the aggravated case of the corporation who have rails in Fulton Street. The impossibility of properly doing anything until the authorities had power to control the action of corporations who get grants from the Legislature or Aldermen, and consequently ignored the authorities, was made manifest. A member of the committee urged that all pavements when opened by corporations or other parties, should be restored by the department's own servants, the cost of such restoration to be paid by the parties having the streets opened; the amount of space to open at one time, duration of time of such opening, where traffic should be suspended, should also be subject to the discretion of the Commissioner of Public Works. On this point the Mayor said that the Corporation Counsel would be asked to see if further legislation was necessary to give such power to the Commissioner. Pending the securing of the needed legislation suggested, the Commissioner of Public Works proposed to submit a list of the streets that most imperatively needed pavement to the members of the committee, and expressed himself as desirous of expending the $500,000 appropriated by the Board of Estimate and Apportionment on such streets as the committee should agree most needed it. After assuring the Mayor and the Commissioner that the Chamber of Commerce would heartily co-operate with them in a movement to secure better pavements for this city, the meeting adjourned for further conference when the data promised by General Newton should be prepared.

A MOVE TO PROTECT OUR PAVEMENTS.*

We are heartily glad to at last see a move made toward stopping the irresponsible tearing up of New York streets. The Park Board has adopted a resolution under which all gas and electric companies, sewer builders and the like, applying for permission to tear up the streets for any purpose, will be compelled to deposit a sum of money equal to the original cost of the pavement, to be held until the roadway is restored to good condition.

This is precisely what we have been urging for years, and the soundness of the principle and the necessity for action cannot be questioned. All other city authorities should promptly follow the example, which, however, does not go far enough, since the work of restoration should, as we have before stated, be done by the city's own employees and under the direction of its engineers, the cost to be paid out of this deposit, which should be more than original cost of pavement. It is easier to get good work that way than to wrangle with the contractor of the parties who rip up the streets.

MAINTENANCE OF PAVEMENTS IN LONDON.†

St. George's Parish.—From a report by Mr. George Livingston, C. E., Surveyor of St. George's, Hanover Square, London, we gather the following :

There are forty-two miles of paved streets in the parish. With the exception of Piccadilly and two miles paved with other materials, all of these are macadamized, the road metal consisting either of granite, flints, or gravel. The labor of maintenance is performed by what is called a "lifting" gang, the men being paid fourpence per hour. The engineer and stoker of the road-roller receive more. This gang is supplemented for a part of the year by men called "district roadmen."

The average annual *cost of maintenance of the macadamized streets*, based on the traffic and cost of the past five years, is estimated at 2s. 6d.

---

* Ed. xx, 268. † xvii, 84.

per yard for the future, and the estimate for granite, including renewal in thirty years, is a trifle more than one-half as much. The addition of cost of cleansing in each case would make the disparity still greater.

Systems of Maintenance.—Four systems of maintenance and repair are discussed:

First—By contract under fixed prices for labor and materials. The difficulties of determining the exact condition the streets shall be in at the expiration of the contract, and of getting efficient contractors is dwelt upon. The convenience of a large staff of workmen is also mentioned.

Second—By piece-work or part contract. The Vestry would by this supply the materials and the contractor the labor and tools.

Third—By a superior class of men in the employment of the parish. This is essentially the present plan, except that by offering a higher price for labor better men would be obtained.

Fourth—By the substitution of a different description of pavement.

Street Cleaning.—Street cleaning is considered one of the most important matters to deal with. It is done in part by contract and in part by the parish staff. There are in the employ of the parish in the "in wards" twenty-two men, one foreman, and a sweeping-machine. Men are paid 2s. 6d., foreman 3s. 6d. per day, beginning at 4 o'clock in the morning in summer and 6 o'clock in the winter. Sweepings are removed by carts as soon as possible. The sweepers are mostly paupers and cost more on account of small quantity of work done than a better class of labor. Having no water-proof clothing furnished them, no work is done in wet weather. In the out wards there are thirty-two districts, in each of which a roadman sweeps the crossings and collects the refuse by barrow for removal by the carts. He also does all repairing needed. The total annual cost for cleaning the streets is about £10,000.

"Notwithstanding this large outlay, the streets of the parish (in the opinion of the committee) continue to be most inefficiently cleansed. The nuisance most complained of is, that the mud is swept from the centre and deposited in large heaps at the sides of the road in or near the channel, where it is left for an unreasonable time to the great inconvenience of the public. The simple cause of this is that the streets which have been swept yield a greater quantity of mud than can, by the present system, be removed. This is due chiefly to an insufficient staff of horses and carts to remove the slop and the difficulty attending its ultimate disposal. Any increase, therefore, in the staff necessary to perform this work more rapidly must entail additional expense.

"*The difficulties* in the way of a speedy cleansing of the streets, however, under existing circumstances are numerous. The chief are:

"*First*—The nature and condition of the pavement to be cleansed.

"*Second*—The imperfect method of sweeping.

"*Third*—The amount of mud to be removed.

"*Fourth*—The objectionable and dilatory method of removal.

The cleansing of the streets, both as regards its cost and efficiency, depends, to a very great extent, upon the description of pavement of which they are composed; and it cannot be disputed that *macadamized streets* (such as most of the thoroughfares of this parish), subject to heavy traffic, yield a much greater quantity of mud than any other pavement. For example, the average quantity of slop removed from the macadamized portion of Piccadilly (previous to its being paved with wood), between Engine Street and Hyde Park Corner, comprising an area of about 8,600 yards, was seldom less than from 25 to 30, and frequently as much as 40 cart-loads a day in wet weather; notwithstanding this, the roadway was never in a satisfactory condition. Whereas, since it has been paved, the average quantity of slop removed, under equal conditions, from the entire length of the roadway—nearly three-quarters of a mile—representing an area of about 20,000 yards, does not exceed 12 cart-loads a day, the greater portion of which is a valuable manure. Every one who has observed the roadway under the two conditions of pavement must be sensible of the vast improvement that has been effected. I have no hesitation in saying that Piccadilly, which was formerly one of the worst kept and filthiest roads in the parish, is now, considering the labor employed, one of the handsomest, cleanest, and best kept thoroughfares in the metropolis.

The surface of streets, under certain conditions, is frequently covered with a thick, sticky, greasy mud, due to various circumstances, the state of the atmosphere, the situation of the street, and the influence of wind and sun, etc., which the present system of sweeping, especially on macadam, utterly fails to remove, and which is the occasion of serious accidents. According to some returns of the Registrar-General, the average annual number of persons killed in the streets of London is about 200, and the number injured by accident about 2,000. The greasy and slippery state of the streets is often the cause of these casualties.

The macadamized roads, also, at times, are "permeated with the solutions of the surface dung deposits, and become excrement sodden," the dry decomposition of which neither the broom-sweeping nor the scraper can wholly remove; and the moist exhalations produced by street-watering (during the process of evaporation), especially in hot weather, are most offensive. As a matter of public health and safety, therefore, the perfect cleansing of the roads and streets is most essential. The quantity of mud and slop to be removed from the streets depends entirely upon the condition of their surface and the nature of the weather; and the accumulation swept from the streets of this parish, which averages about 30,000 loads a year, is so great that its speedy removal from all the streets is, under present arrangements, impossible. This is, in great measure, due to the fact that the entire bulk, after having been scooped up in shovelfuls, has to be removed by horses and carts. A more filthy, more costly, and less speedy method of street cleansing cannot be well imagined, and there are many who regard it as discreditable to the sanitary arrangements of this capital.

If the cleansing of the streets is continued to be done on the present principle, there is an obvious remedy for the evils complained of, provided the necessary expense is incurred. Suppose, for instance, it is required to have the whole of the streets in the parish cleansed by a certain hour in the morning (say 10 o'clock), all that would be necessary would be to employ a sufficient number of men to sweep the mud and dirt from the streets before that time, and to make it imperative upon the contractor to remove it; but the cost of providing such a staff would be serious. In cleansing thoroughfares, especially those of great traffic, speed and efficiency are of the first importance; any scheme, therefore, to be proposed for the better cleansing of the streets must have regard to such results. One method which I venture to suggest, by which this might be effected would be by fixing in each of the principal streets of the parish a number of hydrants, at convenient distances apart, to which suitable lengths of hose-pipe could be attached. By this arrangement, the whole length of a street could be thoroughly cleansed in a comparatively short time, and after the street had been washed by the application of water in this manner, and swept by an India-rubber broom or squeegee, made on the principle of the present horse-scraper, so as to remove any wet or moisture from the surface, it would be rendered perfectly dry and free from every description of filth.

Street-Washing.—Under existing arrangements, the sweepings of the streets (as before stated) are removed by contract, and it is naturally the object of the contractor in performing this work to avoid as much cartage as possible; consequently, in wet weather, the mud and slop are often either left at the sides of the streets, or are, as is frequently the case, when opportunity offers, swept into the gullies. The adoption of the hose-pipe, however, would dispense almost entirely with scavengers, and also with the annoyance and labor of cartage; for, by the application of water, the sweepings of the streets would be put in such condition as to cause them to pass off immediately into the sewers.

The passing of this liquefied mud through the sewers might be considered objectionable, as being likely to choke them up. I am of opinion, however, that, so far from creating any stoppage in the sewers, the operation would be most beneficial, as it would have the effect of flushing them. This hose or jet system is not new; and Mr. Heywood, the City Engineer, who tried it, states with reference to the effect it had upon the sewers that "during the experiment in street-washing I had the gullies and sewers within the city carefully examined from time to time, when they were found to be not only as clean as they had previously been, but, if anything, cleaner; and, indeed, I think that if the surfaces of the city pavements were cleansed by water alone, both gullies and sewers would be cleaner than at the present time, for, as before stated, much dust or dirt is now swept into them in such condition that the usual current does not readily move it, whereas, if the streets were daily washed, nothing would go into the sewers excepting that which found its way there by reason of its fluidity."

Where the streets are paved as in the city, no reasonable objection could be raised to the adoption of such a system; with macadamized roads, however, it might be different. Under any circumstances, the hose-pipe would have the effect of thoroughly cleansing the streets of all their impurities, no matter what might be the nature of the pavement or the condition of its surface; and it would be equally applicable and effective in all weathers, frost excepted. Under the present arrangements, and during dry weather, it is at times quite impossible to remove the dust and dung, etc., which accumulate in the streets, and which, as I have already stated, under certain conditions are the cause of offensive smells. To in part mitigate this nuisance, I have caused the streets to be swept during the early hours of the morning; and although this has generally speaking been, to some extent, successful, it cannot at all times be performed with good effect, for in certain seasons the ground is during the early morning so damp from heavy dews that sweeping alone fails to remove the dry dust and other offensive matter from the surface of the streets, into which they have been firmly trodden. For the past two summers I have had the principal thoroughfares washed twice a week, and the result has proved so beneficial that during the excessively hot weather of last summer I had not one single complaint. Such a system of street cleansing would be the most successful method applicable, as cleansing by water produces a perfect state of cleanliness, by the removal of all decomposing matter.

Receptacles for Street Mud.—Failing this scheme, the only other method I can suggest is by making provision in the streets for the reception of the sweepings during the wet weather, which is quite practicable. The quantity of slop or mud swept from the macadamized roads on a wet day averages in this district from four to five loads per mile. In certain streets, of course, the quantity to be removed would be greatly in excess of this; but these would be the exception. All that would be required, therefore, would be to construct in each street a receptacle of sufficient capacity to contain the slop swept from its surface, and as the quantity to be allowed for is small compared with the entire length of the street, this might be effected by a very simple arrangement—namely, the construction of a sufficient number of troughs in each street, at convenient distances apart, placed alongside the curb, and underneath the channel way into which the mud on the streets would be at once swept, and whence it could afterwards be carted. The great advantage of this would be that in sloppy weather the whole of the mud and dirt swept from the surface of the streets would be immediately removed out of sight. The receptacles might be made fixed or movable as required, and constructed so as to present no unsightly appearance; in fact, they would in appearance exactly resemble a lengthened street gully, except that provision would be made to prevent the entrance of storm-waters. In a thoroughfare like Park Lane, for instance, which on a wet day yields about thirty loads of slop, thirty of these receptacles, placed at suitable intervals, each seven feet long by two feet wide and two feet deep, would be ample for the reception of the accumulated slop of the entire length of roadway. The contents would be removed in carts at leisure, either during the day or at night. Under existing arrangements the time occupied in filling a cart varies from twenty minutes to half an hour according to circumstances. By the proposed arrangement a cart could easily be filled in half the shortest period now required.

Sweeping Machines.—Failing both of these suggestions the only other alternative is to increase the number of horse-sweeping machines, and make it binding under heavy fines upon the contractor to remove the whole of the mud as quickly as it is swept from the surface of the streets. A man with a broom will sweep about 3,000 square yards a day; a man and horse with rotary sweeping-machine will sweep an area ten times as great—namely, say 30,000 yards.

In addition to and in connection with either of these systems the erection of "orderly bins" along the sides of the principal streets, to collect the horse-dung and other matter which accumulates during the day, would be of much use and greatly assist the work of cleansing.

Practice in Paris.—In Paris the sweepings of the streets are not removed in carts as is the system here. There the London "slop" cart is unknown, the refuse and mud is not permitted to remain on the surface of the streets, but is, with the aid of water, swept directly into specially constructed sewers.

Conclusions.—The conclusions resulting from the foregoing may be recapitulated as follows :

First.—The work of repairing and maintaining the parish roads would be more economically and efficiently performed by having ample accommodation for the storage of road material; by the employment of able-bodied laborers; and by discontinuing the use of macadam in all leading thoroughfares.

Second.—Cleansing would be more rapidly effected by such methods as the employment of hydrants, with receptacles for sweeping at the street sides; employing more horse sweeping-machines; obliging contractors to remove mud as swept up; increasing the number of parish horses and carts, etc., and discontinuing macadam in leading thoroughfares.

*Maintenance in Chelsea.**—The following items are from the annual report of Mr. George R. Strachan, Surveyor to the Parish of Chelsea, London.

Complaint is made of an insufficient number of street basins, and that the gutter stones, from not being laid on concrete, have settled under heavy loads, so as to form depressions which hold the water. He states the axiom, that "a street on which water gathers is a dirty street, and a dirty street is an expensive street." The use of flints on macadam roads is condemned as being "dusty in summer and dirty in winter" under heavy traffic. Good granite is recommended, but not such as breaks into sharp, razor-shaped pieces. Large stones are a positive evil, and it requires much care to insure that the stones do not exceed two inches in the largest dimension. The paragraph on the wastefulness of macadam roads for streets carrying heavy traffic is worth quoting entire:

Cost of Maintenance.—Broken granite at its best is very wasteful as a material for maintaining carriageways where heavy traffic exists. A cubic yard of Guernsey granite costs the Vestry 12s. 6d. on the wharf this year. The cartage of it to the carriageway, spreading, rolling, and consolidating it costs 9s. 6d. more, or a total of 22s. on the finished carriageway. It is there ground to dust in dry weather and to mud in wet weather, and becomes, taking an average of all weathers and of the traffic weights in Chelsea, four cubic yards of slop or mud. These cost 10s. 4d. for sweeping to one side of the carriageway, 6s. for carting to the wharf, and then 4s. 5d. for their disposal. From first to last the cubic yard of Guernsey granite has cost £2 2s. 9d. In roads like King's Road one year's wear is the utmost that is obtained from it. The average life in the home district is only four years. The temporary use of the material at the large cost named brings vividly to the mind the unscientific and wasteful character of macadam roads in heavy traffic, and points to the necessity of using a better and different class of material for carriageways in main thoroughfares.

Steam Rollers.—As to the value of steam road-rollers the report states that, after a contest in the courts, the right to use them has been maintained, the gas company having sought to restrain them on account of damage to pipes. One of twelve tons and one of ten tons are now in use, at a cost for each of 10s. 6d. per day, including driver, fuel, oil, etc. By the use of one 300 square yards

---

* xvii, 116.

can be finished per day, with a great saving to tradesmen on a street
by the shortening of time occupied in repairs. Repairs by simply
throwing stone loosely on the surface and depending on traffic to
pack them is considered a cruelty to horses and an inconvenience to
the public. Two gas-mains have been cracked by the rollers, but
the insignificant cost of their repair bears no comparison to the
saving by the use of the rollers.

Cost of Repairing.—The following statement is given of the
actual cost of repairing Harrow Road, the area repaired being 9,522
square yards:

	£	s.	d.
Estimated cost...	390	0	6

ACTUAL COST.

	£	s.	d.
477 scores..	35	15	6
403 tons granite.	211	14	1
216 yards hogging..........	52	4	0
13,500 gallons water................................	10	1½	
76¼ days' horse hire..............................	34	6	3
Labor..	48	4	10

Roller.

	£	s.	d.
Maintenance (1 per cent. on £375).....................	3	15	0
3,500 gallons water...........................	2	7½	
Half ton coal.....................................	9	3½	
125 bundles wood	4	4½	
Six chaldrons coke...............	2	14	0
	£390	0	1
Below estimate.......................................	0	5	

Cost per square yard; 9.82*d.*
Average area rolled per day of 10 hours, 423½ square yards.
Average thickness of granite, about 1.38 inches.

Wood Pavements.—As to wood pavements, some which have
borne a traffic of 550 tons per yard in width during sixteen hours
per day for seven years are good for some time to come. These
pavements wear most at the sides of the road where the water lies
on its way to street-basins; the surface also wears uneven in the
older portions. Hard wood has not proved satisfactory; when one
block is defective it wears more rapidly than adjacent ones, and cup-
shaped depressions are formed, instead of saucer-shaped as in the
softer woods.

Asphalt Pavements.—Of asphalt pavements Mr. Strachan says:
"I venture to state that there is not a pavement in use so economical, healthy and clean, and with so many advantages as asphalt. Its
one disadvantage of slipperiness does not, in my opinion, outweigh
its advantages."

The various carriageways now paved with it are in better condition than when first completed. " They make no mud, and are always wholesome. A heavy rain washes them clean, and an extra watering attains the same end."

The footways paved with asphalt mastic one inch thick are spoken of in the highest terms, although compressed asphalt is undoubtedly most durable.

The trouble from cracks in the latter and methods of overcoming were treated of in previous articles.

Flag-stone footways allow water to percolate through the joints into cellars unless they are laid on concrete.

Cleaning.—The main streets are swept daily, others three times and twice a week according to their importance. Six gangs of men are engaged in this work, and are assisted by four sweeping-machines and three scraping machines whenever practicable. The wood pavement is washed during dry weather, and the slop occasioned thereby is removed as quickly as the present system allows of. Street orderly men are regularly employed on the wood pavement to remove the horse-manure and the refuse. During the year 19,695 cubic yards of slop and street sweepings have been swept from the streets, and have been carted to the wharf at 2s. 3d. per load. These were shot into barges, and removed at 1s. 8d. per load.

Removal of Refuse.—By contract, the carts are to call at every house once a week at least, and the men are to ask for the dust. The system followed when an application is made for the removal of dust is for an inquiry to be made as to why it was necessary. If the contractor is at fault, he is informed of the fact; if the servants have refused to allow the men to take away the dust when they called, their master is informed of that fact; and if the householder has been averse to having it removed weekly, he is urged, on the ground of health, to allow it to be taken each time the men call; and by these means it is endeavored to secure a regular and weekly removal of all the household refuse.

The collecting contractor is instructed not to collect trade refuse as dust, and a strict watch is kept on what he delivers to the wharves.

The public grumbles at the local authority for its neglect in the removal of dust, yet they put out of the reach of the local authority the only means of doing the work well. The custom of tips is the curse of local effort. It demoralizes the men, it causes them to act dishonestly in removing rubbish that ought not to be otherwise disposed of, and it inflicts a great hardship on the poor.

Street Watering.—For street-watering the price paid for water is 9d. per 1,000 gallons, and the cost has been £65 19s. 10d. per mile watered.

Street Opening by Gas and Water Companies.—Street openings by gas and water companies cause damage to the roads and are an annoyance to the public. They may be necessary for the business of the companies as at present conducted, but they are an injury to the roads. I am convinced that sooner or later the local authorities of the metropolis will have to take united action for their own protection against the powers and customs of these bodies. I have already referred to the steam-roller case where the law has subordinated the rights of the public in the repair of the streets to the rights of these commercial undertakings. At the close of the year the law has been held to make the public liable for personal injury under the following circumstances: A water company placed an iron box in a York stone footway under their statutory powers. In time the stone wore below the level of the iron box and an unfortunate pedestrian tripped up over it and sustained serious injury. He brought an action for compensation, and it was held that the local authority was liable. In this case the local authority cannot successfully refuse to have the iron boxes fixed in their pavements, and yet they are required to keep the pavements up to the boxes. If the boxes were not there the pavement would wear equally and no injury would be caused. This is a serious liability. I have not objected to the Chelsea Water-Works Company inserting such works and apparatus as they deem necessary in the streets, but if these liabilities are to be thrown upon the Vestry, it will be my duty to draw the attention of the Vestry to the question. It is no light matter to engage in lawsuits with these companies, when one of them can spend £100,000 in law, and then pay 10 per cent. to its shareholders out of the annual revenue.

MAINTAINING MACADAM ROADS IN LIVERPOOL.*

On the subject of pavement in cities of the Continent of Europe and Great Britain, Mr. Clement Dunscombe, City Engineer of Liverpool, writes :

In the area referred to there are innumerable descriptions of pavements, some good and some bad, and in many of the large cities and towns the same remark applies. Some of these cities and towns are, as a whole, extremely well paved, but in no city or town is the pavement laid down so thoroughly well constructed, so durable, and so sanitary as in Liverpool ; and what, to some, may appear extravagant in the methods here pursued can be proved by demonstration to be the most economical over a series of years. The cost of the maintenance of the carriageway pavements and footways in London is nominal, and a large proportion of the amount expended under this head is due to the retention of macadam roads, which are maintained as such owing to the exigencies due to local circumstances. The low cost of maintenance is obtained by laying down the pavements in the best possible manner, and making all repairs at once when needed, so that the pavements are always in a superior condition, and the cost of maintenance is reduced thereby to a minimum.

THE CARE OF PAVEMENTS.†

Professor J. S. Newberry also gives some good advice in a paper on "The Street Pavements of New York," published in the *School of Mines Quarterly* for July, in which he clearly points out what is so little appreciated here, that even the best pavements require constant and prompt attention, saying :

In a recent visit to Washington I found some of the streets in a bad condition. Even on Fourteenth Street, in front of the Treasury building, the asphalt pavement is full of holes, and the condition of this great thoroughfare has led to an opinion which I found quite prevalent, that asphalt was only adapted to streets where traffic was not great and the

---

* xv, 485. † xx, 155.

vehicles were light. This is a mistake, however. There is no street in America, or elsewhere in the world, that has as much traffic as Cheapside, London, and among the vehicles which pass through it are omnibuses, loaded with passengers inside and on top, carts of all descriptions, and, heaviest of all, the trucks of the great brewers, with their enormous horses and tons of ale and porter. And yet Cheapside is paved with asphalt, and is as smooth as a house-floor. The secret of its perfection is the thorough manner in which the pavement is laid and the incessant care given to it. In nothing is the axiom truer than in the asphalt pavements, "that a stitch in time saves nine." The material has little hardness, and if from irregular settling of the road-bed or local violence a break occurs, the passing wheels rapidly shear off the sides of the hole, and it soon assumes formidable dimensions. In London this is prevented by constant watchfulness; persons are employed to traverse the street with a light repairing outfit, and wherever a defect is observed, this is patched at once, and so effectually that the spot cannot be distinguished. The contractors who lay the pavements agree to keep them in order for fifteen years, at a price which does not average more than a few cents a square yard.

Our people seem to think that no pavement is a good one unless when once laid it will for ever take care of itself; but there is no such pavement. Even our rough stone roadways would pay excellent interest on the expenditure necessary for constant inspection and repairs promptly made when needed.

Such articles as these, and others like that of Clemens Herschel on "The Science of Road Making," should go far to direct this awakened public interest into intelligent and effective lines of action and insure, for instance, that the $3,000,000 recently appropriated for improving the pavements of New York City shall be expended to the best possible advantage. Such an object-lesson to the rest of the country as a well-paved metropolis could not fail to be of incalculable value, and it is earnestly to be hoped that the opportunity now presented to teach this lesson may not be lost through the ignorance or indifference of any one in authority.

We do not believe that it will be, and we find our strongest reason for this belief in what certainly seems to be an aroused and enlightened public sentiment in favor of good pavement and plenty of it.

CLEANING PAVEMENTS IN THE CITY OF LONDON.*

From the last report of Mr. William Haywood, Engineer of the city, which, it must be remembered, is a small portion only of the great metropolis, we gather some interesting items as to the present ideas of the English engineers on pavements.

The carriageways on portions of three streets have been laid with *Val de Travers compressed asphalt.* Another street which had been paved with Henson's wood pavement was also relaid in the same asphalt, the wood having become in bad condition.

---

* xvi, 347.

Four streets had been laid with *Limmer compressed asphalt*. Two others which had been paved with wood, and were in bad condition, were also relaid with Limmer asphalt.

Wood pavement was renewed in three streets and parts of streets.

Granite was used in two streets, but the "relaying of streets with granite diminishes yearly, wood and asphalt (principally the latter) gradually being substituted for granite in the city."

For *footways* Val de Travers compressed asphalt was used in four streets, Limmer asphalt in six streets, and stone in five streets.

The whole of the carriageways are *swept once daily*, the main thoroughfares, in wet weather, swept a second time during the day.

In addition to the general cleansing of the whole surface, the *street orderly system* was in operation on the carriageways of all the main thoroughfares, about 150 boys being employed for that purpose. The great advantage of this street orderly system is most apparent in wet weather, the thoroughfares where they are employed being at such times in a much greater state of cleanliness than they are elsewhere.

The *work of cleansing* the main thoroughfares often begins as early as two, three or four o'clock in the morning. The street orderlies then take up the work at about 7.30 A. M., and cease work at about 4.30 P. M.; late in the evening, when the weather permits, or the condition of the surface renders it necessary, the carriageway pavements are washed. The cleansing of the carriageways in the main streets is, therefore, almost continuously going on.

The *quantity of water used* during the year 1886 for washing the streets and courts was about 2,247,790 gallons. The number of nights when the water was used was 88.

The *footway pavements* are swept by the Commission whenever it is necessary, and in wet weather those in the main thoroughfares are cleansed with squeegees during the day. It is a statutory obligation on the part of the occupiers of property to keep the footways clean in front of their premises, an obligation but little attended to by the inhabitants in any part of the city.

The *courts and alleys* inhabited by the poorer classes were washed with jet and hose twice a week between May and the end of October. A very few places are washed nightly throughout the year for special reasons.

The street orderlies and scavengers have for some years been occupied at times in *strewing sand and gravel upon the streets* which have the greatest traffic. The average quantity so strewn during the last three years has been about 650 cubic yards, and the tendency is for it to increase. The extra work in removing this quantity of material when converted into mud is, in some states of the weather, very great indeed, especially from wood pavements. This circumstance also adds to the difficulty of keeping the streets clean, as well as to the cost of cleansing.

The engineer complains of the trouble arising from the collection of house refuse, owing to the abuse of the privilege allowed of placing their dust and refuse on the footways in receptacles of insufficient size, or with none at all.

CHAPTER VII.

NOTES.

EXPERIENCE WITH ASPHALT AND OTHER PAVEMENTS IN THE CITY OF LONDON.*

Continuing our account of experience in England, we come next to a series of reports on the pavements in the "City of London" proper, where the travel is probably more per unit of width than in any other city in the world. We are indebted for the information to William Haywood, Esq., Engineer and Surveyor to the Commissioners of Sewers of the city, and the reports date from 1886 back to 1853. The first *wood pavement* seems to have been laid there in 1839. One laid in 1842 up to 1853 had had two renewals, making the average total cost (including $4\frac{1}{2}d$. per annum for repairs) to be $1s$. $11d$. per year. *Granite* for the same time cost $9\frac{3}{4}d$. total per year. To the cost above for wood should be added about $1\frac{1}{2}d$. for sanding to prevent slipperiness. The duration of granite in this statement is taken at twenty-five years.

The practice is to lay the new granite in the streets carrying the heaviest traffic, then after a certain amount of wear to remove it, redress it as required, and lay in second-class streets, and, finally, after a second removal, it is placed in streets with least traffic on them. The dressings are used in macadam, and finally the worn-out stones take the same course.

Statistics are given of a *granite pavement* on London Bridge, where it lasted twelve years under this enormous traffic before renewal. The stones, originally 9 inches deep and 6 wide, had worn down an average of 2 inches. The cost per year was $4\frac{1}{4}d$. for maintenance and $1s$. $10\frac{3}{4}d$. for laying, or $2s$. $3d$ per year.

In the Poultry, another very busy street, the wear in six years was $1\frac{1}{2}$ inches.

From a table given we learn that the first squared *granite blocks* were laid down in 1828 and were not renewed for from sixteen to twenty-five years. These were 6 inches square by 9 inches deep. The 3 and 4 inch granite cubes were introduced in 1844, and bore the very heavy traffic of Cheapside and similar streets for seven to nine years before relaying, many of the blocks being still fit for use .

---

* xvii, 150.

Up to 1852, of 21 *wood pavements*, the time down before renewal ran from twenty-three months up to eight and one-half years, the larger number lasting less than four years.

There are about 51 miles of public ways in the "city," containing 441,250 square yards of carriageway and over 300,000 square yards footway.

The next report in the series dates 1870, and contains an account of a pavement called "*McDonnell's Patent Adamantean*," composed of broken stones not over 3 inches in diameter imbedded in asphalt, and laid in blocks 18x12 inches and 6 inches deep, with ½-inch joints filled with asphalt.

An interesting case is reported where suit was brought by the owner of a building against the Commissioners for damages caused by *raising the grade* of the street so that he had to use steps to go down to his property. The case was decided on final appeal in favor of the Commissioners, on the ground that it was done in the interest of public traffic.

The report states that since 1867 the Commission has removed all dust, ashes and trade refuse.

The street "orderly" system, and erection of iron bins along the sidewalks for the reception of the manure collected, was begun at that time and has been continued to the present.

The average amount of material removed per week, not including snow, was 972½ loads.

The usual requirements as to *maintenance* are as follows: In the case of *asphalt pavements*, the contractors for the pavement when new agree to maintain it in good repair for two years without cost to the city, and after this for fifteen years more at a price per square yard per annum named at the time of bidding and based on the total area paved. The pavement to be in good condition (to the satisfaction of the engineer) at the end of seventeen years, and to weigh at that time not less than a given amount per square yard.

For *wood pavements* the agreement is entirely similar except the provision as to weight.

The report of 1871 mentions the first laying of *granite blocks* with wide joints, and filling with small pebbles, and a composition of pitch and creosote oil poured hot.

Observations made that year show that there were fewer *falls of horses* from slipping on pavements of compressed asphalt than on the ordinary granite pavement.

Experiments were made that year on melting snow from the streets by steam and gas.

On account of disturbance of the streets for laying pipes of various kinds subways are pointed out as the only remedy. They were then in use in four streets.

It is interesting to notice in all the reports thus far, frequent mention of street widenings, which have done so much towards relieving the street traffic during recent years.

In 1872 mention is made of the *failure of a pavement consisting of blocks of compressed asphalt* laid on concrete and grouted with bitumen. These were 13x6½ inches and 2½ inches deep. Much consideration was given to asphalt pavements this year.

The first specimen of the regular *Nicholson wood pavement* was laid also, and as asphalt was considered unsuitable for grades steeper than one in sixty, wood is suggested as a possible substitute in such localities, taking the place of the asphalt.

Complaint is still made of the disturbance of the streets for laying of pipes.

In 1873, the *Adamantean pavement* laid three years before, is reported upon. It showed serious wear at the end of one year, at the end of eighteen months received extensive repairs, and in two years was so bad that further payments upon it were withheld.

A pavement of Trinidad bitumen, broken stone, chalk, etc., laid hot, was proven to be a failure; also, one consisting of a mixture of certain oils, caustic lime, pitch, sawdust, etc.

A pavement of compressed asphalt from Seyssel rock (Société Francaise des Asphalté), also from Montrotier asphalt, proved enduring.

Experiments also showed that such pavements would not aid in spreading fire.

Patent wood pavements of various kinds began to be experimented with.

A footway of Portland cement laid under pressure was also put down.

Seven street-sweeping machines are reported in use.

The *washing down of streets* paved with asphalt begun in 1866, and was still under trial as an experiment.

It would seem from a statement made here that the city retains full control of the subways under the streets.

More streets were torn up for pipes of the Gas Company and Hydraulic Pressure Company.

Conflicts of authority are noted between the Commission and Metropolitan Board of Works as to control of certain streets, etc.

A separate *report* was made this year *on asphalt pavements*. The tests of thickness by pieces cut out, weighed for compressed asphalt from about 16½ to 22 pounds per square foot; the original thickness being about 2 to 2½ inches and reduced thickness $\frac{3}{16}$ to $\frac{3}{4}$ of an inch less. It would seem to indicate less care in compression in some cases than in others.

In 1874 a patent "asphalt" pavement, consisting of tar, cement, sand, and sawdust, laid 2¼ inches thick, while hot, under a pressure of 112 pounds per square inch, on concrete foundation, was tried, and taken up in two months.

Another, called a metallic asphalt, consisting of blocks 2x2 feet and 4 inches thick, made up of manufactured "asphalt" and burnt ballast, also failed in the same time.

This year the "Court of Common Council" resolved, "That in their opinion tramways in the 'City' of London will occasion greater inconvenience to the general public and to the traffic within the city than they will conduce to the convenience of the public."

Comparative Advantages of Wood and Asphalt.—A special report made this year on asphalt and wood pavements gives conclusions as follows: Asphalt is less noisy than granite and wood less than asphalt. Asphalt needs close attention to repairs or is speedily knocked to pieces. Wood is in time unequally worn, and causes more jolting and noise than when new; disturbing those inside of carriages, however, the most. Large blocks with wide joints wear more unequally than small blocks with close joints. Asphalt is smoother, cleaner, and drier. Water remains longer on wood and dirt remains in the joints. The smell mentioned in connection with wood has caused no complaints, and, on the contrary, people living on line of streets paved with it are anxious to have it continued. Asphalt is most pleasing to the eye, and, on the whole, most pleasant to travel upon. Both should be kept perfectly clean, but this is most difficult with wood. Washing is the best method for all pavements. This makes asphalt slippery, but does not affect wood. One can be laid about as fast as the other. In good weather 125 to 129 yards laid per day. Asphalt is easily repaired and wood less so, and not so permanently.

As to *slipperiness*, it was found in Paris that 1 horse in 1,308 slipped on granite, and one in 1,409 on asphalt. An extended series of observations for fifty days in London, with weather mostly dry and cold, the asphalt sanded, when wet, gave one fall for each horse traveling 191 miles on asphalt and 330 on wood. On wet days this became respectively 192 and 432 miles. The complete falls were in the ratio of about 1 wood to 4 asphalt. The wood gives more chance to the horse to save himself, while a small quantity of mud makes asphalt very slippery. It is more difficult for the same reason for a horse to get up from the latter.

Wood, on the contrary, seems to be more slippery when frost and snow prevail. It is not so safe at ordinary times to drive fast on asphalt as on wood ; and a horse can be stopped quicker on the latter, except in frost.

Compressed asphalt in all cases proved more durable than other asphalts. The opinion is expressed that no asphalt will last more than four to six years without much repairs, and that their entire surface must be renewed in from six to ten years. The cost reported on eight compressed asphalts at the date of the report for whole cost and maintenance for the seventeen-year term was, per year, from 1s. 4½d. to 2s. 4½d., the average being 1s. 8d.

At the same date *wood pavements* were reported, under a heavy traffic, as lasting nine to eleven and one-quarter years, the total surface having been relaid at least once, and additional blocks inserted from time to time, the mean total cost per annum being 2s. 7½d. for those under heaviest traffic and 2s. 4½d. for lighter traffic.

In 1875 Ludgate Circus was completed, giving a diameter of 160 feet, and greatly relieving the consequent traffic at that point.

A large number of patent wood pavements laid down for experiment, replacing granite blocks.

Cleansing of the footways begun by the Commission in 1872 was continued, and a method of sprinkling streets by means of perforated pipes laid along the curbs put in operation as an experiment.

The slipperiness of asphalt had become a subject of complaint, and sand recommended as a remedy, sparingly used.

In 1876 the granite sets with "asphalt" joints, laid three years before, were found to be in very bad condition, as it was impracticable to remove blocks for repairs without considerable expense, the smallest repair requiring a boiler and special heating apparatus. The pavement was a noisy one also.

The *apparatus for removing snow* by melting was used, and the cost of melting a cubic yard found to be 11¼d., 192 feet of gas melting one yard. Salt was then used, and then forced towards the gullies by hand labor.

This year another patent apparatus for washing streets from pipes along the centre was put in place for trial.

A number of new "resting-places," or refuges for pedestrians in crossing public streets, were constructed. This very desirable improvement is worthy of imitation in many of our crowded cities.

In 1887 another of the patent pavements, called "*Barnett's iron asphalt*," proved a failure, and the company threw up its contracts for maintenance.

The compressed asphalt seemed steadily growing in favor, and also certain styles of wood.

This year a *pavement of blocks composed of clay highly compressed and hard-burned* was laid. The blocks were 3¼x8 inches and 6 inches deep, weighing 12 pounds. These were laid on "ordinary ballast," with ¼-inch joints filled with ballast. The removal of granite block pavements continued.

The *use of salt* on the streets for removing snow ("coupled with quick and careful sweeping so soon as the salt has done its work ") was continued as the most effective means.

In 1879 *Davison's patent iron and asphalt pavement* was laid, consisting of iron frames with projecting iron studs, set in mastic asphalt on a concrete foundation. A similar one with lead disks proved a failure in three months. Up to this date thirteen kinds of asphalt pavements and eleven of wood had been laid, so that nearly every portion of the great arterial thoroughfares were paved with one of these.

In 1880 the iron and asphalt pavement and one of the mastic asphalts proved failures, also a "noiseless granite" pavement laid on felt, etc.

This year was laid a sidewalk pavement consisting of slabs made of pulverized granite sifted to the fineness of sand, mixed with Portland cement and water, and after setting, dipped in a silicate of soda. A nuisance of steam discharged into sewers, and escaping into streets, etc., was abated by the Commission forcing the construction of a special sewer by the offenders.

In 1880 another subway for pipes, etc., was constructed by parties owning property on each side of a newly-opened street. A specimen of the so-called granolithic pavement, made of crushed granite and lime or cement, was laid. *Wood and asphalt* had now almost entirely *replaced granite* in the main thoroughfares and many of the minor streets, asphalt predominating. Three more "resting-places" were introduced.

. The construction of subways was again forcibly urged as the only means by which to prevent tearing up of the streets.

In 1882 the *washing of streets* by a hose and jets having been in use two years was reported upon favorably. In the same year a special report gives the annual *cost of maintenance* only of various pavements as follows :

The average for 16 of Val de Travers compressed asphalt was 1s. per yard ; of 7 Limmer compressed asphalt, 9½d.; of 5 Société Francais des Asphalté's compressed asphalt, 11d.; of 9 improved wood, 12½d.; of 18 of all kinds of wood, 13½d.; and of 5 granite block, 5¼d.

In 1883, the reports states that "the most favorable period for using water (for cleansing streets) is when the surfaces of the wood and asphalt are in a moist and greasy condition," the mud being then most easily removed. *Street washing* was done on 187 nights.

In 1884 the so-called *International pavement*, consisting of large blocks of asphaltum and crushed limestone, highly compressed, and laid on a concrete foundation, of which a specimen was laid three years before, proved a failure. The Asphaltic Wood Pavement Company also failed. The large amount of *sand* required *on the asphalt pavements* and the removal of the mud formed is reported as requiring 50 per cent. more laborers to care for the streets.

In 1885 complaints were made of *smell* arising *from a wood pavement*. The street was one on which the sun rarely shines, and not being well adapted for wood it was replaced by asphalt. The Henson Wood Paving Company failed.

In 1886 the engineer reports all the *wood pavement companies* as either bankrupts or as having failed to fulfill their contracts but three.

This ends the lists of reports received to the present time. There are many other matters treated in them ; and being the summary of the experience of such a dense population, the results are especially valuable to city engineers everywhere.

TRAFFIC IN PARIS.*

With a view to testing the comparative durability of pavement in Paris, statistics have been prepared showing the number of vehicles and horses passing every twenty-four hours through the principal streets of Paris. Every twenty-four hours there pass through the Rue de Rivoli, 23,233 vehicles and 42,035 horses; Avenue de l'Opera, 29,460 vehicles; Boulevard Sebastopol, 46,318; Boulevard des Italiens, 30,125; Boulevard des Madeleine, 17,524; Place de l'Etoile, 18,311; Rue de Pont-Neuf, 20,082; Rue Lafayette, 12,210.

OBSERVATIONS ON STREET TRAFFIC AND STREET PAVEMENTS.†

A paper of much interest was read by Captain F. V. Greene of the U. S. Engineers, at the meeting of the American Society of Civil Engineers in this city on December 16, 1885. Under the title "An account of Some Observations of Street Traffic," it gave the results of very extended observations on the street traffic in the United States, with reference to weight of vehicles and loads, character of pavements, width of streets, etc. These observations were made by a corps of observers in ten cities, who reported to Captain Greene. The cities were New York, Philadelphia, Chicago, Boston, St. Louis, New Orleans, Buffalo, Louisville and Omaha, and included notes on thirty-six streets.

---

* vi, 475.   † xiii, 81.

The observations were made between 7 A. M. and 2 P. M., on six consecutive days at each place. They were first recorded by one or more observers by the punch and "trip-slip" method in use by car conductors, slips of different colors being used for vehicles of different tonnage. A summary of each hour's observations was then transferred to a daily report made on a printed form. These were then sent to the author by whom the computation of total tonnage, tonnage by vehicle, and tonnage per foot of width were made. The accidents to horses were also observed and recorded in three classes —viz., falls on knees, falls on haunches, and complete falls. Record was also made of the width and grade of the street, state of the weather, temperature of the air, average speed of the vehicles and length of street on which accidents were observed and recorded. On the back of the printed form of report full instructions were given for the observers, especially in regard to the weight of different kinds of vehicles, and every precaution was taken to have the observations made on a uniform system, and to eliminate the personal equation of the observer and leave as little as possible to his judgment.

Each day's report was subscribed and sworn to by the observers before a notary public ; these reports were submitted to the members of the society present for their examination. The pavements on the streets observed were of granite, asphalt and wood. In compiling the results as to traffic the standard for comparison was the average daily tonnage per foot of width. This was found to range from 273 tons on Broadway, New York, to seven tons on Olive Street in St. Louis. The average weight per vehicle varied from .68 ton on Fifth Avenue in New York to 2.08 tons on a part of Wabash Avenue in Chicago. Several remarkable instances were given of the manner in which smooth asphalt streets have drawn the traffic away from stone pavements—the most noticeable instance being in St. Louis, where, on Olive Street, paved with granite, the daily traffic per foot of width is only seven tons, while on Locust Street it is 103 tons, the two streets being parallel, adjacent and similar in every respect and paved at the same time. Comparison was also made with the London observations, showing a very much larger traffic (as high as 422 tons on one street) in the latter city, and the cause of the difference was stated to be the absence of street cars and the large number of cabs and omnibuses thus rendered necessary. Reference was also made to the observations of Messrs. Deacon and Stayton, in England, on the comparative wear of different classes of pavement under different amounts of traffic as actually observed; and the desirability of keeping such records by municipal engineers in this country was pointed out.

In regard to accidents, the standard for comparison was the distance traveled before an accident occurred. This was obtained by multiplying the distance over which the observations were recorded by the number of horses, and dividing the product by the number of accidents. The general result was : on asphalt, 583 miles ; granite, 413 miles ; wood 272 miles. There were only three sets of observations on wood, and hence as to this class of pavement the results are inconclusive. As to granite and asphalt, the large number of observations (over 800,000 horses and 81,000 miles traveled) justified the conclusion that fewer accidents occur on asphalt than on granite. This agreed with the result of similar observations in London, although in the latter city, from local causes, the accidents on each class of pavements were about three times as numerous as in the cities observed in America. In classifying the accidents it was found that the falls on knees were much more numerous on granite pavements than on asphalt. The cause of this was pointed out, the granite blocks being too wide (about 5 inches instead of 3) to afford a proper footing for the horses ; and the falls on haunches on asphalt being due to sudden pulling up and turning of the horses as noted in nearly every case in the reports.

From Capt. Greene's tables it is evident that the heaviest traffic, both in this country and in Europe, seeks the asphalt pavement in preference to either granite or wood ; and although there are as yet no trustworthy statistics regarding first cost and maintenance, Capt. Greene favors asphalt as the universal pavement.

IRON WHEELWAYS IN NEW YORK STREETS.[*]

General Roy Stone, who was for some time General Inspector of Street Pavements when General John Newton was Commissioner ot Public Works in New York City, urges that a pair of steel rails 12 inches wide be laid for wagon tracks on one or more of our heavy trucking streets ; the rails to be grooved and corrugated to prevent slipping.

The plan of providing a smooth surface for the wheels of vehicles is not entirely new ; it was used at Pompeii, and is still in general use in Northern Italy, where two parallel lines of granite slabs, 24 inches wide, are laid 28 inches apart. In England Mr. Walker laid two lines of granite tramways, on Commercial Road, London, many years ago, of stones 16 inches wide, 12 inches thick and 5 or 6 feet long, and the same arrangement was to be found in Liverpool until within seven or eight years. It is now thought better to make all of the street pavement smooth enough to permit its legitimate use as a roadway, and it is doubtful if any granite tramways are now to be found in any of the larger English cities.

---

[*] Ed. xx, 85.

If we can never emancipate our chief city from the ignorant influences that prescribed the noisy granite pavement for Fifth Avenue, which disgraces that street, General Stone's project might be adopted as a makeshift. It seems high time, however, that the provincialism which has characterized the views of its citizens as to pavements should be cast aside, and that they should unite in an effort to have New York at least as well paved as other cities of approximate wealth and importance. There is no street in New York with a third of the vehicular traffic of Cheapside, London, yet that has been paved for twelve or fourteen years with compressed asphalt. It has no street with much more than half the traffic of the Strand, but that thoroughfare has for several years been paved with wood. Both of these pavements are nearly noiseless.

The shameful neglect, against which *The Engineering and Building Record* protested, that allowed the small piece of wood pavement on Fifth Avenue to go to ruin in the course of four or five years, when such pavements on the Strand, in London, last nine years, should not be urged as an argument against a pavement that would be such a luxury for residence streets of suitable width.

If we are to be condemned in perpetuity to noisy granite pavements the width of the stones should be reduced to three inches, as is the practice in England, and the joints, which should be narrowed proportionally, should be thoroughly calked. so that they could not be depositories for street filth, and the streets could then be better cleaned.

EFFECT ON HEALTH OF NOISY STREET PAVEMENTS.*

What are the effects, if any, upon the health of those constantly exposed to it, of the continuous noise and roar of the business streets of a great city? We are not now referring to the mere inconvenience which is felt in localities where conversation in an ordinary tone of voice is made difficult by the clangor of machinery or by the rumble of street traffic, but to the effects produced upon an average healthy man by long exposure to such influences as to his ability to work and as to his enjoyment of life

That noise often has a bad effect upon a sick person is well known ; the desirability of quiet to permit of restful sleep to the weary brain is admitted by all. Even in cases of sickness, however, there is a great difference between the effects produced by a comparatively steady, continuous rumble or roar, such as that of a water-fall, or of the sea on the beach, or even of the ordinary passage of vehicles over a stone pavement, and a sudden intermittent interruption of silence such as is produced by the clangor of

* Ed. xiv, 7.

bells or the passage of a train on the elevated railway. The former, by its monotony, may be actually soothing and lulling in its effects, so much so that a person accustomed to it may at first find it difficult to sleep in a place that is absolutely quiet, while the latter always cause more or less shock to the person enfeebled by disease, though habit may do much to lessen the perception of them.

If continued noise produces any effect on the healthy man it must be chiefly, if not entirely, due to the increased concentration of attention to what is being said which it makes necessary—to increase of strain, in other words. Clerks and accountants employed in noisy localities do not seem to find any special difficulty in writing, in keeping books, etc., so long as they are in good health, but if they come to the office a little below par—say with a slight headache—then they sometimes feel that the noise makes it more difficult to add up a long column of figures or to write an important business letter with clearness and accuracy, that it increases the strain on attention to the matter in hand very much as it does in the case of conversation.

In a previous editorial we have referred to the increasing prevalence of affections of the nervous system among business men, and it is claimed by some that this is in part due to the almost incessant noise to which they are exposed.

As it is not possible to diminish street traffic, the only remedy for the evil is to provide a street surface which shall be as little resonant as is consistent with durability and safety. All engineers who have written on the subject of street pavements call attention to noiselessness as one of the conditions which it is desirable to secure, yet it may well be doubted whether in actual practice sufficient attention is paid to this requirement.

Certainly there are many streets in New York City, especially in those quarters occupied by residences of the better class, which have unnecessarily noisy pavements. In this respect Washington is much superior to any city in this country, and the work of its Engineer Department in this direction for the last ten years is worthy of special commendation and of imitation elsewhere.

It is true that these smooth, noiseless pavements are not suited to the requirements of heavy traffic, being slippery in wet weather, and not being sufficiently durable under such circumstances to meet the requirements of a reasonable economy ; but there are many streets in all cities where the traffic is light, and these considerations do not apply. We are not so badly off as Philadelphia and Baltimore, with their cobble-stone pavements, but there is plenty of room for improvement here, and we believe that public opinion will be strongly in favor of the man or men who set to work to bring this improvement about.

FIRST COST AND MAINTENANCE.*

Cedar-block pavement costs in Milwaukee, laid on boards, 85 cents per square yard. In St. Paul, creosoted and laid after the English method, $3.40. In London, England, the *treating* of the blocks by chemicals cost $2.50 to $4.83 per yard, and cost of maintenance each year 20 per cent. of cost. In Leavenworth and Kansas City, on six inches of concrete, the first cost is $2.30 to $2.63 with an average life of the blocks of four to six years, when the blocks can be renewed on the old base.

In Omaha, Colorado sandstone pavement costs $2.61 to $2.67 for blocks eight inches deep on eight inches of sand, and Sioux Falls granite $2.53 on sand and $3.09 on broken stone and sand foundation. Asphalt on a 6-inch concrete base cost $2.98, with a money guarantee to keep it in perfect order for five years.

In Washington granite cost, in 1884, $2.50 to $2.65.

In St. Joseph, the cost for asphalt, with the five years' guarantee, was $2.63 in 1886.

In Topeka the estimate is $3, which would include grading of about nine inches of the street.

TESTS OF DURABILITY.†

It may not be amiss here to recall the experiments so carefully made nine years ago in St. Louis. These consisted in rolling a two-wheeled cart forward and back over two strips of the pavement to be tested which were each 22 inches wide. The tires were each $2\frac{1}{2}$ inches wide, and loaded to 800 pounds per inch, or two tons total load. The heaviest traffic in St. Louis was 75 tons per day per foot of width, but the average for business streets was 35 tons. Estimating the effect of horses' hoofs at one-third of this, 50 tons per foot was taken as a standard. The total number of revolutions shown by the counter attached to the propelling crank corresponds to the total number of double trips (across the samples and back) borne by the specimens:

If R = number of such revolutions, or 2 R = number of trips,

L = total load in tons on machine.

W = sum of widths of the two samples tested in inches,

$\dfrac{2\,R \times L}{W}$ = total tons carried per inch of width of specimen,

$\frac{50}{12}$ = tons per *day* per inch width on the streets.

and $\dfrac{2\,R \times L}{W} \div \frac{50}{12}$ = the number of days equivalent wear on the streets that the specimen has borne.

---

* xv, 375. (Report to Common Council of Topeka, Kan.) † xv, 658.

The samples were weighed before and after testing, and under this load of 800 pounds per inch width of tire, borne for a period equal to eight years and six months on the streets, and 2½ years of about 400 pounds per inch, the total abrasion of the fire-brick pavement was 9 per cent. of its weight, or a depth of ¾ inch. The asphaltum block pavement wore under same traffic 14 per cent. but the bricks in the pavement tested were about one-half of them broken and but one of the asphaltum blocks. Granite blocks after a test equal to eight years and seven months traffic had worn scarcely appreciably, while limestone had lost 10 per cent., and in actual traffic would have lost still more by action of the elements. As a result of all the tests granite blocks were chosen for use on all streets having heavy traffic.

These tests seem to bear out the claims that are made in favor of brick pavements for light traffic.

PAVEMENTS IN AMERICAN CITIES.*

Alderman M. S. Chipman, of Waterbury, Conn., has been investigating street pavements in other cities with a view to ascertain the best method for improving the condition of the streets of Waterbury. In November, 1888, he sent out to various cities circulars asking information about the methods in vogue. Some of the questions were as follows : What kind of street pavements do you use ? Have you any method of assessing for pavements laid ? Is the assessment confined to the streets paved ? Is this basis of assesment the cost of the pavement ? How much are horse railroads assessed if tracks are in streets paved ?

The following is a tabulated summary, taken from the *Morning Telegraph* of New London, Conn., of the replies received from the following cities : Albany, Auburn, Boston, Buffalo, Cambridge, Camden, Elmira, Hartford, Hoboken, Holyoke, Lowell, Meriden, Middletown, New Bedford, New Haven, New London, Newton, Norwich, Oswego, Paterson, Pawtucket, Providence, Rochester, Springfield, Syracuse, Troy, Utica and Worcester :

Of twenty-eight cities where pavements are laid, twenty-four use stone blocks, some use granite exclusively, while others use sand-stone, granite or trap rock. Five cities use some asphalt, either in blocks or in sheet. In seventeen cities the whole cost of paving falls upon the city at large. In eight cities the whole cost of paving is assessed upon the abutting property. In three cities the cost is divided between the abutting property and the city at large in the proportion of two-thirds to the abutting property and one third upon the city at large, these last named being New Haven, Conn., Oswego, N. Y., and Utica, N. Y. In all cases where the property is assessed the assessment is upon the frontage of the property, or approximately on the frontage, the location or value of the property not governing

---

the amount at all. The horse railroad companies are assessed directly, or are required to pave in seventeen cities; are not assessed in six, and in the others they either do their own paving or maintain and keep in repair the space between their rails and from one and one-half to two feet outside the rails. In twenty-one cities the cost of paving is paid from the general tax; four only issue bonds or improvement certificates bearing interest. Comparing by States we find that in eight Massachusetts cities, seven use granite blocks, two Rhode Island cities use granite blocks. In none of the Massachusetts cities is there any assessment on the abutting property. In the six Connecticut cities three use the granite blocks. In the six Connecticut cities five put the whole cost upon the city at large, while one places the cost upon both city and abutters. In the nine New York cities all use stone blocks. In six the whole cost falls on the abutting property. In two the cost is upon both city and abutters. In one the whole cost falls on the city at large. The three New Jersey cities all use stone blocks; two of them place the cost on the abutters; one places the cost on the city at large.

PAVEMENTS IN OMAHA.*

As the object of these articles is to give the results of experience under all conditions of practice, we shall in the present article quote from the experience at Omaha as detailed by Mr. Rosewater in the report for 1886. As is well known, the temperature ranges here are extreme, and therefore most trying on any material used for a covering.

Material Used.—Until 1883 the material used for both curbing and paving was Nebraska limestone. This disintegrates rapidly under the action of frost, and was also high in price owing to lack of competition.

Colorado sandstone is much better, and the larger part of the curbing now in use is of this material. To prevent a monopoly, however, Mankato limestone and Berea sandstone are admitted, but they are softer and not so durable.

The result of competition has, however, been a reduction in price in the Colorado sandstone of at least 16 per cent.

Up to 1882 but one street had been paved. This had a Telford macadam 12 inches deep of Nebraska limestone. It soon wore into ruts and became covered with mud. In the fall of that year, as the result of an investigation in other cities, "stone and sheet-asphaltum" were recommended for pavements "on all but steep thoroughfares, and upon grades where stone blocks would not afford sufficient footing to use macadam."

Payment.—The charter provided that street railways were to pay for paving within their lines, the city paid for street intersections, and property owners for pavements opposite their lots. Funds

---

* xvi, 236.

were raised on twenty-year bonds at 6 per cent. for intersections. The property owners paid in five installments, at respectively sixty days, one, two, three, and four years, the amount being at first paid by "district bonds." Later on the time of payment for property holders was extended to ten annual payments. The majority of the property holders determine on the material to be used.

A bitter controversy was brought about by the acceptance of a bid to lay Sioux Falls stone under proposals calling for granite. The courts finally decided that it was a granite. The result has been that Colorado stone has fallen from $3.49 in 1884 to $2.60 in 1886, and Sioux Falls stone to $2.60 as against $4.35. Competition has reduced all rates except that for asphaltum, so that wood and stone are the most recent favorites.

Forms of Cross-Section.—Two widely different forms of cross-section have been adopted for streets, depending on whether they were provided with storm-water sewers or not. Where these *are* provided the gutters are shallow, and the centre of the street is about at the level of the curbs.

In other streets a deep, flat, V-shaped gutter is used, and the street centre is one to six inches below the curbs.

Foundations.—As to *foundations* for pavements the experience seems to classify them in order of merit as follows : (1) concrete, (2) broken stone and sand, and (3) sand. Even in this frigid climate a concrete foundation six inches thick seems to be ample for all purposes. This statement is made with the reservation that the concrete shall be put down under competent inspection. No great difference in result was found from the use of the various sands of the neighborhood.

Asphalt.—As to surface material for pavements sheet-asphaltum is given first place for cleanliness, lightness of traction, and non-absorbent qualities. It requires specialists on the other hand for its maintenance, and where wind and moisture collect on flat grades it rapidly disintegrates. For this reason it is not now used in gutters. Along street-car tracks it is absolutely necessary to provide a toothing of stone to prevent formation of ruts. As to resisting a climate ranging from 35° Fahr. below zero to 120° Fahr. above, four years' experience decides in its favor. As in all forms of block paving cracks are developed by extreme cold, but they close up closely in summer. It does not seem adapted to grades of over five per cent.

Stone.—Sioux Falls stone is a quartzite, close-grained, non-absorbent, and frost-proof. It does not break as evenly as granite or sandstone, and as it will probably wear slippery it is recommended that the blocks be limited to three inches thickness. It withstands the heaviest traffic.

Colorado sandstone affords a good footing, and seems to withstand the elements well, but its durability under wear has not been fully tested. Wood can be recommended for its cheapness ; compared with the other materials for durability it is not so good.

Maintenance.—As to maintenance of paving, the usual difficulties of a constant breaking and opening of streets for all sorts of purposes have been experienced. It is recommended that a price be fixed by ordinance for each such opening, and that before the permit shall issue a deposit be made of funds sufficient to cover the price of restoring and keeping the street surface in repair. This price to be fixed by a contract to be made with some responsible man who shall do all such work.

SHOULD PAVING CONTRACTS BE GUARANTEED.[*]

An ordinance was lately introduced in the Common Council of Cleveland providing that, in all future contracts for paving in that city, the contractor should guarantee the durability of his pavement for a period of ten years, and agree to keep it in good condition during that period and leave it in good condition at its close, the guarantee to be secured by the retention of ten per cent. of the contract price, which was to be invested in interest-bearing securities mutually satisfactory to the city and contractor.

· This ordinance was referred to a special committee of the Board of Improvements, who made a report signed by City Civil Engineer Rice, Mayor Babcock and Alderman Randell. This report, with comments in parallel-column form, as given below, has been somewhat widely circulated.

REPORT.	COMMENTS.
To the Honorable Board of Improvements : GENTLEMEN: The attached ordinance, No. 544, on careful investigation, does not bear out the fair promises on its face. I believe the same to be wrong in principle, for the following reasons : . To attempt to group all kinds of pavements and all conditions under one rule is absurd.	The city of Philadelphia, June 13, 1884, adopted an ordinance containing the following section : "SECTION 1. The Select and Common Councils of the city of Philadelphia do ordain, That, for the purpose of obtaining professional advice on the subject of paving the streets of Philadelphia, the Mayor is authorized to obtain from three engineers, *distinguished for their knowledge and experience of pavements,* a written opinion and report concerning the subject of pavements in Philadelphia."

---

[*] Ed. xviii, 256.

REPORT.

COMMENTS.

One of the recommendations contained in the report made by the Committee of Experts is as follows :

" All contracts for pavements (excluding macadam) should contain a clause by which the contractor guarantees to keep the pavement in good repair for a period of not less than five years from the date of its acceptance by the city, and to insure a compliance with this guarantee, the city should retain ten per centum of its cost until the expiration of the five years term."

Report signed by Q. A. Gillmore, F. V. Greene, and Edward P. North.

These gentlemen represented the highest order of engineering talent the United States has yet possessed, and not one of them was ever under suspicion of abuse of his professional integrity. Every one of them had had long and varied experience with street paving and was recognized as an authority in the business.

A distinction should be made between pavements laid under the city's specifications and those laid under the specifications of a contractor. A distinction should be made between pavements composed of materials purchasable in open market and open to competition, and pavements where the purchase of material and competition is barred by patents and other restrictions.

As no pavement is laid without the city's approval of the specifications, all such specifications are from necessity the city's, as all are also the contractor's when he accepts them. What distinction in guaranteeing shall be made on account of material that may or may not be purchasable in unrestricted or open market? What has that to do with the fact that dishonest or incompetent contractors may, regardless of material, do bad work—bad work that a guarantee will prevent?

It is probable that a guarantee of any number of years would insure better work for which the property would pay, not the contractor.

This is self-evident, and unless the paragraph carries by implication the opinion that property owners ought not to pay for what they get, calls for no comment. In such case it is too absurd for notice.

REPORT.

The fact that block stone pavements are good for ten years is patent to every observer, and to pay an additional amount to redemonstrate this fact is not economy or wisdom on the part of the city.

Such faults as this class of pavements possess are due to foundation, not the wear of the material.

The same necessity for a foundation does not exist where the soil is sand or gravel, but if this fault is considered vital, and the time has, in your opinion, arrived for adding costly foundations, it is a simple matter.

In the case of sheet pavements a foundation is a necessity, no matter what the character of the soil is.

A contractor may do a business of $100,000 on a capital of $20,000, and if you retain 10 per cent. on this amount of work done, you retain an amount equal to half his capital, or all or more of his profits for the year. This would aggregate or retire such a sum in ten years of active contract work, as to, in all probability, swamp the strongest paving firm that has ever attempted to lay pavements in this city. All contractors with moderate means would be driven from the field inside of two years, and the work forced into combinations and trusts. This would do away with the small competition we now have and run up prices. In the case of contractors who have done work for this city, this 10 per cent. would amount in one instance to the abstraction of $20,000 in four years, and in another to the enormous sum of $100,000 in ten years.

The city authorities have the power and the remedy of this matter in their own hands without the passage of an ordinance of such a dubious character. Guar-

COMMENTS.

As no block stone pavement laid in the city of Cleveland has lasted ten years in good condition, and as most of it is in wretchedly bad condition in less than five years, it must be patent that the claim made cannot be sustained. And if a guarantee will be the means of securing good foundations or anything else, it is entitled to a trial.

This reference to the opinion of the Board is worthy of careful attention, and the conclusion they reach will, it is hoped, be announced before any more pavement is laid *without a foundation*.

True; and contractors laying asphalt sheet pavement always advocate the best practice known to paving construction.

This consideration for the welfare of contractors is charitable, but is not business. An inspection of the pavements of the city does not establish a special claim to consideration; particularly so if the result is to be still further depletion of the taxpayers, from whom the money is to be taken for the repair of pavements. But why not take good bonds to protect a guarantee, and so wipe out the hardship so mournfully set forth?

If the power and remedies exist, as stated, there must have been official delinquency on the part of particular officials, whose duties have been seriously neglected, and

REPORT.

COMMENTS.

antees may be exacted for all new and untried material whenever or wherever necessary. Foundations may be put in whenever or wherever necessary, and proper steps to encourage competition may be taken.

the question of the hour is, what are they doing about it? The many miles of unevenly settled pavements, both old and new, found in the streets of Cleveland, are positive evidence of the need of better foundations for block stone pavements, and the legitimate question is, are better foundations to be provided for in the specifications for such work?

Respectfully submitted, Walter P. Rice, City Civil Engineer; B. D. Babcock, Mayor; Frank Randell, Alderman.

The practice of requiring the contractor to guarantee the performance of his work is not uncommon, nearly all the sheet asphalt pavements laid in this country have been laid with a guarantee for their repair through a specified term of years, but it does not seem fair to a contractor to require him to make all repairs without compensation. It would be disastrous, for instance, in New York City where so little control is exercised over those who are daily permitted for one cause or another to open the streets. The European practice often is to specify that the contractor shall keep the pavement in good repair for two or three years, free from all charges to the Municipality, and for fifteen years after that, at a fixed annual price per square yard, giving to the Municipality at the end of seventeen or eighteen years from its acceptance a pavement as good in all respects as when originally laid. Payment per square yard for maintenance is not estimated on the surface actually repaired, but on the total area paved by the contractor, and the precise amount of subsidence or wear permitted before repairs are required is carefully specified. Some instances are given in E. P. North's paper on the "Construction and Maintenance of Roads" in the Transactions of the American Society of Civil Engineers for 1879.

Any such provision in this country should be modified in favor of the contractor, by making due allowance for the frequent tearing up of the streets by plumbers and pipe-line men, a practice which is not so frequent in European cities as it unfortunately is here.

PART II.

ROADS:

CONSTRUCTION; MAINTENANCE; LEGISLATION.

THE REPAIR AND MAINTENANCE OF ROADS.*

From the *L. A. W. Bulletin* we abstract from an article, especially written for the National Cyclists' Union, some very pertinent remarks on the subject of road repairing and maintenance. The roads in many parts of our country are a disgrace to civilization, and it is undoubtedly true that an immense saving in money would result if their management would be left to the intelligent direction of educated engineers, instead of the ignorant roadmasters who consider a road made when it is heaped up in the middle :

Although railways may have altered the character of the traffic on the common roads, the actual number of vehicles and amount of produce conveyed over them is no doubt greater than it ever was. The railways, by facilitating traffic, have greatly increased the resources of the country, but every ton of goods conveyed by railway must first pass over the roads leading to or from the place of production.

Economy in Good Roads.—For economical and convenient traction, roads should be maintained in thoroughly good order. This, however, is too seldom the case, and many roads are kept in such a condition as to be a disgrace to the country.

This is the more inexcusable as it is a fact well known to all who have had experience in the matter, that roads well maintained and kept in good order cost less than bad roads. It is plain that if by keeping roads in good order, four horses are enabled to do the work of five, or three or four (by no means an unreasonable supposition), the economy of horse labor and wear and tear of vehicles and harness must be considerable, but economy in the actual cost of maintenance generally follows as well. Experience proves that a road with sufficient strength, good surface, and thorough drainage can be kept in first-rate order with a much smaller quantity of materials than an inferior, ill-kept road requires, and, though a greater amount of manual labor may be necessary, a good road on the whole is generally more cheaply maintained than a bad one, especially when there is any considerable amount of traffic. The indirect saving in the cost of traction and wear and tear of vehicles and horses, which would result from better roads, would probably far exceed any direct saving in expenditure on the roads.

* xvi, 97.   By W. H. Wheeler, M. Inst, C. E.

It is estimated that the saving in England by improving the roads (and they are now greatly better than in this country) would be $100,000,000 per year.　This is on the supposition that three horses would be able to do the work of four.　It is no exaggeration to say that on many of the roads in *this* country one horse would do the work of two were the roads put in good condition.　The author, referring to English experience, quotes a case where the tolls on a turnpike were insufficient to pay expenses, but by a reorganization of the labor system and the selection of good materials, not only was the roadbed so much improved that the loads hauled were *doubled*, but the income paid off the debt.

Supervision.—To accomplish this, however, requires the supervision of a skilled engineer, and this is the rock on which authorities here are sure to come to grief, since every roadmaster is his own engineer.

There is no point upon which a more decided coincidence of opinion exists among all those who profess what may now be called the science of roadmaking, than that the first effectual step toward general improvement must be the employment of persons of superior ability and experience as superintending surveyors.　The duties of surveyor demand suitable education and talents, and some skill in the science of an engineer should also be regarded as a valuable qualification, and these qualifications must be fairly remunerated.

In the case of roads in town and suburban districts, these are generally under the control of a board having efficient officers, and if the roads are not properly kept it cannot be from want of proper machinery to effect this object.　In many of these cases, however, the fact does not seem to be fully realized that efficiency in management means economy in cost, and that efficiency and cost are invariably in reverse ratio.

Vehicles.—The growing practice of using narrow wheels for vehicles carrying heavy loads is very destructive to the roads, the width of surface covered by the narrow wheel not being sufficient to bear the heavy weight imposed on it.　It has been proved by repeated and careful experiments that wheels with two-and-a-half-inch tires cause double the wear on a road than those of which have four-and-a-half-inch tires,　On good roads there is no great advantage in a greater width than four and a half inches.　Telford's rule was one inch of tire for every 500 weight on the wheel.　By this a cart weighing with its load two tons would require four-inch tires.　The following proportions have been advised by competent authority :

	WIDTH OF TIRE.	
	Without springs.	With springs.
	Inches.	Inches.
¼ to ½ ton on each wheel........................	3	1
½ to 1 ton on each wheel.	4	3
1 to 1½ tons on each wheel......................	6	4

It has even been recommended by some that all carts weighing over 600 pounds should be restricted to wheels of not less width than four inches, and the vehicles drawn by more than three horses should have wheels of not less than six inches.

Wheels of a large diameter also do less harm to a road than small ones. Stones will be pushed and moved out of their places by the small fore wheels of a wagon, whereas the larger hind wheels will pass over without displacing them. Large and broad wheels not only do less harm to the roads, but cause less draught for the horses.

Requirements of a Good Road.—A perfectly good road should have a firm and unyielding foundation, good drainage, a hard and compact surface free from all ruts, hollows, or depressions. The surface neither too flat to allow water to stand, nor too convex to be inconvenient to the traffic; free from loose stones, the fresh material being put on whenever practicable in winter in such a manner as to inconvenience the traffic as little as possible; all mud scraped off the surface as soon as it arises and not left in heaps on the road, the longitudinal inclination not to be greater than such that a horse may trot down hill with a vehicle behind it without danger.

Foundation.—With the exception of the first and last, all these conditions may be obtained in our existing highways. The great bulk of the roads have been made without any sufficient foundation, and the coating of surface material is so thin that unless constant and vigilant attention is given, the wheels rapidly cut through the crust in wet or frosty weather. The rain thus gains a way into the under foundation, and large quantities of new material are required to be put on before the road can be restored. It is impossible now to provide these roads with good foundations, but it is practicable, notwithstanding, to convert them by proper management into very good roads.

Gradients.—In hilly districts it is not possible to reduce the inclination to such a gradient that a horse can safely trot down, but there exist numerous cases where, at a comparatively small expenditure, by taking off the crest of a hill and filling the hollow, dangerous portions of roads might be improved and convenience of the traffic greatly increased. A gradient having a rise of one foot in every 40 may be considered perfectly safe for a horse to trot down with a light vehicle. An inclination of one in 20 is too steep for convenient traction, and should, if possible, never be exceeded. The latter gradient is too steep, and is the cause of great inconvenience, especially in level districts, it being frequently impossible for a horse to take the load that he can well draw along the level road up such an inclination.

Ease of Traction.—The power required to move vehicles along a road varies with its conditions. The harder and more level the surface the less tractive power is required. On a good paved road a force of 33 pounds is required to move a ton; on a well-constructed road maintained in good order, a force equal to 46 pounds, and on an ordinary country road, fairly well kept, 65 pounds is required; whereas, on a badly kept road, with loose surface, the force mounts up to double and even treble the last figure.

The result of experiments made on a macadam road by the author with a dynamometer attached to a wagon having 3-inch wheels, the total load being two tons, showed that on a road repaired with hard gravel picked off the land and broken by hand, a tractive force varying from 224 pounds up to 280 pounds was required to move the wagon. The same road, repaired with granite broken into cubes to pass through a two and a half inch ring, and on which more labor and attention had been bestowed, required exactly half the force, or 112 pounds.

Construction.—Although it is seldom that new roads forming main lines of communication have now to be made, yet it is desirable briefly to state the method of construction adopted by the best roadmakers.

The plan adopted by Telford, the greatest road engineer probably since the time of the Romans, was first to level and drain the site of the proposed road, then to lay upon it a solid pavement of large stones, and on this a layer of stones carefully broken so as to pass through a two and a half inch gauge, and no stone weighing more than six ounces, and over all gravel or other fine material in sufficient quantity to hide the stones. Great attention was paid to the surface until it became thoroughly solidified, and then it would stand for several years with a very little repair, one of the roads constructed in this way requiring nothing to be done, beyond cleaning the dirt off, for six years after its construction.

Macadam, who earned his reputation more as a converter of bad roads into good ones than as a constructor of new roads, differed from Telford in his system. Instead of the first coat of rough pitching, he substituted a layer of hard stones broken into angular fragments. This was watched, and as ruts or inequalities formed they were raked and leveled and fresh material added until a hard and level surface was obtained. The material used for coating was the hardest he could obtain, preference being given to granite, greenstone, or basalt.

The great art in roadmaking is so to construct the surface that no wet can penetrate from the top, and so that the dirt cannot work up from the bottom. The foundation must be of such materials as to form a uniform and unyielding body.

Burnt clay is dry and porous, and, if well burnt and thickly covered, makes a good material, but unless these conditions are complied with, it is liable to rapid deterioration.

With the exception of some of the principal thoroughfares, few roads in this country have been formed with a proper foundation. Generally a quantity of such material as could most easily be procured in the neighborhood has been shot down, leveled, and then left to become consolidated by the traffic. This has gradually been worn away by the action of the weather and the wear and tear of the traffic, till only a very slight coating of hard material remains. In wet and frosty weather the wheels rapidly cut through this thin skin, allowing the water to get below the surface level and the mud to work up from below.

A visible track once made in a road, all the vehicles will follow in the same course, each cutting into the road and forming two deep ruts. The hardest-paved road will wear away if the traffic be all kept in one track. If drivers would only vary their track a few inches, one set of wheels would counteract the effect of the others, and the road remain uninjured. The advantages of this is proved by the fact that wherever there is a turn in the road, however deep the ruts may be on the straight part, they disappear at the turn, because at the turning the horses coming from different directions naturally vary their course round the corners, and one wheel obliterates the track of the other. The experience of all roadmen can testify to the fact that at these points less material and labor is required than at any other part of the road.

One material point in the maintenance of roads is the transverse form. Roads that are too flat hold the water and keep the surface wet. A road that is too round is not only inconvenient for the traffic, but wears badly.

On a road that is too round there is a necessity for the traffic all to follow in the same track along the centre of the road, that being the only part where a vehicle can run upright. The continual tread of the horses' feet in one track in the centre soon forms a depression which holds the water, so that such a road is not so dry, and wears more unevenly than one of a flatter section in which the traffic is more evenly distributed over the whole width. If the surface be allowed to wear into tracks and ruts, the roundness will not clear it of water. The best form for the transverse section of a road is that of an ellipse, the surface having a very small inclination in the middle and a steeper fall at the sides toward the gutter.

Maintenance.—In dealing with the ordinary repair of roads they must be divided into two classes, the first being those which have to endure the constant and heavy traffic of towns and their immediate neighborhood, and the second, those subject only to provincial and agricultural traffic.

The attrition caused by the wheels of the thousands of vehicles which pass over town roads, many of them carrying weights of several tons, produces an immense amount of wear and tear and necessitates frequent and expensive repairs. The great advantage and economy of paved surfaces in towns need not be dealt with here. Where there is much traffic it is impossible to prevent macadamized roads from wearing into dust in summer and mud in winter. The thickness of the material, thus converted from the hardest known rocks into mud and dust, has been proved to amount to as much in one season as 4 inches over the part of the surface exposed to the most wear and 2 inches over the whole surface. The difficulty of replacing this material and combining effectiveness of repair, with consideration for the users of the road, has been a matter of much discussion. The practice of laying a thick coating of stone over the road and leaving the traffic to grind it down to a level surface is barbarous to the horses, destructive to all light vehicles, and wasteful of the stone. To enable broken stone to bond together and form an even surface, some substance must be used that will allow each stone to keep undisturbed with its neighbor. Unless some filling-in material is supplied, from one-third to one-fifth of the whole mass must be ground down to fill up the interstices before the stone can be bedded. In the process the angles are rubbed off the stones until their surface becomes rounded, rendering them liable to be constantly loosened. On the other hand, all fine material added, especially if it be not of good character, is liable to work into dust in summer and mud in winter. If the filling material be of an adhesive character in certain states of the weather it clogs the wheels and tends to the destruction of the road. After a sharp frost, when the thaw begins, the effect may frequently be seen of the material of which the surface is composed combining with the moisture and horse-dung to work into a sticky paste, which adheres to the wheels of vehicles and draws out the stones from the roads. Some road surveyors have adopted the plan of mixing the filling-in material with tar before it is placed on the coating of stone. A road treated in this way if well rolled before being used has a surface almost impervious to wet, and, owing to its elasticity, of a very enduring character. Granite chips or the screenings of the broken slag are an excellent material for mixing with tar. The heap of material after mixing should be allowed to remain for several days before use, otherwise it will be too sticky.

Constant watering in dry weather and cleansing of the surface is absolutely necessary to maintain the surface of a macadam road in good condition. The method of repair observed by the best road surveyors is to have the roads repaired in winter. The material used is the hardest and toughest to be procured, broken to an even and regular size, the fragments not larger than will pass through a two and a half inch ring. The surface is covered with sharp material of the same character as the stone ground in a stone-breaker to a size known as chips. After the stone and filling in has been evenly spread, the surface is well watered and rolled with a steam roller weighing not more than ten tons. A greater weight than this is apt to damage the gas and water pipes. After continual rolling the surface becomes thoroughly consolidated and even and fit for the traffic.

Steam rolling is in every way better and more economical than horse rolling, and that the advantages arising from the use of these rollers are economy, facility of perfect construction, comfort to persons and horses using the roads, improved surface, diminishing the wear and tear of vehicles, saving of material amounting to as much as 25 to 50 per cent.

In Paris the method of repairing macadamized roads consists in having the stones broken to such a size that no stone weighs more than five ounces. The interstices are filled with sharp sand. After the road is completed it is maintained in order by constant watering and sweeping with a machine—the sweeping following the watering, and the surface being made thoroughly wet, so that the machine will work easily. This is done early in the morning, before the traffic is about. In warm, dry weather the road is frequently watered during the day with hose from stand-pipes, the surface being only moistened : and not made thoroughly wet. In many towns in this country water is poured on to the surface of a road covered with dirt, so that a sea of mud has to be waded through by persons having to cross the road, and their boots become covered with mud on the brightest summer day. The mud soon after dries, and is carried in clouds of dust over the goods in the shops and the clothes of the passengers.

The repair of country roads is a much easier task than that of town roads. The amount of traffic is so small that with ordinary skill, care and attention there ought to be no difficulty in keeping the highways of this country in excellent order at a small cost.

The main points to be observed are : The constant care on the part of the roadman never to let his road get out of order ; in no case does the old proverb, "A stitch in time saves nine," apply more forcibly than to road repair ; the use of good materials ; proper attention to surface drainage ; cleaning the mud off in wet weather ; removing all loose stones in dry weather.

The roadman should never lose sight of the fact that loose stones are annoying to every user of the road, and always more or less dangerous.

One very essential matter that requires the attention of the roadman is the scraping and cleaning the mud arising from the wearing away of the materials and the droppings of the horses. It is impossible for the wet to evaporate quickly when the surface is covered with mud. In winter it acts like a wet blanket, preventing the sun and wind from drying the road, and in summer it causes clouds of dust. A road that has not been scraped will remain wet and dirty several days after an adjoining road, which has had the mud removed, has become quite dry. No road surface can be kept hard, nor can ruts be prevented where mud is allowed to accumulate and remain on the road. To prevent mud forming in wet weather the dirt should be removed in summer. It should be borne in mind that the summer dust makes winter mud, and that it requires much less labor to remove the dust from the roads in summer than the mud in winter.

At the sides of all roads maintained in good order a gutter is formed, and these should be kept clear and free from grass and weeds. They should always be laid out by a competent person, and suitable outfalls made for the water.

Material.—The chief requisites of a material for repairing roads are, that it should not only be hard, but tough, and that its composition should be such as that it shall not easily be affected by the weather. For the best road the fragments should be angular and cubical, so as to bond well together, each stone should fit against and be wedged compactly up to its neighbors, as in a piece of mosaic-work, and no space left for water to penetrate.

The materials used must, to a certain extent, depend upon the locality, but it is far more economical to use good material, although it has to be brought from a distance, than inferior local stone procured at less cost. No material is so efficient and really economical as granite or some of the hard volcanic rocks. Even in first cost it is cheaper than gravel, limestone or slag. There is less labor and carting in putting a durable and lasting material on roads than one that is more perishable and requires frequently renewing.

Some of the harder limestones and sandstones make fairly good roads. In damp weather in summer they make pleasant roads, but the former especially, although hard and tough, becomes disintegrated rapidly by frost, and both wear into dust in dry weather, and make roads dirty in wet and dusty in dry weather.

The softer kinds of limestone are utterly unfitted for repairing good roads.

Flint gravel makes a clean road, and if properly attended to gives a firm surface, but not a very even one. If the gravel is picked off the land or taken from the bed of a river it consists of stones of different degrees of hardness which wear unevenly, and all gravel is more or less round. Even if thoroughly broken there is left one part of the stone which has a rounded surface. Sea shingle and gravel, consisting of rounded pebbles, although clean and hard, never consolidate. Round gravel is one of the worst and most dangerous materials that can be used for roads. If the gravel becomes consolidated in winter it is certain to work loose in the following summer. The constant tendency of all stones having rounded surfaces is to roll round one another when weight is brought on them, and so work loose.

The materials to be avoided are those that are quickly converted into mud or dust and that work loose, and preference should be given to those that, avoiding these evils, make as even a surface as practicable and seldom require repair. The question of first cost, within reasonable limits, need not be considered, as the better the material, the less of it is required, and the less the labor and carting.

With regard to the size to which the fragments should be broken, opinions vary considerably. The old rule was that no stone that would not pass through a two and a half inch gauge should be used for the surface repairs. More recently smaller sizes have been advocated, and the element of weight as a standard advocated. The following standard has been recommended :

	Maximum Weight.	Minimum Weight.
Granite and similar rocks......................	3½ oz.	½ oz.
Flint and similar stones........................	5 "	¾ "
Limestone and similar stones........	6 "	1 "

One-half of the total quantity to be of the maximum weight, one-eighth of the minimum weight, the remainder to be composed of stones varying between these.

There is no doubt that the hard and tough rocks should be broken into smaller fragments than those that are softer. For ordinary country roads the cleaner and more even the size of the stone the better, as there is always in the road plenty of loose material for bedding the new stones. The author's experience leads him to give as his opinion that for the repair of a main road in rural districts, where granite or similar hard material has not been long used, the size to pass through a two and one-half inch gauge is not too large. If the fragments are smaller than this they are squeezed into and lost in the softer material, their size not being large enough to give sufficient bearing surface to withstand the weight of the wheels of heavy vehicles.

After granite has been used for some time and the road has become fairly coated with it, so as to give a good resisting surface, a size to pass through a one and three-quarter inch gauge is sufficiently large to fill up the depressions. If a smaller size than this is used it will not stand the crushing of the wheels, and soon becomes dust.

All stone should, when practicable, be carted and placed in heaps at the sides of the road during the summer, when the roads are hard. The material is then ready as wanted, and there is less wear and tear on the road by carting than if it be done in winter. All stones harden by exposure to the weather. Stones that have been lying during the summer in heaps will wear longer than those brought directly from the quarry.

As far as practicable materials for repairs should be placed on the roads in the early part of winter to allow of its being thoroughly bedded and incorporated with the old material. As a matter of experience it has been found that stone put on in the early spring, although apparently well bedded and set, will work out again much sooner than that put on in November and December. If it is necessary, as will sometimes be the case, to place the stones on the road at other times than in winter in order to fill up depressions or to prevent the formation of ruts, the surface should be loosened with the pick, the stone evenly spread and covered with road-scrapings. If this be carefully done the place so repaired will soon become as firm and as level as the rest of the road. For repairs of this kind only the smallest kind of material should be used. Many surveyors keep a supply of granite chips purposely for the repair of hollows and defective places in summer time.

Cost of Repairs.—It may be taken as a fair estimate for the main roads of this country, excluding the strictly urban portions, that a mile of road will require 40 tons of granite to repair it, and that one man can keep in order four miles of road. Putting the granite at $2.50 per ton, and the wages $1.25 a day, this, with an allowance for carting, extra labor, and incidental expenses, would give $175 a year as the cost per mile. After a road has been thoroughly well coated with granite, the quantity of material required will decrease considerably. On the other hand, roads which have previously been repaired with soft and inferior material will require a larger quantity of material and more labor.

THE ADMINISTRATION OF COMMON ROADS IN FRANCE.[*]

· Under the title "Local Government and County Councils in France," M. Waddington, the French Ambassador at London, has some very interesting notes on the administration of the public roads of France.

France is divided into departments, each department into arrondissements, each arrondissement into cantons, and the administration unit is the *canton*, which may be composed of several rural parishes, a town and adjacent parishes, or a portion of a city. An *arrondissement* consists of five or six cantons, while the *departments*, of which there are 86 in France, are generally composed of 35 or 40 cantons, but vary from a minimum of 17 to a maximum of 62, and from the Department of the Seine, with nearly 3,000,000 inhabitants, to that of Belfort with 80,000. M. Waddington compares these departments to the English county, and uses the word county in speaking of them.

Throughout France the roads are divided into five classes as follows :

1. *Routes Nationales.*—These great highways, which preceded the railroads, and parallel to which the railroads were generally located, lead from Paris towards the frontiers, join large towns like Lyons and Bordeaux, or connect the different fortified posts along the frontier. At the beginning of the century they were the main arteries of traffic through France and the only roads kept in good repair. They are now much less used, but are kept in good order and are entirely maintained by the State under the direction of the *Ingenieurs des Ponts et Chaussees.*

2. *Routes Departmentales.*—These connect the different towns of a department with each other and with those in a neighboring department. They are not quite as broad as the "Routes Nationales," but have more traffic than any other class of roads and are kept in admirable order entirely out of the funds of the department, or "county rates," and are under the direct management of the "*conseil general,*" an elective council of one member from each canton.

3. *Chemins de Grande Communication.*—These are almost equivalent to those last mentioned and the two are now often completely amalgamated. The "Routes Departmentales" were, while maintained out of county rates, managed by State engineers, and the "Chemins" were constructed and maintained partly by county fines and partly by contributions of the different parishes, and managed by county officials.

4. *Chemins d' Intérêt Common.*—These are country roads for local circulation of less width and less solid construction than the preceding, and

*xviii, 135.

are mainly maintained by parochial contributions, with the aid of variable annual grants from the county.

5. *Chemins Vicinaux Ordinaires.*—Purely parochial roads, connecting the villages or hamlets of the parish. They are maintained out of the resources of the "commune," and under the supervision of the Mayors of the commune, but county officials lend their assistance or advice when requested, and make plans for their construction, to which both the State and county contribute aid in the case of the poorer parishes after the commune has furnished proof that it is able to maintain the road when once constructed.

The construction and maintenance of these roads, other than the first mentioned, is managed by the "conseils généraux," and though in some instances the service is managed by engineers of the "Ponts et Chaussées," who in that case receive an extra pay from the department, they are, in general, in charge of "agents voyers." An "agent voyer en chef" for the department, "agents voyers d'arrondissement," and in each canton one or two "agents voyers cantonaux." The latter have charge of the "cantonniers," who are permanently employed, with a month's holiday in harvest time, and execute the current repairs and maintenance, aided by workmen temporarily employed in case of emergency.

These "agents voyers" now form a very considerable body of highly-trained men, supporting a monthly review which deals with questions of interest to them.

A special budget for the roads of the department is prepared by the "agent voyer en chef" and laid before the "conseil général" each year. It is divided into two sections, the first relating to maintenance and the second dealing with the reconstruction and improvement of old roads or the construction of new roads.

The resources are (1) "Prestations en naturi," (2) "Subventions industrielles," and (3) Contributions from the county rates. Every taxpayer is bound to furnish for the maintenance of the roads in his parish and vicinity three days' labor, called "journées de prestation," not only for himself and all laborers permanently in his employment, but of all horses, donkeys, mules, draught oxen, and carts in his possession, or commute such services at a rate established every year; or the prestations may be converted into piece work.

The "subventions industrielles" are levied on the principle that certain industries and manufactures, such as beet-sugar works, distilleries, etc., which cart heavy loads, cause an abnormal wear of the roads, the repairing of which cannot be fairly charged to the general taxpayers. The amount of this subvention, which must always be spent on certain specified roads, is arranged with the "agents voyers" under the sanction of the "conseil général," or, if the parties cannot agree, the question is referred to the administrative tribunal of the department, with whom the final decision lies.

The old *corve*, or warning peasants out to work on the roads once or twice a year, with the consequence that the roads were never in good order, has long been abolished in France, and the system of continuous maintenance substituted, has given that country the best roads of any except Italy, where a system of continuous maintenance under the charge of permanent officers is also in use. The system in this country does not require either description or characterization.

ROADS AND ROAD-MAKING.[*]

The progress of civilization has everywhere been marked by good roads. It may even be said to be largely due to them. Ancient Rome was not only famous for its own roads, but it carried the art of road-making into all its conquered provinces. As its civilization disappeared in the degeneracy of the Dark Ages, good roads ceased to exist, and they only reappeared when modern nations began to emerge from the Middle Ages. It is often said that the test of civilization in any country is the consumption of iron ; but this is true only because railroads are the chief consumers of iron, and they are but one form of roadway.

It is an undeniable fact, that while the United States has the finest railway system in the world—the most perfectly adapted to the work it has to do, and the cheapest in charges for transportation—yet its roads and its city streets are far inferior to those of France, England, Germany, Austria and Italy. Doubtless, the admirable character of its railways is itself the cause of its bad roads and streets, for the railways serve their purpose so well that there is less apparent need of good carriage roads. All the other countries above named had reached a high degree of civilization before the advent of railways, about fifty years ago, whereas, about three-fourths of the present area of the United States have been settled and populated during the railway era. The rapid advance in wealth and population of the principal countries of Europe during the latter part of the eighteenth and early part of the nineteenth century would have been impossible without a corresponding and simultaneous improvement in the quality of their roads. The still more rapid advance of America during this century has been accomplished chiefly through the instrumentality of railways, and these have so thoroughly intersected the country in every direction, bringing the merchant and manufacturer at one end, and the farmer and miner at the other, into such close communication, that the necessity for good roads has been overlooked. The opinion is now gaining ground, however, that notwithstanding the excellent and cheap service of the railways, there is a great loss in the unnecessary cost of transportation in hauling merchandise through the mud to reach the railroad, and again over rough cobblestones when it leaves the cars at its destination. And independent of the commercial aspect of the question, there is still to be considered the comfort and convenience of those who use roads and streets for pleasure riding and driving, and to whom good road surfaces are absolutely necessary. During the last few years there has been a constant increase in the attention and thought devoted to the question of roads both without and within cities, and the object of this article is to give briefly such information as to the history and present condition of the art of road-making as may be useful in this discussion.

Roman Roads.—The much-quoted Roman roads were, in reality, far inferior to the best roads of modern Europe, and were much more costly. Hence they may be dismissed in a few words. They were stone pavements with a very thick concrete foundation ; or, as described by another writer, they were " masonry walls laid on their sides." The most famous of them was the Appian Way, constructed about 313 B. C., from Rome to Capua, and

* xx, 145. Abstract of a paper by Captain Francis V. Greene, in *Harper's Weekly* of August 10, 1889.

subsequently extended to Brundusium (Brindisi). The foundation consisted of one or two courses of large, flat stones, laid in lime mortar ; next came a layer of concrete made of one part of lime and three of broken stone, thoroughly mixed and consolidated by ramming ; on this was spread a thin layer of mortar, in which the stones forming the top course were bedded. These stones were of basaltic lava, about 12 or 14 inches in width, with smooth upper surfaces, but irregular sides, and when carefully jointed together they formed a large mosaic. The total thickness of the road was about three feet, and its width varied from 12 to 20 feet. On either side were raised footways, paved with stone, and at frequent intervals were stepping-stones for mounting horses. It was also marked by mile-stones, indicating the distance from the forum at Rome.

This road was certainly durable, as is proved by the fact that although it had to be rebuilt by Trajan, at the end of the first century A. D., parts of it are still in existence, 2,200 years after it was first constructed ; but it was deficient in the other qualities of a good road. Horace is authority for the statement that it was " less fatiguing to people who travel slowly."

Similar roads were built in Gaul, in Great Britain, during the Roman occupation, and in Thrace by the Emperor Trajan.

With the decline of Rome, road-making shared the fate of the other mechanical arts, and for the time was forgotten. Good roads were unknown again in Europe until the middle of the eighteenth century. They were revived almost simultaneously in France and England, and soon afterward in the other chief countries of Europe.

Macadam Roads.—Among English-speaking races the perfection of modern roads is generally attributed to two Englishmen, Macadam and Telford, who rebuilt nearly all the English roads in the early part of this century. Telford was a distinguished engineer, while Macadam prided himself on being nothing but a road-maker. It is also generally believed that to Macadam is due the principle of using small angular fragments of clean stone, which, under traffic, unite into a solid mass. The distinctive feature of Telford's roads was a layer of irregular stone, from six to eight inches in size, carefully placed on the ground as a foundation for the smaller stones, technically called the road metal. The chief object of this foundation was to afford good drainage, and prevent the metal from being pushed into the ground in places where it was soft ; but Macadam always denied its utility or necessity, and engineers are still divided on this question. In regard to the size of the metal, Telford specified that the stones should be as nearly as possible uniform in size, the largest of which should pass, in its longest dimensions, through a ring two and one-half inches in diameter. Macadam preferred the test of weight, and insisted that no stone should weigh more than six ounces—which is the weight of a cube of one and one-half inches of hard, compact limestone. His overseers were provided with a small pair of scales and a 6-ounce weight, in order to test larger stones.

It is a fact, however, that the *correct principles of modern road-building* are not *due to* either Macadam or Telford, but to a French engineer, *Trésaguet*, who anticipated them in every detail by about thirty years. In a memoir prepared in 1775 Trésaguet advocated the small angular fragments of broken stone of Macadam, and the rough paving foundation of Telford. He built the high roads from Paris to Toulouse, and from Paris to the Spanish frontier. His views were adopted by all French engineers at the end of the last century, and it was in accordance with them that the Simplon and other great roads over the Alps, as well as the principal roads of France, were built under Napoleon.

The excellence of broken stone roads caused their universal adoption in the first half of this century, and in only two particulars have any improvements been made upon them to the present day. The first is in regard to the manner of *breaking the stone.* Macadam caused the stone to be broken by hand on the side of the road, the size and weight of the hammer being carefully specified. Now they are much more quickly and cheaply broken by machine. Two classes of stone-crushers have been devised for this purpose. The first, usually known as the Blake, consists essentially of a strong iron frame, near one end of which is a movable jaw of iron. By means of a toggle-joint and an eccentric this jaw is moved back and forward a slight distance from the frame. As the jaw recedes, the opening increases and the stone descends; as it approaches the frame the stone is crushed. The second class is known as the Gates, and consists of a solid mass of iron shaped somewhat like a bell, which is supported within an iron cone. By means of an eccentric shaft a rocking and rotary motion is given to the bell, so that each point of its surface is successively brought near to and removed from the surface of the cone, which causes the stone to descend and be crushed as before. These machines are driven by steam engines and are of various sizes, capable of crushing from ten to two hundred tons per day. By regulating the width of opening between the jaws, or within the cone, the size to which the stone can be broken is correspondingly regulated; and by the use of revolving screens with openings of various sizes, the stones of different sizes can be separated and delivered in separate piles of one-half inch, one inch, and two and one-half inches, etc.

The other improvement is in the *use of rollers* to consolidate the road and give a smooth, uniform surface, instead of allowing this work to be slowly and painfully performed by the vehicles using it. Horse-rollers were introduced about 1834, and steam-rollers about 1860. There was for some time a discussion as to the relative economy and merits of the two kinds of rollers, but this has now been settled in favor of the steam-rollers. Of steam-rollers there are three principal varieties, the first known as the Ballaison, or Gellerat, designed in France, the second known as the Aveling and Porter, designed in England, and the third known as the Lindelof, designed in America. Of these the Aveling and Porter is the best for Macadam roads, and the Lindelof for rolling plastic pavements.

Construction.—Macadam roads are now everywhere constructed on substantially the same principles. The ground is first cleared and leveled of the prescribed width, and, if necessary, excavated to the depth of the road-covering. All roots of trees, and soft and spongy spaces not affording a firm bearing, are removed and their places filled with good gravel or broken stone. The surface is then rolled with a heavy roller in order thoroughly to compact it. If the Telford foundation is used, it is placed on the rolled earth in the form of irregular stones, from six to eight inches in size, carefully placed in position and forming a rough pavement, on which the Macadam metal is placed. If the Telford foundation is not used, the metal is placed directly on the earth, in a uniform layer not exceeding six inches in depth. This is then thoroughly compacted by rolling with a heavy roller for several hours, until the metal will not yield under the roller. Another layer of broken stone of the same depth is then placed on

the first and compacted in the same manner. Finally a layer of from one
to two inches in depth of very fine broken stone or gravel, not exceeding
three-quarters of an inch in largest dimensions, is spread on the surface,
and this in turn is compacted by rolling. The road is then ready for use.
The rolling is greatly facilitated and the compactness of the road increased
by thoroughly sprinkling each layer in connection with the rolling. In

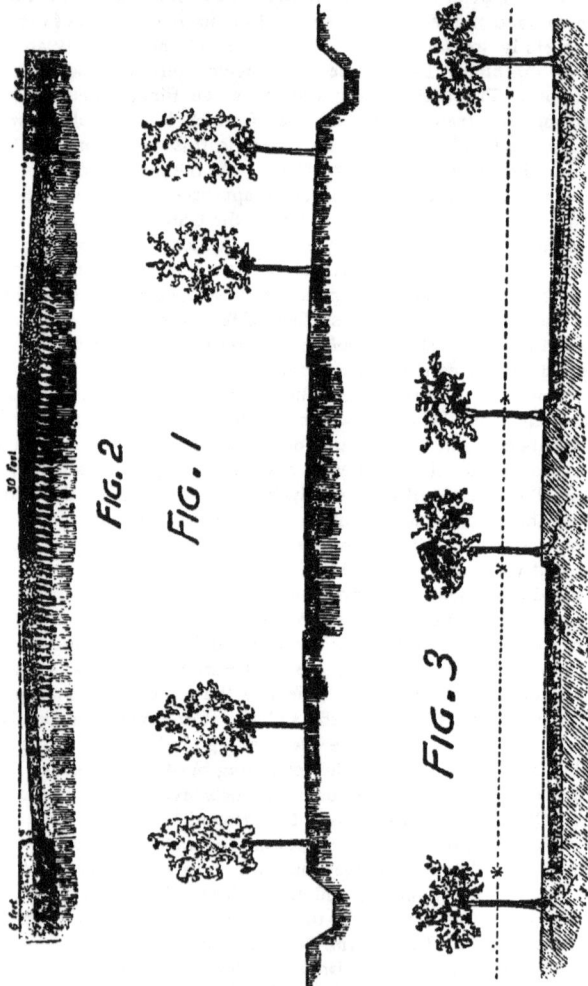

many cases the total thickness of the Macadam is only eight inches, instead
of twelve to thirteen inches, as above described.

Figure 1 is a section of the high-road or *chaussée* from Paris to Cher-
bourg. Figure 2 is a section of Telford's Holyhead Road ; and Figure 3 is a
section of the Western Boulevard of New York.

The cost of such roads depends chiefly on two factors, the price of labor and the price of broken stone. In addition to this is the cost of culverts and bridges, which must be provided for any road, whatever the road surface may be. The price of broken stone varies from 70 cents to $2 per ton, depending on the character of the stone and the distance which it has to be hauled. For a road 30 feet wide and 9 inches thick, about 5,500 tons are required for each mile in length. The cost of the road surface alone is about $12,000 per mile. The cost of embankment, excavation, culverts, drains, stone gutters, etc., may carry the cost up to $70,000 per mile.

These figures might even be increased in the case of roads traversing a mountainous district, where expensive embankments, cuttings in rock and earth, retaining walls, etc., would be necessary. The laying out of such roads calls for the same surveys and the same engineering skill as in the laying out of railways.

The shape or cross-section to be given to the road has been the subject of much discussion in the past. Roads which are much rounded in the centre shed the water very easily, but, on the other hand, they are very uncomfortable for vehicles. There has also been much dispute as to whether the cross shape of the road should be a curve, or should consist of two straight lines meeting at the centre. It is now generally conceded that the cross section should be a curve, and that the height of the road should be about one-sixtieth of its width—*i. e.*, in roads 30 feet wide the centre should be 6 inches higher than the sides, in roads 40 feet wide it should be 8 inches, and so on in proportion to its width.

The great cost of *macadam roads*, and the comparative lack of necessity for them in consequence of the enormous development of railways, has prevented their construction to any great extent *in America*. The National Road, which was intended to form the great highway across the Alleghanies from the Potomac to the Ohio, was begun under authority of Congress about sixty years ago, but it had only progressed a short distance beyond Cumberland, Md., when its construction was abandoned, in consequence of the building of railways for the same purpose. Macadamized roads have, therefore, been confined to city or suburban streets, and to a few of the older States in the East. Even the turnpikes, or toll-roads, originally built by corporations which made their profit by levying toll on each passing horse or vehicle, were macadamized only for a small portion of their width in the centre, leaving earth roads on each side. These latter were habitually used in summer, leaving the hard central portion, whose surface was seldom kept smooth for use during the rains and mud of winter.

With the exception of these few turnpike high-roads, American roads have been built of whatever material was nearest to hand. Frequently, if not generally, they were made by simply plowing a ditch on each side and throwing the earth into a mound in the centre. An improvement on this was to spread a layer of bank gravel containing a large proportion of clay over the road; and on the New England coast, where a rocky soil and clean gravel or beach shingle were everywhere available, these materials were used, and formed a comparatively hard and durable road surface. Through the swamps and clay soils of the South, where stone and gravel were not available, the corduroy road was much used. This consisted in felling trees, stripping the branches, and placing the trunks across the road; and it was probably the most inhuman device ever suggested as a means of communication. In central New York, and in some parts of the West, plank roads were at one time constructed, but their lack of durability caused this system to be soon abandoned.

Maintenance.—The condition of a road depends not only upon the manner in which it is constructed, but upon the manner in which it is maintained. The best of roads are being constantly worn by traffic, and if they are not quickly repaired whenever any defects appear, they are soon destroyed. Macadam's reputation was made not in building new roads, but in repairing old roads and *keeping them always in good order*. In order to accomplish this result incessant attention is necessary, so as to fill up any ruts or holes the moment they appear, and prevent them from being enlarged by travel and rain. The road thus gradually wears down, but always presents a uniform and smooth surface; and when its thickness is reduced to about five inches it is necessary to make general repairs by covering it with a new coating of stone. The amount of wear is proportional to the volume of traffic. On some of the heavily traveled macadam streets of London and Paris it has been as much as four inches in a year, but on high roads between cities it is often as low as one-half inch in a year.

Nowhere is the art of road-making and maintenance carried to such perfection as in France, where the necessity of constant supervision and prompt repairs is fully appreciated. Her roads have a length of about 200,000 miles, of which more than 120,000 miles are macadamized. They have cost nearly $600,000,000 for construction, and the sum of $18,000,000 (or about three per cent. of first cost) is annually spent for their maintenance. Until we are prepared to expend the necessary sums for solid construction and incessant maintenance, we cannot have good roads. With an area of 204,000 square miles, and a population of 38,000,000 inhabitants, France has about one mile of road to every square mile of territory, and to every 190 inhabitants; its roads have cost about $3,000 for each square mile, and about $18 for each inhabitant; their maintenance costs annually $90 for each square mile, and 48 cents for each inhabitant.

The State of New York has an area of 47,000 square miles, and a population of about 6,500,000, the number of inhabitants per square mile being about three-fourths the number in France. On the basis of area, in order that its roads should be equal to those of France, their length should be 46,000 miles; the first cost would be $138,000,000, and the annual cost of maintenance would be $4,140,000, or 64 cents for each inhabitant. The railroads of this State have cost nearly $900,000,000, and the annual expense of maintaining their road-beds is fully 6 per cent. of their first cost. It is evident that it would not be an impossible task to create a system of roads corresponding in excellence to the railroads whenever the necessity for them is fully recognized; and it would not be difficult to prove that the benefits derived in cheapening the cost of transportation to the railroads, of which the roads would act as feeders, would be more than an equivalent for the expense. Nor would the cost in reality be anything like the large sums above named, for many of the existing roads contain an abundance of stone, which could be taken up, broken and relaid, after the manner in which Macadam rebuilt the roads of England, the cost of which is stated in his memoir to have been as low as $600 per mile. Owing to the increase in the cost of labor since Macadam's time, the cost would now be about $2,500 per mile.

It is worth while to note the manner in which France maintains these splendid roads. The data is all available in the ninth volume of *Debauve's Manual for the Engineers of the Ponts et Chaussees.* While we have no such large body of trained engineers in the public service, and while our political organization does not permit the adoption of the system as a whole, yet there are many of its features which are not only applicable to us, but are essential to any satisfactory method of road maintenance.

The roads in each department in France are under the general supervision of the Prefect of the department, and their construction and repair are entrusted to the engineers of the ponts et chaussées. The necessary funds for this purpose are allotted to each department by the Minister of Public Works. The high roads are divided into two classes—national roads, running through two or more departments and connecting the chief cities, and departmental roads, connecting the principal cities within a single department. The local roads are divided into three classes—the important local roads, the ordinary local roads, and the by roads. Each road is thus classified according to its use and the traffic upon it, as determined by actual count at stated periods. The construction and the maintenance are varied according to the use and the volume of traffic. Some of the national roads are paved with stone blocks, like city streets, for long distances; others are macadamized; and the local roads are of gravel. The engineer-in-chief has charge of all the roads in the department; under him are engineers having charge of certain districts, and under each of these are superintendents and overseers, each in charge of a certain length of road, and with a certain force of laborers and the necessary materials for keeping the road always in good order. It is, in short, the same system of constant inspection, maintenance and repair which is in use on every one of our principal railroads, but which is never applied to our roads.

The fundamental principles of maintenance, as laid down in the *Manual of Instruction*, are only two in number—viz.: 1. The removal of the daily wear of the road, whether in the form of mud or dust; 2. The prompt replacement of this wear by new materials.

Each road is divided into sections called *cantons* ; on heavily traveled roads a canton may be only 100 yards long, on light roads it may be a mile; and to each canton there is a workman known as a *cantonnier*, who is responsible for the condition of the road in his canton. He lives in the immediate vicinity, and is obliged to be on the road from 5 A. M. to 7 P. M. in summer, and from sunrise to sunset in winter; he can rest two hours for his noonday meal, but with this exception he must be always at work between the hours above stated. He has the following tools—viz.: wheelbarrow, iron shovel, wooden shovel, pick, iron scraper, wooden scraper, broom, iron rake, crowbar, hammer and tape-line. His duties are: 1. To keep the gutters clear so that the water can run off freely. 2. To scrape off the mud in wet weather and sweep off the dust in dry weather, so as to keep his canton always clean. 3. To clean off the snow as far as possible, and break up the ice on the surface of the road and in the gutters during the winter. 4. To pick up all loose stones, break them, and pile them in regularly shaped piles on the side of the road, ready for use in repairing ruts and holes. 5. To keep the mile-posts in good order. 6. To take care of the trees bordering the road.

The six adjacent cantonniers form a squad called a *brigade*, which is under a foreman known as a *cantonnier-chef*, and forms the unit of working force. Several brigades are placed under the charge of a *conducteur*, or superintendent, who has charge of a section of forty or fifty miles of road, for the good order of which he is responsible, and every part of which he must inspect and report upon twice a month. Several sections are placed under an engineer, who has charge of all the roads in an *arrondissement*, or township, and must inspect every part of them once in three months. Finally, the engineer-in-chief has charge of all the roads in the department or province, eighty-seven of which constitute the territory of France.

During the winter, when the repairs are heavy, and whenever a general resurfacing of the road is undertaken, the regular cantonniers are assisted by auxiliary labor hired for the time being. The broken stone required for such work is furnished by contract,

It should be borne in mind that this is not a mere paper organization, or code of forgotten statutes, but an actual working system in full operation to-day. It is the result of 120 years of thought and labor devoted to an important subject by some of the best minds in France, and the result is the most superb system of roads to be found anywhere in the world. The cost is surprisingly small, considering what is accomplished. The actual cost per mile of maintaining the national roads (all macadamized) is given in Debauve's *Manual* for each of the eighty-seven departments. It varies from $60 to $500 per mile, with an average of $150, of which about half is for labor and half for materials. For maintaining less important roads the average cost per mile is as follows : departmental roads, $135 ; important local roads, $92 ; ordinary local roads, $57 ; by-roads, $42.

It would seem as if a somewhat analogous system might be devised in America, by which the roads in each State might be placed in charge of the State Engineer, the repairs in each county to be made by the county survey, or according to the instructions of the State Engineer, a uniform road tax of, say, five mills to be levied throughout the State, but the amount of taxes raised in each county to be expended in that county. With an estimated valuation in the State of New York of $1,200,000,000 (exclusive of city property) for the census of 1890, such a tax would yield $6,000,000 per annum for the roads of the entire State ; and this sum judiciously expended, according to well-digested plans and under competent supervision, would in a few years rebuild nearly all our important roads and maintain them in good order.* The present system of independent action or inaction by each Board of County Commissioners is known to be a complete failure. What it costs for the entire State cannot be stated, for there are no statistics on the subject. Possibly, if the statistics were available, it would be found that the total cost is fully as great as the sum above stated,

---

* A bill of a somewhat similar character is now pending in the Pennsylvania Legislature. It provides for a uniform road tax of seven and a half mills, to be raised in each county by a board of road commissioners, and expended under their direction by a county engineer, provided that not less than forty per cent. of the road tax shall be expended in macadamizing or other permanent improvement. The act further provides that the County Engineer shall be appointed by the Court of Common Pleas, that the roads shall be classified into highways, roads, and lanes, that the county shall be subdivided into districts, each in charge of a supervisor, and that he shall make plans and specifications for all work upon roads, and report at stated periods concerning the same. The only defect in the proposed plan is its failure to provide some central supervision for the entire State, so that the roads should be constructed and maintained on a uniform system in the different counties.

although the result is almost nothing. As for toll roads, and compulsory labor or a tax in lieu thereof, they are both out of date at the end of the nineteenth century.

In brief, then, the only system for good country roads, as shown by universal experience, is a bed of stone, broken into small angular fragments and thoroughly rolled, and maintained in good order by a small force of laborers, under proper organization and supervision, constantly at work summer and winter in cleaning off the road and repairing any defects the moment they appear ; to which must be added, from time to time, according to the amount of traffic and resulting wear, a general renewal of the road surface with the same materials.

Pavements.—City streets are simply roads of very heavy traffic, and the problem of paving is road-making designed to meet certain special conditions. A vast amount of ingenuity has been expended in the effort to make pavements that would be indestructible, but the effort is entirely futile. In the constant attrition of wheels and pavement something must be worn, and if the pavement is indestructible the vehicles will soon be destroyed. That pavement is the cheapest which affords the least wear to its own surface and to the vehicles combined. A good pavement should be durable, smooth, cleanly, as nearly noiseless as possible, and afford a good foot-hold for horses. Every form of construction material—iron, brick, stone,* and wood—has been tried in every conceivable manner of application during the last fifty years. The results of this large experience—as to cost and durability, ease of traction and cleanliness, noiselessness and slipperiness—have been carefully studied by French and English engineers, and to a certain, though much less extent by American engineers. While it cannot be said that the exact amount of wear in terms of the traffic has been fully determined, nor that the effect of different pavements upon the wear of vehicles and the cost of transportation has been mathematically demonstrated, yet certain fundamental principles are now generally admitted by all who have given careful thought to the matter—viz.: 1. A foundation is necessary, which constitutes the real pavement, and which is indestructible. 2. On this foundation a suitable wearing surface should be laid, and renewed from time to time. 3. The only suitable wearing surfaces are stone blocks, asphalt, and wood.

In reality these principles are only a development of the macadam road. Since the surface of macadam is worn too rapidly by heavy traffic, it must be protected with a renewable surface, leaving the body of broken stone as the permanent road-bed. As broken stone and cement mixed with sand will acquire in a few days the solidity that macadam will attain only after several months or years, the bed of macadam metal has naturally given place to a bed of concrete. This is universally conceded to be the proper foundation for any good city pavement. A thickness of six inches has been found by experience to be amply sufficient ; in cases of exceptionally heavy traffic it should be made of Portland cement, but in all other cases the ordinary American cements are quite strong enough.

In selecting the wearing surface, due regard should be had to the gradient, the traffic and the climate. Stone blocks are the most durable, but they are the most expensive, the most noisy, and offer the greatest

---

* Asphalt pavements are really a form of stone. The asphalt which they contain acts as a cement to hold together the limestone or sand which forms the body of the material, being from 85 per cent. to 95 per cent. of its weight.

resistance to traffic. Asphalt is the smoothest and cleanest, but it should not be used on grades of more than 4½ in 100. Wood is the least durable, but it is smooth and noiseless. Among different kinds of stone, sandstone and limestone are not sufficiently durable, and trap is so hard that it polishes and becomes very slippery under traffic. Hence, granite is considered the best stone to use. Of asphalt there are two varieties, the natural bituminous limestone of France and the artificial bituminous sandstone, made by mixing sand with pure asphalt, which is largely used in many American cities. Of wood many varieties, both hard and soft, have been used, but the best wooden pavements of London and Paris are made of Baltic fir.

Acting on these general principles, engineers have usually recommended granite blocks in streets of heavy traffic or steep grades, and asphalt or wood for residence streets. They have for many years condemned macadam as a city pavement on account of its lack of durability and because it cannot be kept clean, being always muddy when watered and dusty when dry. There are still large areas of macadam in the cities of Europe as well as of New England, but the expense of maintaining them is so great that they are being replaced as rapidly as possible. The wood pavement on a concrete foundation has not been popular in America on account of its lack of durability, the wood surface requiring renewal every five or six years; but it is largely used in London and Paris.

The granite block surface has been used more largely than any other, an undue importance having been attributed to the element of durability, regardless of all other qualities. But of late years the questions of noiselessness, cleanliness and ease of traction have been more fully considered, and the result has been a large development of smooth-surface pavements —i. e., asphalt and wood.

The limits of this article do not admit an exhaustive statement of the relative merits of the different kinds of road surfaces, but certain facts in relation to them may be briefly stated.

1. *As to Durability.*—The average life of granite blocks under heavy traffic in London is fifteen years, during which time the wear is about two inches, and the edges become so rounded that the pavement is as rough as cobblestones. They can then be taken up, redressed, and laid on streets of lighter traffic, where they will last for twenty years more, during which time the wear is another two inches. The blocks are then so worn that they have not a sufficient depth for a pavement surface, but can be sent to the crusher and broken up for concrete.

The average life of asphalt as laid in London and Paris is seventeen years. Cheapside was paved with asphalt in 1871, and after sustaining the heaviest traffic for seventeen years, it had worn down about one inch when it was resurfaced in 1878. The life of asphalt as laid in America is not yet fully determined. The first good asphalt pavement was laid on Pennsylvania Avenue, in Washington, in 1876, and it is reported that it will be resurfaced this year, after thirteen years' use. On the other hand, several streets laid in Washington in 1879 are in perfect order, and do not show any apparent diminution in thickness after ten years' use. It is probable that the average life will prove to be about fifteen years.

The average life of wood in London and Paris is from six to seven years, as shown by the experience of large numbers of streets.

2. *Ease of Traction.*—Elaborate experiments have been made by Morin, MacNeil, Rumford, Gordon, and others, to determine the force required to draw a given load on various surfaces. The results agree fairly well, and show that the force is from $\frac{1}{10}$ to $\frac{1}{118}$ of the load, depending on the surface. The result of all the experiments, as regards the relative value of different surfaces, is as follows :

Force required to draw one ton.

Iron...	10 pounds.
Asphalt..	15 "
Wood..	21 "
Best stone blocks...	33 "
Inferior stone blocks......	50 "
Average cobble-stone	90 "
Macadam.......	100 "
Earth..........	200 "

—*i. e.*, if a certain amount of force is necessary to draw one ton on iron rails on level ground, it will require additional force in the proportions above stated to draw the same load on the other surfaces. The importance of these facts is but little realized, and in the absence of accurate statistics as to the number of vehicles, the amount of tonnage, and the distance traveled in large cities, it is impossible accurately to demonstrate their effect ; but it can be approximately estimated. For instance, in the city of New York it is estimated that there are 12,000 trucks, carrying an average load of 1½ tons for 12 miles on each of 300 days in the year, at an average daily cost of $4 for each truck. The result is about 65,000,000 tons transported one mile in every year, at a total cost of $14,400,000, or at the rate of over 22 cents per ton-mile. The excessive nature of this charge is seen when it is remembered that the same goods are now carried by rail at $\frac{1}{10}$ of 1 cent per mile. On asphalt or wood pavements the same horses could transport a load three times as heavy as on the present rough stone pavements. If the saving in transportation is proportional to the load carried it would amount to nearly $10,000,000 per annum. It is safe to say that at least one-half of this amount would be saved by substituting smooth pavements for those now in use in New York.

3. *Cleanliness.*—The joints of a block pavement are receptacles for manure, urine, and all other street filth, and these joints can never be perfectly cleaned. The only remedy is to make the joints as small as possible. This is easily accomplished in wooden pavements where the blocks are sawed to exact shape. In stone pavements it is more difficult, but the dirt spaces are reduced to a minimum by filling the joints with gravel and hot tar, which renders them water-proof, and fills them up flush with the surface. When this work is carefully done with proper materials the filling is very durable, and remains in place for many years. It can easily be replaced when worn or broken by travel, by raking out the joints and refilling them. On asphalt pavements there are no joints, the surface being continuous, and for this reason the asphalt is the cleanest of all pavements.

There are two methods of cleaning streets. The cheapest, and the one most commonly used, is to clean the pavements (preferably at night, and after being sprinkled to lay the dust) by revolving brooms attached to carts. The broom is set at an angle, and revolved by cog-wheels connecting with the main wheels. The dirt is thus brushed into the gutter, where it is collected into piles and removed by carts. The other method consists in removing by hand every particle of manure or dirt the instant it is placed on the street. Boys or men are stationed on every block, and provided with a broom and dust-pan, or canvas bag, into which they brush the dirt, and deposit it in a receptacle placed on the sidewalk, whence it is removed every few hours by carts. Broadway between Seventeenth and Twenty-third Streets, and Fourteenth and Twenty-third Streets between Fifth and Sixth Avenues, in front of the large dry goods stores, are thus cleaned by private enterprise. In London this work is done at public expense by large numbers of boys between ten and fourteen years of age, whose dexterity in darting between the horses and wheels in the most crowded thoroughfares is quite remarkable. Iron boxes are placed on the curbstones at intervals of about 150 feet, into which they empty the contents of their dust-pans, and the boxes are in turn emptied into carts, and hauled away every few hours. The expense of this hand labor is much greater than a daily sweeping with machines, but it is very much more effective. Where the streets are not properly cleaned, sprinkling is resorted to in order to lay the dust, and the result is only to substitute one evil for another, for the sprinkling turns the dust into mud, and renders all pavements very slippery. Pavements of all kinds should be kept dry and perfectly clean.

4. *Noise.*—The asphalt and wood pavements have a great superiority over stone in the matter of noise. Wood is probably the most noiseless of all, as the only sound is a low rumbling, due to the wheels passing over the joints of the blocks. On asphalt there is a click of the horses' feet, but no noise from the wheels; this is hardly noticeable in summer, but is observed in winter, when the pavement becomes harder. But both the rumbling and the click are insignificant in comparison with the roar caused by the mingling of countless blows of iron shoes and wheel tires on stone blocks. Several eminent physicians have expressed the opinion that this incessant noise is the chief cause of the nervous diseases which have come to be such a feature of modern city life.

5. *Foot-hold.*—The opinion generally prevails that granite block pavements are less slippery than smooth pavements, but careful observations show that this is not the fact. The best foot-hold for a horse is afforded by the soft, dry soil of a race-track; next to this is a gravel road, and then macadam. But all of these surfaces are out of the question on heavily traveled streets. Exhaustive experiments, conducted by Colonel Haywood in London, showed that the relative proportion of falls of horses on different pavements, under the average conditions of weather, was as follows: On asphalt, 1; on granite, 1.47; on wood, 0. Similar observations in American cities established the following: On asphalt, 1; on granite, 1.40; on wood, 0.

It is thus evident that under ordinary conditions, such as exist on probably three hundred and fifty days in a year, the number of accidents to horses is much greater on stone pavements than on either asphalt or wood. In fact, the surface of granite, or of any stone sufficiently hard for use on streets, polishes under traffic and becomes very slippery. The only foot-hold afforded to the horses is in the joints between the blocks. On the other hand, under certain conditions, such as a light, dry snow, or a fine rain on a dirty surface, asphalt and wood are more slippery than stone. The surface of these materials is not so slippery even under these circumstances as the stone, but they have no joints to prevent the horse from completely losing his footing. The number of accidents on stone pavements, under the circumstances named, is very great, but not so great as on the smoother pavements. But when kept dry and clean, both asphalt and wood afford a perfectly good foot-hold for horses, if reasonable care is exercised in turning corners. It is the practice in London and Paris to sprinkle sand on the smooth pavements, when the conditions are unfavorable, and the same practice is followed daily under all circumstances by the street-car companies in New York on the stone pavements used by their horses.

6. *Cost.*—The prices of labor and materials differ so widely in various cities, and at times in the same city, the conditions of traffic and cleanliness are so different on different streets, and the character of the maintenance is so different, that it is extremely difficult to form comparative tables of cost of the different road surfaces that can be relied upon as accurate. It is evident at a glance that the cost of construction is only one factor in the problem, and not the most important one. The main question to be determined is the cost of construction and interest on the same added to cost of maintenance during a long term of years. And by maintenance is meant maintaining the surface in a condition practically as good as when first laid. Of course if stone blocks are placed upon a street and become full of ruts and depressions at the end of five years (as has happened on Broadway between Seventeenth and Twenty-third Streets), these defects will not become very much worse in another twenty or even thirty years, even if no repairs are made. The cost of maintenance under such circumstances would be very different from the figures obtained from the experience of Paris, London, Manchester, or Liverpool, where the surface is always kept in good order. In the following statements the comparison is made between different pavements laid in the best manner, with concrete foundations, and maintained at all times in a condition substantially as good as when first put down.

Cost.—In Law & Clark's *Treatise on Roads* are given a great number of tables of first cost and maintenance of pavements in English cities, and in Debauve's *Manual* and the notes of the engineers accompanying the annual budgets of Paris are given similar data in regard to French cities. They differ widely, according to varying circumstances; but all agree in showing the excessive cost of macadam under city traffic, which ranges from fifty cents to over two dollars per yard in every year. They also agree in the general statement that of pavements proper the granite is the cheapest, asphalt next, and wood the most expensive. The only scientific attempt to reduce these varying data to a uniform basis of cost for a given traffic is that made by Mr. Deacon in a paper read before the Institution of Civil Engineers in 1879, and since widely quoted. He had extensive statistics of cost and traffic in several English cities, and he reduced them to a uniform standard of 100,000 tons of traffic per year on each yard of width of the pavement. This is equivalent to about one hundred tons per day on each foot of width, and would be produced on a street forty feet wide by about 5,000 vehicles of an average weight, including load, of 1,800 pounds each. This is substantially the traffic of Fifth Avenue, in New York. He counted interest at three per cent., sinking-fund at fifty years, and maintenance at actual cost. His figures are as follows:

For 100,000 tons annual traffic per yard of width: granite blocks, 26 cents per yard per year; bituminous concrete, 45 cents; wood, 52 cents; macadam, 71 cents.

The "bituminous concrete" referred to in his tables was a mixture of coal-tar and gravel used in Liverpool. Data now available for asphalt streets would place their cost about midway been granite blocks and bituminous concrete—*i. e.*, about 35 cents.

In America, owing to the absence of accurate statistics on the cost of maintaining granite, it is difficult to give exact figures, but it is believed, from present experience, that the relative expense of the granite block pavement on Fifth Avenue and the asphalt pavement on Madison Avenue in fifty years will be as follows, per square yard per annum :

	Granite.	Asphalt.
Cost of construction	$4.60	$3.75
Interest at 3 per cent., and sinking-fund at 50 years......	.27	.22
Annual maintenance...	.04	.10
Three renewals of surface at $2.50 $\dfrac{2.50 \times 3}{50}$ —		.15
Four renewals of surface at $2 25 $\dfrac{2.25 \times 4}{50}$ —		.18
Total per year...............	.46	.50

In the above statement the cost of renewing granite surface is taken at $2 per yard, The actual cost, on the basis of the contract price for Fifth Avenue, would be $3.75, from which should be deducted the value of the old stones, estimated at $1.25, which would be available for redressing and use on the lighter streets.

There are no statistics in America as to the expense during a term of years of wood pavement on a concrete foundation. In Paris the current contracts run for eighteen years, and the entire cost, both of construction and maintenance, is paid in annual installments during that period of 89 cents per yard for each year. It is stipulated that the surface is to be renewed every six years.

In brief, then, of the three wearing surfaces granite block is the cheapest, but at the same time the noisiest, the most destructive to vehicles, and the most expensive for transportation. Asphalt is the smoothest and cleanest, and is slightly more expensive than granite; wood is the most noiseless, is quite smooth, but is the most expensive.

There are various other pavements, such as brick, wooden blocks on plank, macadam, etc., which are useful in villages and small towns, but are incapable of standing the traffic of large cities, and hence are not discussed here.

Street Railways.—One of the principal features of road-making within city limits is the construction of street railways. It is less than forty years since they were introduced, and their greatest development has taken place during the last fifteen years. Their use is not fully comprehended in Europe, which still adheres in the main to omnibuses and cabs as means of public transport. London alone has more than 15,000 cabs, and while these, in connection with smooth pavements, are a great luxury for the rich, they afford a very inadequate service for the poor. In America the street cars are universally employed in cities of all sizes, and with the final perfection of the mechanical motors in place of horses, they will have completely solved the problem of passenger traffic in cities. No idea of restricting their use would be tolerated for a moment. Their utility is so great that until recently the managers have been allowed to adopt the form of construction which they considered cheapest and most suitable for their cars, regardless of the rights of others who make a common use of the streets. Ordinarily they have built their roads with cross-ties, on which were placed longitudinal sleepers, and to these were spiked a rail with a broad base and without any vertical web or girder. The ordinary forms are the New York rail (Fig. 4) and the Philadelphia rail (Fig. 5). The

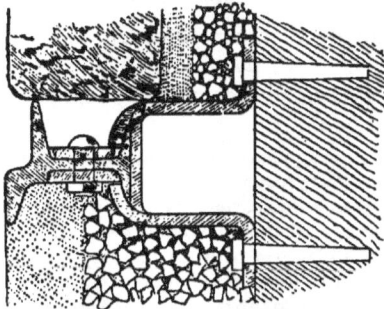

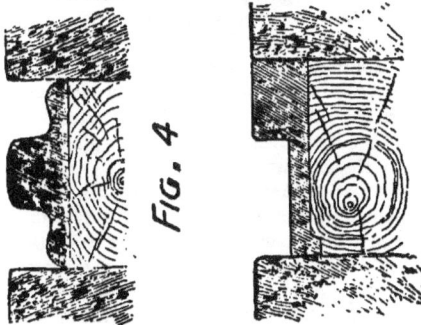

longitudinal sleepers quickly rot, the spikes pull out, leaving the rail loose, and the rail itself has a high shoulder or center-piece, which twists the wheels of every passing vehicle, and not infrequently breaks the axle. Within the last few years a more permanent construction has been devised by the use of a girder rail whose vertical web gives the necessary stiffness and does away with the necessity of the longitudinal sleeper. The rail rests

upon chairs securely fastened to the cross-ties, and no spikes are necessary. If the cross-ties are embedded in concrete, the construction is very
solid and permanent. The form of the head of the rail is, however, the
most important feature as regards the obstruction to street traffic. The
center-bearing rail is shown in Fig. 4, the side-bearing in Figs. 5 and 6, and
the flat grooved rail in Fig. 7. It needs only a glance at these sections to see
the great superiority of the last form, by which the obstruction to vehicles
is almost wholly removed. Wheels pass over it at any angle without catching in the groove. This form of rail has been in use in European cities for
many years, and it has lately been introduced in Washington and in Boston.
Objection is made to it by the street car companies on the ground that the
groove becomes filled with dirt in summer and with ice in winter ; but these
objections are of the same specious character as those so long used by elec-
trical companies in opposition to placing their wires underground, and do
not, in fact, constitute any practical difficulty. While the center-bearing
rail is undoubtedly the best for the interests of the street-car companies,
yet it is an intolerable nuisance for all others who use the streets. And this
nuisance is wholly unjustifiable when it can be avoided by the use of another
form of rail.

THE COMMON ROADS*

We are permitted to present here an abstract of Professor
Shaler's article in *Scribner's Magazine*. We have already alluded
editorially to the appearing of such articles in our popular magazines
as a hopeful sign of the times.

I propose in the following pages to take up the most important ways of
commerce—viz., the ordinary roads.

The United States, as a whole, remains less provided with such means
of communication than any other area of equal general culture in the world.

Toll Roads.—The difference between the road-making motive of the
New Englander, accustomed to the strong government of the town system,
and of those from the Virginia group of States, who are bred under the weak
communal system of the county organization, is perhaps better shown in
the matter of roadways than in any other feature of the social life. At the
present time in New England there is scarce a single toll road, except it be
where, as in the White Mountains, ways have been constructed for pleasure
traveling alone. On the other hand, in Kentucky and the other States
which have inherited their theories of life from Virginia, there are no good
ways which are really the property of the public.

There can be no question that the toll-road district of the United States
has before it a problem of a far more menacing nature than that afforded
by our railway system. The turnpikes should be made free.

Construction and Maintenance.—In most rural districts of the United
States the common roads are built and maintained in the most ignorant and
inefficient manner. In no other phase of public duties does the American
citizen appear to such disadvantage as in the construction of roads. The
voting part of the population is summoned each year to give one or two
days to working out the road tax. The busy people and those who are
forehanded may pay their assessment in money ; but the most of the population finds it more convenient to attend the annual road-making in person.

* xx, 333.

Theoretically the gangs of men are under the supervision of a road-master. More commonly some elder of the multitude is by common consent absolved from personal labor and made superintendent of operations.

Arriving on the ground long after the usual time of beginning work, the road-makers proceed to discuss the general question of road-making and other matters of public concern, until slow-acting conscience convinces them that they should be about their task. They then with much delibera-tion take the mud out of the roadside ditches, if, indeed the way is ditched at all, and plaster the same on the centre of the road. A plow is brought into requisition, which destroys the best part of the road, that which is partly grassed and bush-grown, and the soft mass is heaped up in the cen-tral parts of the way. The sloughs or cradle-holes are filled with this mate-rial, or perhaps a little brush may be cut and heaped in, making a very frail support for the wheels. An hour or two is consumed at noonday by lunch and a further discussion of public and private affairs. A little work is done in the afternoon, and at the end of the day the road-making is abandoned until the next year.

Cost.—If we take the misapplied expenses of our country ways ; if we count at the same time the mere social disadvantages which they bring to the people, it is probable that the sum of the road tax in this country is greater than that of our ordinary taxation. From some data which I have gathered in my personal experience with roads, I am inclined to think that even in New England the cost to the public arising from ineffective road-ways, as well as from the waste of money expended on them, amounts to not less than an average of ten dollars a year on each household. In this reckoning I have included the loss of time and of transporting power of vehicles, the wear and tear of wagons and carriages and the beasts which draw them. It is probable that the expenditure in this direction is greater than that which is incurred for schools or any other single element of public interest. I am inclined to think that it comes near the sum of all our State and Federal taxation together.

Road-Masters.—It would be greatly to the benefit of our system of road management if men could be thoroughly well educated for the duty of road-masters. A well-instructed expert could readily take charge of all the roads in an ordinary county. Bringing to bear the experience which has been gained in the art of road-making, he could greatly diminish the cost of construction and maintenance, and, without any addition to the present expenditure of labor, secure good and permanent ways. No other step seems so likely to advance this element in our policy so effectively as the institution of educated road-masters.

Clay Roads.—Clay roads can only be made into satisfactory ways by means of effective drainage. Deep side ditches are absolutely necessary for such roads, and the narrower the roadway the more effective will be this drainage work. It is a great mistake in such roads to have any more width than is imperatively necessary for the uses of the structure. If the ditches extend to a depth which would maintain the crown of the road two feet above the water level, and the roadway is of the least possible width, the problem of protection against mud is most easily solved.

To effect any satisfactory solution of the difficulties which beset such roads it is necessary, however, either to construct an artificial surface of timber or of stone, which is always a matter of great cost, or to mingle some binding materials with the clay. If gravelly materials, or, what is better, shingly waste, such as is often produced by frost action on slaty stones, can be commingled in the proportion of one-half with the clay, a firm road-bed can commonly be secured, provided the road is well ditched. This commingled gravel or other solid substance must extend at least for a foot below the surface in order to withstand any heavy carriages. In many cases an equally good result can be accomplished by covering the surface with repeated coatings of any shrubby vegetable matter. In Northern Minnesota I have seen the material known as "excelsior"—*i. e.*, strip-like shavings, cut by machinery from blocks of wood, serve admirably to prevent the motion of the clay, and I am of the opinion that it would, in clay countries where stone cannot readily be obtained, but where timber is plenty, be an admirable device to have a machine for making excelsior to be used as a road material. On the surface such woody matter rapidly decays, but when worked by the wheels into the clay it may last for several seasons. At no great cost the material might be saturated with creosote, and thus rendered much more resisting to decay. The finest branches of trees, the leaves of pines, even rushes, may serve the need, if they can be cheaply applied.

Sandy Roads.—In sandy countries the problem of road maintenance is very much simpler than it is in the regions underlaid by clay. The aim here should be to have the roadway as narrow and well defined as possible. In most cases it is desirable to have all the vehicles run in the same track, with an abundant growth of vegetation either side of the rut, for by this means the shearing of the sands is in a great measure avoided.

Macadamized Roads.—We come now to the type of roadway which should be constructed wherever the culture and condition of the country permit the expenditure of a considerable amount of money on its main carriageways—a construction commonly known as the turnpike. The essential feature of all such ways consists in the substitution of a compact mass of stony matter in place of one of ordinary soil. When properly built they so far spare the expenses of reconstruction as in many cases to be, in the long run, more economical than clay roads. All macadamized roads should be double ; on one side covered with stone, on the other having the ordinary foundation of the soil. If the soilway is kept in fair repair it will be preferred by sensible teamsters for more than half the year in all regions, and in many sections of the country for more than three-fourths of the time. In preparing such a way care should be taken, where possible, to remove the whole of the soil proper in order to secure a foundation on the subsoil, which, having escaped in the main the action of frost, as well as the disorganizing effect of roots, is firmer than the over soil. Founded on hard pan or subsoil, it is commonly possible to make a tolerably permanent road by placing upon the bed a layer of from eight to twelve inches of broken limestone, or, what is better, a less thickness of broken shale. The fragments should, if possible, in all cases be of a somewhat limy nature, for in such materials a process of natural cementation goes on whereby the mass soon becomes very firm. If possible, the inter-spaces should also be filled with powdered or finely broken limestone, not with sand, which usually does not add much to the firmness of the way. Where the underlying layer of soil is not very compact, it is in almost all cases advantageous to lay a floor of flat stones, like a loose pavement, and upon this to place the true macadam or broken bits of rock material.

Laying out a Road.—It is in the study of the road project, as regards location, that we find the greatest difficulty in our American system of rural ways. My much experience with this problem has convinced me that an educated road-master can do much for our people in bettering the placing of our roadways. I have frequently to traverse a road three miles in length which crosses two deep valleys, the declivities making it very difficult, if not impossible, to maintain the ways in fair condition. The difficulties might have been avoided, and a nearly horizontal way secured, by a slight deflection from the present line.

Vehicles.—The character of the vehicles which are used upon a road-way has a great influence upon its endurance to the beat of the wheels. With the same burden a two-wheeled cart does far more damage to the road than one of four wheels, and this because of the suddenness in the motion of the wheels and their irregular twisting movement in the track-way. The greatest defect in our American carriages is that for a given weight of carriage and burden the tires of the wheels are extremely narrow. It is true that on ill-conditioned and muddy roads a narrow wheel-tread is advantageous for the reason that the thick mud has a less extended hold when it wraps around the felloes and spokes ; but with this arrange-ment the interests of the roadway are sacrificed to the convenience of the individual who drives upon it. These narrow wheels, with tires often not more than an inch in diameter, cut like knives into the roadbed and so deepen the ruts. If we could require that no vehicle should have a tire less than an inch and a half in diameter, and that all springless carriages should have tires at least two inches in diameter, increasing in width with the burden, we would secure our ways against a considerable part of the evils from which they suffer.

IMPROVING COUNTRY ROADS.*

The subject of the improvement of county roads is one which is attracting wide-spread attention. The fact exists that our highways in the rural districts are, as a general rule, in an unsatisfactory condition, many of them being almost impassable without great discomfort during large por-tions of the year, while few are kept in a proper state of repair. They are far inferior to those throughout England and several other countries in Europe, while the public roads in the New England States are conspicuously better than ours.

This situation may have arisen because of our vast expanse of territory, the effort to maintain too many highways, the large expense involved in their proper care, inattention and indifference on the part of the people, or possibly from a defective system of highway laws. Whatever may be the real cause, it cannot be denied that our highways are deteriorating, and that some adequate remedy should be devised. It is apparent that they are not constructed with any special skill, little or no engineering talent being employed, and the matter of culverts and drainage being largely overlooked. It is asserted by some that the present system of permitting each freeholder to " work out " his road taxes operates badly, and, being a relic of old times, might be essentially modified with beneficial results. There seems to be a lack of official responsibility and competent supervision.

* xxi, 87. From the annual message of Gov. David B. Hill, of New York, January 7, 1890.

Neither commissioners of highways nor pathmasters are always selected for their especial fitness for the discharge of the important duties involved in the proper construction of suitable highways or in their care and maintenance. In some of the Western States these duties devolve upon a county civil engineer, who has entire charge, and whose functions are performed for the benefit of the whole county, freed from local influences and interests.

When highways are once properly built, the inexpensiveness of their proper maintenance is not generally understood ; but the principal difficulty in the past has arisen from their originally defective construction. Country highways running through a town should not be regarded as principally for the benefit of that town. That may be their primary object, but they serve a broader and more comprehensive purpose in affording means of communication for all the people, and should be viewed as a part of a great general system. The burdens imposed upon the taxpayers in the country are conceded to be onerous and various, and it cannot be reasonably expected that they will manifest an unusual interest or a large degree of pride in the maintenance of superb and expensive highways to an extent beyond the actual needs of the immediate neighborhood.

But the required improvement of our highways should not be considered in any narrow or selfish spirit, nor should local interests alone be consulted. The interests of the whole State are involved. The aspect of the question naturally presents the inquiry whether the State itself should not take the lead in the matter of so pressing and desirable an improvement.

It has been suggested that the State should proceed to construct through every county two highways, running in different directions, and intersecting each other in about the center of the county, such roads to form a part of a complete general system, those in each county to connect with those of adjoining counties, and to be known everywhere as State roads, constructed, cared for, and maintained at the expense of the State at large, under the direction and supervision of the State Engineer and Surveyor or other competent authority to be designated. This system when once completed, would enable a person to start from New York City, Albany, or any other point, on foot or in carriage, and visit every county in the State without once leaving the State roads, thus insuring comfort, convenience, pleasure and speed. These roads should be macadamized or constructed of crushed stone or other suitable material, with proper culverts, good bridges, adequate drainage, watering troughs, and sign boards, so as to compare favorably with the best country roads in other countries ; and existing highways could be utilized for this purpose so far as feasible.

EDUCATION IN ROAD-MAKING.*

ELIZABETH, N. J., January 16, 1889.

SIR : You have in past volumes discussed quite fully the subject of road-making. Now that the matter of better roads is being agitated in a number of localities, and there is a probability of radical changes in the road laws of some of the States, it would seem that a presentation in your columns of such information as you can gather as to the latest enactments of the several States might be of great value to the public.

F. COLLINGWOOD.

We thank Mr. Collingwood for his suggestion. It is, as he says, in line of what we have done much of in the past and shall do even more in the future, as there is nothing more important, from an engineering point of view, to the comfort and commerce of a country than good highways. We are glad to see that more interest is being taken in this matter. At a recent meeting of the Board of Trade of Elizabeth, N. J., in discussing the appointment of a committee to adopt means for the improvement of country roads, Mr. Collingwood said:

What the country needs is éducation in road-making. The amount of waste by reason of bad roads is enormous. We can't carry the loads we ought to. If with good roads we can carry twice the loads we can on poor ones then they save half the labor of hauling.

In the far West, twenty years ago, first-class roads were built in many troublesome places as a measure of true economy, while here, ten miles from New York, we have roads so horrible that teams are hardly able to haul empty wagons over them.

He pointed out also the cause of the evil and its remedy, saying that not one in ten thousand of the roadmasters knew how to build good roads, and that the whole work should be done under the direction of competent engineers.

It is obvious that the question of one organization and control of the road-making force is primary and fundamental, and that until we have a system of administration for our highways equal to, though not necessarily the same as that of France, we cannot expect to have as good roads as the French rejoice in. It is then at present more a matter of legislation than of specifications, and it is to be hoped that our law-makers may see its importance in season. Meetings of citizens, such as that held in Elizabeth, may do much to clarify the legislative vision.

COLLINGWOOD ON ROAD-MAKING.*

The following letter addressed by F. Collingwood to the Elizabeth, (N. J.) *Daily Journal* refers to the meeting of the Board of Trade of that city to consider the improvement of the county roads, to which brief reference was made in our last issue. Mr. Collingwood has given special attention to the subject of road-making, and his letter can be read with profit by any one who has to do with such work.

ELIZABETH, N. J., January 17, 1889.

MR. EDITOR: My purely chance attendance at the meeting of the Board of Trade seems to have given me a notoriety that I do not covet. I cannot, however, pass by a statement made mention of by you without explaining

---

* xix, 120.

myself a little more fully. I refer to the idea held by not a few that a thin covering of say three inches of road metal (broken stone) will make a road.

The philosophy of road making is very simple, and is identically of the same character as the obtaining of a suitable foundation for a building. A tyro in engineering soon learns that if he would found on a yielding soil, he must spread the base of his foundation if he would prevent immoderate settlement, and the softer the earth strata, the wider or larger must be the base.

Now in the case of a wagon wheel on a roadbed saturated with water, the material directly under the wheel is both compressed and moved to one side and the wheel sinks until the resistance beneath is equal to the load upon it.

If we place a comparatively unyielding mass (such as macadam stone) of sufficient thickness and sufficiently large area, upon this surface, side movement will be prevented and the compression reduced practically to zero. The pressure from the wheel will be transmitted to the soil through a cone of the underlying mass, which will have its apex at the wheel, and its base at the surface of the soil. The diameter of the base of the cone will increase as the depth increases, but the area of the base, which is the import- ant element in its bearing power, will increase in proportion to this depth multiplied by itself, or as mathematicians say, "as the square of the depth." If then the base for one inch of depth would bear some given load, three inches of depth would bear nine times as much; nine inches of depth would bear eighty-one times, and twelve inches one hundred and forty-four times as much. The facts shown by experience are that while three inches depth would last during good weather, it will cut through when a heavily loaded wagon passes after rain of any magnitude, and the road will soon be full of deep ruts.

The latest practice among engineers recognizes the fact that well- drained roads of sharp descent which can hardly get much softened by water, do not need as thick covering as flat roads from which the water does not flow so quickly. The thinnest covering I have ever heard mentioned, however, was six inches, and that not on roads having very heavy traffic. For country macadam roads the limits would be between six inches and twelve inches, but the judgment of the engineer is needed to decide each case. For cities, where the loads are heavy, even more is desirable.

The explanation made above shows at once one other fact—viz.: in order that the road covering may act as described, there must be a proper bond- ing and interlocking of the various layers of stone, which can only be secured by careful work, under intelligent and competent inspection. A large stone in the base of the pavement furnishes a smooth surface on its top for the covering stone to slide upon, and will be the first place to invite a break in the surface. There are such to be seen near our own doors.

I have said nothing about the very important matters of drainage, or rolling, which are also essentials.

Before such work is contracted for all matters of drainage, grades, etc., should be arranged for and full and explicit specification prepared by a com- petent engineer, and the work must be watched through all its stages to see that the specifications are carried out.

I think we shall go to a useless expense, if the work to be done is left to the unscientific methods which have prevailed for the last hundred years. This should be provided against in the laws that it is proposed to ask for,

F. COLLINGWOOD.

ROAD SAVING—VEHICLES AND TRACTION.*

SYRACUSE, N. Y., January 31, 1889.

SIR: In your issue of January 26 I note "Road Making," by F. Collingwood. Allow me to make a suggestion as to road saving. Have the State pass a law making it imperative on all two-horse farm wagons, and all others that carry over one ton, to use 4-inch width of tires, and to have the hind wheels set eight inches further apart than the forward ones, and that all carriages also shall have the hind wheels the width of their tires further apart than the forward ones. With this law in force road making will be reduced 66 per cent. When wagons are thus made they will not track and you have no ruts. No one ever saw ruts around the corners of any country road. Each wheel rolls down and does not cut the road ; and it is absolutely untrue that the wagon draws harder ; it is absolutely true that it draws easier.   W. A. SWEET.

WASTEFUL ROAD MAKING.†

The Minneapolis Society of Civil Engineers held a meeting on March 6, at which George E. Crary read a paper on "Permanent Improvements on Highways." He made the point that $800,000 was being practically thrown away every year in the State by the present method of road-making. He would have the State portioned into three or five districts, with a competent superintendent in charge of each, and responsible for the condition of the roads under his jurisdiction.

CLASSIFICATION AND MAINTENANCE OF ROADS. ‡

The Engineers' Society of Western Pennsylvania, at its last meeting in Pittsburg, Pa., adopted the report of the Committee on Improvement of the Country Roads. It calls for a division of the roads of the State into highways, roads and lanes, defining highways as what are commonly called county roads, leading entirely through the county, or from one important centre to another ; roads, as what are now commonly called township roads. Lanes were defined as private roads, leading only to small hamlets or private properties. The draft of the law to be submitted to the Pennsylvania State Legislature provides for road commissioners who shall appoint county engineers to have charge and control of the improvement of the county roads, in shortening distances, lowering grades and furnishing a good road surface. The committee was instructed to use every means to have this bill adopted by the State Legislature. A full report was read upon the advantages of good country roads to the community in general.

---

* xix, 134.   † xix, 207.   ‡ xix, 236.

ROAD IMPROVEMENT IN THE SOUTH.*

Interest in the improvement of the country's roads is not confined to any one section of the country. *Bradstreet's* says :

We have seen what attention the subject has attracted in the North and East ; it is apparently attracting no less attention in the South. For example, a proposition to hold a congress to consider the best means for improving the roads of the State has been brought forward in Georgia, and, judging from the comments of the State press, the proposal is being favorably received. The congress will be held in Atlanta in May next, and it is hoped that a practical scheme of reform in road administration will be adopted. The question is one which addresses itself with peculiar force to the commercial and agricultural interests of the country. The merchant and the agriculturist alike find their account in a ready exchange of products, an easy marketing of crops, and it need not be said that such a desirable consummation may be very much retarded by bad roads. It is plain, from indications such as that to which attention has just been directed, that interest in the question of road improvement has by no means culminated yet.

SPECIFICATIONS FOR GRAVEL ROADWAY.†

The specifications for a gravel roadway from which the following extract is made, were prepared by Dr. D. W. Mead, City Engineer of Rockford, Ill., who finds that a gravel roadway gives much better satisfaction in that town than macadam made with the local stone.

A half mile improved two years ago under these specifications he reports has given good results.

Extracts from specifications :

" The streets shall be excavated or filled by the city to the grade given by the City Engineer, and the surface prepared by said city for receiving the rubble.

"On the surface so prepared a layer of large-sized rubble shall be thrown. This course shall be arranged in a close and compact form, and the interstices of the larger stones shall be filled in with sound stone chippings, all to the satisfaction of the Commissioner of Streets. The stones for the foundation course shall be generally not less than five inches in any dimension.

" Over the foundation course a layer of rubble broken as nearly as possible to a cubical form, not less than one and one-fourth inches, nor more than two and one-half inches in any dimension shall be placed. The rubble shall * * * * be well compacted and rolled with the city roller, to the satisfaction of said Commissioner of Streets.

"Over the rubble a layer of screened gravel mixed with clay shall be placed. The layer shall be when finished at least two inches in thickness at the center and one inch in thickness at the side. This layer shall be well rolled and compacted with the street-roller to the satisfaction of said Commissioners.

" Both before and while rolling the rubble and gravel layers said layers shall be flooded or sprinkled."

---

* xix, 184.   † xviii, 321.

ROAD LEGISLATION.*

The necessity for better roads in America has been many times shown in these columns, and a welcome addition to the literature of the subject comes in a carefully prepared paper by Professor Jeremiah W. Jenks, of Knox College, Galesburg, Ill., upon road legislation, published as a tract by the American Economic Association. The author's personal observation has been in a field where the need of reform is especially great, for the soil of the West is at once generally far from being the best for roads and yet calculated to produce such crops of grain that good highways are very important. The advantages of proper laying out by engineers and better construction, such as might be provided for by law, are urged in detail, and some estimates of the money advantage to be gained can hardly fail to surprise those who have given the subject no particular thought. Thus, it is suggested that proper roads would increase the value of two-thirds of the good farms of the Mississippi Valley $10 an acre, and that, by an estimate which no well-informed man would question, the State of Illinois alone would gain no less than $160,000,000.

Some interesting suggestions as to what should be sought by laws are given, but the chief value of the paper will perhaps be found in an elaborate compilation of laws as they now are. The French system is explained in detail, and those of some other European countries briefly, and then the laws of each of the States are given in tabular form, so that a comparison may be made at a glance. It is enough to say that since hap-hazard road-making, regardless of economical or scientific methods, is so general in this country, it is also true that the laws make no intelligent provision for anything better. Back of inadequacy of the law is, of course, general igno rance of the subject, and we are glad to see leading popular periodicals begin work toward dispelling this. An article in *Harper's Weekly*, by Captain F. V. Green, has already been referred to in these columns, and later there has appeared in *Scribner's Magazine* a paper by N. S. Shaler, which clearly indicates the wastefulness of the present method, and makes some plain and simple suggestions for reform.

GOOD ROADS, THEIR ECONOMIC VALUE AND HOW TO MAKE THEM.†

It is an injury amounting to a calamity to the farmers—the impassable state of the country roads after a day's heavy rain. It is then, when his fields are too wet to work in, that the farmer can, without inconvenience, bring his produce to market. But, unfor-

---

*Ed. xx, 297. †xx, 187. By J. F. Pope, C. E., in the Austin (Tex.) *Statesman*.

tunately, the same rain-fall that prevents work in the fields, equally, from its effect on the roads, precludes his access to market, as the hauling of even an empty wagon is then a wearisome labor to driver and team. The farmer must, therefore, perforce sit idle, unable either to cultivate his land or market his produce ; the latter, perhaps, spoiling on his farm, instead of being profitably sold to his city customers. At length, when a few days of fine weather has dried his land, and rendered the space between two fences called a country road passable, he has to determine which is to him the least harmful, to neglect his land or his marketing. For, it is just when his time should be divided between plowing his land and marketing his produce that the greatest and most frequent rain-falls occur ; and while he is grateful to Providence for the beneficial showers that enrich and invigorate his land, he must think with bitterness of the laws and administration that, by its negligence or inaction, deprives him so unjustly of the use he could then make of his time in the profitable sale of his produce. While the impassable state of the country roads is a calamity to the farmer, it is also a serious injury to the residents of the cities, as, instead of getting their daily supply of vegetables and farm produce from the local farmers and market gardeners, they have to depend on importation from other States to supplement the deficiency due to the impassable state of the country roads ; and for these imported articles they must necessarily pay a higher rate to cover the extra cost and risk of transportation. This in a great measure accounts for the vast importation of country produce into the city from other States. The grocer, or city purveyor of these articles, must always have a supply on hand to meet the demands of his customers, and when, from whatever cause, he cannot depend on the regularity of the local supply, he must otherwise arrange, at even an enhanced cost, to meet the daily demand. He must, therefore, keep up his outer State trade connection, even when it would be to his temporary advantage to purchase from the local producers. Thus, while all farm and garden products are at an extravagant price in the stores, the local farmers and market gardeners cannot always get sale for their produce. The money that should circulate locally to the advantage of the general community is remitted to benefit the citizens of other States. Thus, the interests of both the urban and rural population of the country are seriously affected by the want of means of communication between town and country. In fine, it is a lamentable state of governmental administration when a day's rain paralyzes the internal trade of a State and obliges the citizens of an agricultural country to depend for their food supplies to importation from other States. But, even under the most advantageous conditions of the weather, the farmer is still seriously injured by the rough, crude state of the country roads, badly aligned, badly graded, and full of holes and ruts. Thus, in bad weather the farmer is virtually cut off from town, and in good weather, from the faulty construction and ill-kept conditions of the roads, the value of his time and team are reduced one-half, as on a hard, smooth surface his team will draw a greater load in less time than on the ordinary ill-constructed country road. Moreover, not only is the power of his team decreased, but there is also the great loss by wear and tear on his harness and wagon, due to the plunging and jerking through the holes and ruts. Again, there is, as a rule, no well devised plan or order in the administration of the road fund and the construction and maintenance of the county roads. By fits and starts, without system or method, attempts are made at repairs and improvements, and the money is then lavishly and unintelligently expended. The work is entrusted to an amateur, or theorist, who has as much practical knowledge of road-making as he has of anything else of which he is ignorant, and, more frequently than not, a few months after such expenditure the state of that improved road resembles the state of the relapsed sinner. These remarks apply generally to road management throughout the State. It should be more generally known that there is as much science required in road-making as in any other engineering work.

Macadam Roads in India.—The most perfect system of road-making in the world, perfect not only in the excellent condition of the roads, but also in the economy of their construction, is that on which, with other engineering works, I was engaged during my sixteen years' service under the government of British India. The making of cheap and efficient roads is as familiar to me as plowing to an old farmer. I have during my sixteen years' service either constructed or kept in repair, taking one year with another, over 1,000 miles of macadamized roads. The system and method in vogue there is the result of the experience of nearly 100 years ; and as there have been of late so many amateur and theoretical suggestions regarding the construction of roads I will, as briefly as possible, describe the cheap and efficient method of road-making in India, which is equally applicable to Texas, as there is as much variety of soil in the former country as in the latter.

Alignment.—Having determined the points or places to be connected by a road, the country along the proposed route is reconnoitered and the road then aligned. Great discernment is necessary in this alignment, so as to combine directness of route with the avoidance of swamps, hills, etc., or where these have to be crossed to do so as advantageously as circumstances will permit, and to

secure a firm road-bed and easy gradients. There is not much to
be said on this head with reference to the Texas county roads.
Unfortunately these were aligned in the manner I have described,
and while, for obvious reasons, it would be hardly practicable to
re-align all the county roads, still great improvement might be
effected by a slight deviation here and there to shorten a line, avoid
a bad crossing of a creek, a hill, a swamp, or a stretch of deep sandy
soil. The alignment being made, the first requisite of a good road
is efficient drainage.

Drainage.—Drainage does not consist in digging ditches four
to six feet deep along a road. A scientific scratch in the proper
place may be more effectual than a ditch four feet deep run by the
eye of an amateur road-maker. I will again repeat that the first
and absolutely necessary requirement of a good road is efficient
drainage. The road must not be allowed to get sodden with water.
Fortunately, drainage costs but little when devised by a competent
engineer. As a rule, proper drainage can be effected by trenches,
the earth from which would just suffice to raise the roadbed from six
inches to a foot.

Width of Road.—The land taken up for a county road should
not be less than 60 feet in width, to allow for drainage, trenches,
etc., and the width of the made roadway should be 26 feet, raised not
less than 6 inches above the ground. This 26 feet should be made
up as follows : Eighteen feet of a hard central surface, metaled with
broken stone or gravel, with 4 feet of an unmetaled earthen berm on
each side of the hard macadamized central surface. The side slopes
of embanked roads to be, as a general rule, $1\frac{1}{4}$ feet horizontal to 1
foot vertical; in cuttings the slopes to be according to the tenacity or
hardness of the material cut through, varying from $1\frac{1}{4}$ to 1 in ordinary soil to almost a perpendicular rock. The width of roadway
given above—18 feet and 4 feet side berm—is wide enough for all
practicable purposes. More than this is a mere waste of money. A
roadway of greater width may be necessary in Bourbon County,
Ky., but not in sober, temperance-loving Texas. It would be better
to expend the extra money in keeping the width I mention in good
repair, than to waste it on an unnecessary width of 25 or more feet.
I write from the experience of the commerce of a country having a
population to the square mile sixteen times that of Texas, and where
the Government deems it as much a part of its duty to aid and facilitate
the internal commerce of the country by good roads, as it deems it
a part of its functions to protect the personal rights and property
of its citizens by the just administration of law and justice.

Material.—As I have said, the width of the macadamized or
prepared hard surface of the road should be 18 feet, metaled with

either broken stone or gravel, according to the local supply. If of broken stone, it should be the hardest rock locally procurable, broken to pass through an inch and a-half diameter ring. The thickness of the layer of broken stone or gravel should be, as a general rule, 4½ inches at the edges, and 6½ inches at the centre.* If the material used be broken stone, it must be laid in one layer but if of gravel, it must be laid in two layers, the first layer being partly consolidated before the second is laid on, the consolidation to be effected by iron rollers drawn by horses.

Rollers.—In exceptional cases steam rollers might be used, but they would not be economical away from the immediate vicinity of a city; the iron rollers to be 6½ feet long, weighing 3 tons, and so constructed as to weigh nearly double that weight when filled with earth ; the consolidation to be in part done by the unfilled roller, and completed by it when filled. The surface of the completed road should have a camber from the centre to the sides of 1 inch in 4½ feet.† There is no necessity whatever for the three layers of 6 inches each of flat rock, crushed stone and gravel mentioned in the issue of the *Statesman* of January 1.

Cost.—The cost of such three-layer construction is there said to be $2,640 per mile. No contractor in the State would undertake such work under $9,000 per mile. The cost is an easy calculation, and I have made it with the result that it would not be under $9,000. However, no such expensive construction is necessary. My appeal is to the thousands and thousands of miles of the finest country roads in the world constructed as I have described. The average cost of such road would be $2,100 per mile, a minimum of $1,600 and maximum $2,600. It is, however, not enough to construct a road well in the first instance ; it is also necessary to keep it in repair. The repairs will, of course, vary with the amount of traffic. The cost per mile per annum will vary from $100 for the more remote to as much as $300 for the last couple of miles leading into the city. The partial renewal of the surface will also be necessary every four or five years.

---

* These thicknesses of 4½ to 6½ inches would be permissible only on very firm, or, at least, fairly hard ground. On ordinary ground, especially when softened by a thaw after frost, a thickness of not less than 9 inches would be required to prevent cutting through by the wheels of heavily loaded vehicles.—ED. RECORD.

† This gives a rise of but 2 inches in the centre, which might be sufficient as a final condition on exceptionally good ground after all settlement was over. In ordinary conditions the rise should not be less than 6 inches at first and more if the ground underneath is at all soft. Less than this will allow water to lie in the centre after a very little wear, which is detrimental to durability, and often causes considerable mud to collect.—ED. RECORD.

MACADAM AND GRAVEL ROADS IN CHELSEA, MASS.*

From the annual report of the City Engineer, Mr. William E. McClintock, of Chelsea, for the past year, we gather some interesting items. The street pavement that seems to have been decided upon is macadam, but the stone in the neighborhood is a rotten schisty slate, and unfit for the surface. A crusher has therefore been fitted up, with a specially designed elevator, by which the crushed stone is raised about 15 feet for screening, thus saving considerable manual labor and cheapening the process. The present cost is about $1.50 a cubic yard and 50 cents for hauling. To obtain a suitable rock from Malden, consisting of trap, porphyry, etc., will cost about $1.25 additional, or $3.25 per yard, aside for payment for the quarry. He recommends the soft material for foundation and the imported material for metaling.

Gravel has been used, but he says of it: "We have to let it be worn down by team travel, which may be an economy to the city, but is a great loss in wear and tear to the traveling public, both mentally and morally. This gravel is unsurpassed for paving purposes, but for street purposes *it is as well to use beans.*" In respect to repairs of streets, he says the cost is variously estimated in other cities at from 3 cents to 25 cents per square yard per year, and it is bound to cost as much in Chelsea as elsewhere, the lower limit being the least at which he would place it for the very least repairs that should be made, or a cost for the city of $35,000 annually, and $7,000 was the amount actually spent. Sweepings to the amount of 1,168 loads were removed from the streets during the year.

THE BRIDGEPORT MACADAM ROADS.—A GOSPEL OF THINNESS.†

Two systems of making roads of broken stone have for some time been employed. The first, practiced by Macadam, calls for a light coating of stone broken to a uniform size, on a substratum of good earth. The second, introduced by Telford, calls for a foundation of hand-placed and packed stones, over which the broken stone or road metal, is spread and compacted in the same manner as in the case of macadam.

Unfortunately for that portion of our people interested, in other modes of circulation than those afforded by our railroads, the able engineers who projected and built the roads in the Central Park, of New York City, had not, as a general principle, been in charge of wagon roads, and so turned to books of information. To a mathematical mind, nothing could be clearer than the proposition that the road metal transmits the pressure of the wheels to an area propor-

---

* xv, 545.   † xxi, 53.   By Edward P. North.

tional to the square of its depth. From this the other contention of Telford follows, naturally, that the desired depth could be obtained economically and advantageously by the Telford foundation.

The result of the study of authorities was, that Central Park has roads of indefinite thickness, reposing on a bed of thoroughly well-drained earth, at a cost that is prohibitory to any great extension of such roads. But they are built according to authority, and they have been and still are standards for construction, as they are well maintained. Authorities, however, are reputed to be often wrong, as was exemplified on Fifth Avenue, where an equally well-built Telford road, in lack of maintenance, was for some years a nuisance to those who rode on it and a terror to those who lived by it, and many, if not most, of the engineers trained in the school of the Central Park roads, show by their practice a growing belief that a cheaper road is, on the whole, better.

Possibly no more radical departure from the teachings of Telford has anywhere been made than by Mr. B. D. Pierce, the Street Commissioner of Bridgeport, Conn., who was at one time a foreman on the construction of the Central Park roads. When he took charge of the streets and roads of that city they were notoriously bad, excepting a little pavement, and some Telford road laid down at an expense of about $1.50 per square yard. Mr. Pierce started with the intention of building 4-inch macadam roads; an intention adhered to except in two instances to be mentioned hereafter, and last March he had over 45 miles of good macadam roads from 18 to 20 feet wide. Of this, 38.9 miles had cost, including some grading with the maintenance and repairs following extensive renewals of an old water-pipe service, $115,297.25, or only $56\frac{14}{100}$ cents per lineal foot.

During the present season somewhat over 5 miles more road have been constructed in Bridgeport, and the example of cheap and good roads has spread into neighboring towns, so that in addition to the 60 odd miles of good roads in Bridgeport there are about 14 miles of equally good roads in the immediate vicinity, making between 70 and 80 miles on which an ordinary team can haul 3 tons of net load; or Sunol, if it was not for pounding her heels, might try her speed with safety and comfort to her driver.

Mr. Pierce accepted his position as Street Commissioner in 1885 with the understanding that 4-inch roads would be built, and would prove not only cheaper but at least as good as the Telford roads then existing in the city. So far, judging from their appearance, and from conversation with two of the Board of Public Works, as well as with Mr. Pierce, they are better than these Telford roads, and under equally heavy traffic are at least as easily and cheaply maintained. Two theories are advanced in explanation of this. One that the earth works up, into and through the voids in the Telford foundation. The other that the road metal works down into these same voids. To these I desire to add a third; that the Telford foundation acts as an anvil, between which and the wheels of passing vehicles the road metal is pounded to pieces. This, it is well known, is analogous to the effect of the stone blocks used on the introduction of railroads here; blocks which were all taken out probably as early as 1846 to 1848.

In building these roads, the ground is graded and regulated with a gutter 18 inches deep on each side; the soil is then thoroughly rolled with a 15-ton roller, and the stone spread on the surface so prepared. Three varieties of soil are met with in Bridgeport: a fine "dead" sand, which sometimes cannot be rolled on account of its pushing before the roller, without covering it with coarse, broken stone; this expedient, acting as a pavement, prevents movement; loam, a very fine gravel, and a hard pan with mica disseminated through it. Underdraining has in no case been resorted to, the 18-inch gutters being depended on for drainage. After the broken trap-rock is rolled to a bearing, screenings are added as a binder and the road metal is well and thoroughly filled with them, the whole being rolled until the water flushes on the surface. A strong silicious sand is sometimes used, in part, in place of screenings, and when, in dry weather, the road commences to break up or "ravel," out of easy access by watering carts, sand is spread over the spot, which quickly consolidates the road. No loam or clay is used as a binder or filler in the construction of the roads nor in their repair, except when the surface over a ditch is to be replaced and it is too small a patch to justify bringing the roller; then the broken trap is laid down after being mixed with the proper quantity of screenings and the whole covered with loam; the traffic consolidates it in a short time.

Reference was made to two cases in which the thickness of four inches was increased. One of these, a road near the winter quarters of "the greatest show on earth," was built on sand which, on being thoroughly wet, became quicksand. In dry weather the road stood without injury a net load of 42 gross tons, drawn by 24 horses, but in a wet time the road would wave under net loads of 2½ gross tons, and at last broke through in places; three inches of stone were added, since which the road has stood without further care. The standard loads on this road, hauled by the teams of a copper-works, varied before it was stoned, between 12 and 20 gross hundredweights; 50 hundredweights are now regularly hauled. The other instance of reinforcing the depth occurred over an undrained pocket in rock. The road in this instance also waved under heavy loads for some time and then broke through ; an addition of three inches of broken stone stopped all trouble.

The relation of these two instances to the necessity for large expenses for drainage and to the theorem of the square of the depth, can be seen by any engineer. But it should be noted that in addition to Mr. Pierce's care to have all voids in his road metal thoroughly filled with clean (*i. e.*, free from loam) binding, the trap he uses is one of the strongest known. The thorough filling of the interstices probably prevents dirt from working up into the road and acting as a lubricant, and a weaker stone, not presenting such strong angles, might not give a road of sufficient cohesive strength to stand waving on a soft bottom before breaking through.

That underdraining is unnecessary with a well built road, where the gutters on each side are built to grade and kept clean, receive fresh proof from the fact that the main road from the quarry supplying broken stones to the city is so built through a succession of micaceous hardpan cuts. Between 100 and 200 wagons and carts have passed over the road daily through February and March without injuring it, the wagons carrying from 40 to 60 gross hundredweights, and the two-wheeled carts from 20 to 27 hundredweights. But where springs find vent under the road-bed they should in all instances be drained into the gutter.

The gutters, which are carefully made to an established grade, are kept clean and free from grass and weeds, whether paved or unpaved. The weeds are dug up, and the grass is killed by spreading a thin layer of sand over it on hot days ; entailing, in this point as in all others connected with the roads under discussion, persistent and intelligent maintenance.

It should be noticed, in connection with the low cost mentioned above—about 28 cents per square yard—that Bridgeport, in addition to the possession of particularly good trap-rock, is exceptionally favored in the location of its quarry, almost exactly two miles from the centre of the city ; so that the cost of the stone is 82 cents per gross ton, of 21 or 22 cubic feet, delivered to the wagons, and the cost of hauling varies, depending on the distance, from 50 to 75 cents per ton, or between \$1.32 and \$1.57 per gross ton delivered on the road. The trap-rock is broken to 2-inch size by three 7x10 inch Marsden crushers, placed side by side on a platform, to which cars are drawn from the quarry by a wire rope, wound by the same engine which runs the crushers. The interest on the cost of the roller—an Aveling & Porter, now twenty years old—is not reckoned in the above-mentioned cost.

That the cost of the Bridgeport roads has not been under-esti-
mated, is apparently made certain by the contract price of such work
in the neighboring town of Fairfield, where, with a longer haul,
a four-inch road 20 feet wide was built for 85 cents per lineal foot, or
38.3 cents per square yard. This sum included regulating, some
grading and the use of a roller, as well as the contractor's profit.

Nothing written above is intended to imply, if a road is to be
built, and then, as has been the case with the Telford roads under
the care of the Department of Public Works in this city, receives no
attention from a reasoning being, that a Telford road will not last
in some condition longer than a four-inch macadam road. Nor is it
intended to imply that only the best trap-rock is applicable to four-
inch roads, nor that a less or greater depth of road metal may not in
cases be preferred.

But engineers are urged, in view of the successful effort of Mr.
Pierce to give the citizens of Bridgeport three or four miles of good
road for one, to neglect precedents and written authorities and try
to fit their roads to the necessities of traffic, and the means of the
community employing them and the materials in hand. Remember-
ing that it may be better to have a patch fail in a road three miles
long than to have one mile of road that will not fail, neither road
can be kept in good surface without maintenance. And a thoroughness in building roads which involves so large an expenditure as to
prevent road-building, is not, it is submitted, thoroughly good
engineering.

Mr. Pierce is an enthusiast in the matter of his style of roads,
and will doubtless be willing to meet any engineer who notifies him
of his desire to visit Bridgeport and show him the results he has
attained.

THIN MACADAM ROADS.*

Mr. James Owen, County Engineer of Essex County, N. J.,
under whose direction the admirable macadam roads about Orange
have been built, sends the following discussion on Mr. E. P. North's
article on the Bridgeport macadam roads, which appeared in our
last issue.

We should be glad to supplement this with the views of other
engineers who can speak from their experience. Mr. Owen writes:

I was much interested in reading the article in your last issue on the
macadam roads at Bridgeport, by Mr. North, and the experience there
detailed is very gratifying, and as it adds an iota to the very scanty information available on the construction of hard roads, whether of Telford or
Macadam.

* xxi, 74.

I must demur, however, to the acceptance of the construction of 4-inch roads, as there detailed, in any other way than as an isolated fact; whatever success there was must have been due to the peculiarities of soil and location, and such work should not be taken as a precedent to be followed in any and all cases.

Like Mr. Pierce, of Bridgeport, I obtained my early experience in the sixties in the construction of park roads in Brooklyn, when the thickness of pavement was never less than 16 inches, laid on a bed of 12 inches of sand, and was undoubtedly a Telford pavement. When, however, I had to initiate in New Jersey a more economical system I decided on a depth of 12 inches, 8 inches of pavement and 4 inches of broken stone; between 30 and 40 miles were constructed of this depth in the avenues radiating from Newark through the Oranges and Montclair. They have stood the wear and tear for sixteen years admirably, of course with proper repairs, and in only two or three instances did the foundation ever blow up. These roads were county roads, and really main arteries, but when the local committees decided to build their own roads, the divergence of opinion and lack of crystallized sentiment led to the adoption of anything from 4 to 12 inches, and the result has been in the same ratio as the thickness. The Oranges, East, West and the City, have laid their roads 10 to 12 inches in thickness, and the uniformly good condition of their roads are proverbial. Bloomfield and Montclair have been building theirs 6 inches, and the difference is remarkable ; ruts quickly appear, holes are common, and they look as they are, cheap roads. In these cases, however, there was an attempt at a pavement. In other townships, like Clinton, Millburn and parts of South Orange, no attempt has been made further than to spread 4 to 6 inches of broken stone on the natural soil, and the result shows that it is to a certain extent a waste of money. The only advantage accruing in such a road is to keep the wagons from getting mired, but as means of travel they do not reach to a very high order.

In the red sandstone formation of New Jersey, in which all the roads mentioned are laid, there are critical periods, especially when the frost is coming out, when it seems absolutely necessary to have a foundation of some sort to keep the roads from breaking up, and only in specially favored localities is it possible to keep a 6-inch pavement without rutting unless there is a sharp fall to the roads. In the latest roads built in this section at Belleville I adopted a rule of making the thickness of pavement as follows : For grades flatter than 1 per cent., 10 inches ; between 1 and 4 per cent., 8 inches ; and over 4 per cent., 6 inches. This is the thinnest construction advisable in this locality with any certainty of good permanent results, unless the roads are merely built as a preventative from miring instead of from travel.

The practice in gutter construction, as outlined in the Bridgeport, I have also tried, but with no great results except on steep grades. In conclusion, I wish to deprecate as strongly as possible the idea of doing cheap work in road construction, as in my experience there is more money wasted in these attempts than is generally realized. A community, if educated to a proper standard, will prefer to spend a dollar well than fifty cents in make-shifts, and it should be the duty of every engineer to guide them to that end.

THIN ROADS.*

Sir—I am very much obliged to Mr. Owen for his criticism on the article you did me the honor to print on the 28th ultimo ; as it allows me to say what I evidently did not sufficiently emphasize, and which is, in my opinion, the key to success in macadam or thin road building, that the excellence of Mr. Pierce's roads, after the hard stone he can command, is due to his care to have all voids in his road metal thoroughly filled with clean (*i. e.*, free from loam), binding, and to the use of a steam road-roller of competent weight. The points which it seems I should have emphasized are :

First.—Thoroughly compacting the soil by rolling before the broken stone is applied.

Second.—Clean binding, which, if possible, should always offer as great or greater resistance to crushing as the stone employed, in contradistinction to either filling or lubricating material.

Third.—The use of a steam road-roller of competent weight.

Fourth.—The combination of these two, in connection with sufficient skill, resulting in a wheel-way that is without voids in its lower surface, inviting the percolation of mud from below, that is sufficiently compact to shed rain and has some cohesive strength, while the clean binding, which is strenuously insisted on, prevents movement of the stones among themselves and keeps their angles from wearing off.

These points, I think it may justly be said, were entirely overlooked by Mr. Owen in his communication. The thin roads he describes were, as seen by me, made by putting from four to six inches of broken trap on unrolled earth, the broken stone filled with screenings and clay, and in some instances with clay alone, and either left to be compacted by the traffic or rolled with a horse-roller. In the thicker roads, Telford, clay was not always excluded from the binding and some at least, of them were rolled with horse-rollers.

If roads are to be built in this way they should undoubtedly be massive. They will have no cohesive strength, they will not present surfaces impervious either to the mud from below or to the rain from above ; the clay filling will be utterly without the strength necessary to keep stones from wearing each other out under traffic, except when just passing from dampness to dryness, and with even the most careful horse rolling the angles of the stones will be more or less worn off before the road is compacted.

Such a road, if sufficiently massive, will be a great improvement on the red soil of Eastern New Jersey when frost is coming out of the ground, but it must depend on its mass. The construction of such roads, whether thick or thin, is strongly deprecated by the writer in all cases where skill to properly construct thin roads at from one-half to one-third of their cost can be commanded.

I wish to urge in this communication, as in my first, the necessity for doing cheap work in all cases where cheap work can be made to accomplish the service of more costly work, and also in cases where the cheap work and a renewal will not cost more than the expensive work. Further, I would in general substitute skill in construction for mass, in conformity

---

* xxi, 87.

with American practice. Mr. Owen, I think, will agree with me in this, and would not have objected to the Bridgeport roads or their mode of construction if either he had seen them or I had been more felicitous in their description. My preference for a steam-rolled road is due to the fact that horse's feet, in drawing a roller, always churn the broken stones more or less and wear off their angles, and the roller itself does not have the drawing action of driving wheels, an action which, I believe, speedily places the stones in such a position that they present the minimum of void spaces to be filled with binding; for, however strong the fragment of screenings may be, it will not offer the same resistance to crushing that a 2-inch cube does.

There is a necessity for better roads. A thick road will last longer under both neglectful construction and maintenance than a thin road, but it costs from two to three times as much. Communities in general are not yet educated to spend even fifty cents much less a dollar. And a thoroughness in building roads, which involves so large an expenditure as to prevent road building, is not, it is submitted, thoroughly good engineering, particularly if skillful building and maintenance will keep the thin road in as good order as one thicker and more expensive.   EDWARD P. NORTH.

MACADAM ROADS IN UNION COUNTY, N. J.*

Some 35 miles of Telford macadam roads are being constructed in Union County, N. J., under the direction of F. A. Dunham, the County Engineer, and in view of the recent discussion of the subject in these columns by Messrs. North and Owen, it was believed that some account of this work would be of interest. A representative of THE ENGINEERING AND BUILDING RECORD accordingly called on Mr. Dunham and learned the following facts: Some 15 or 16 miles of road are under construction, but none have yet been completed. The paving is 16 feet wide, and when 12 inches thick and with a haul of material of from a mile to a mile and a-half it costs from 90 cents to $1 per square yard, or from $8,000 to $10,000 per mile. The bottom stone costs, delivered, about $1.15 per ton, and the fine stone somewhat more.

In constructing the road the ground is first leveled off to the proper height, all soft places are carefully dug out, and with other hollows filled to the sub-grade with gravel, cinders or slag.

The surface is then thoroughly rolled with a horse roller weighing from 2 to 5 tons, and the Telford foundation laid on by hand. This is of sound trap rock, and for 12-inch paving the stones are 8 inches deep, not more than 5 inches wide on top, and from 8 to 12 inches long; for 8-inch paving the stones are 5 inches deep, not more than 3 inches wide, and from 5 to 8 inches long. They are set on their broadest side, with their length at right angles to the road, and so as to break joint at least 1 inch. They are then firmly wedged by driving down between them other stones of the same quality, and as near as may be, the same depth. All irregular projections are "napped" off with a light hammer so as to leave a tolerably uniform surface.

On the foundation thus prepared, the broken stone or macadam of sound trap rock is laid in two courses. The stones of the bottom course for the 12-inch paving must pass through a 2½-inch ring and be leveled off to 2 inches below the finished surface, and for the 8-inch paving the stones must pass through a 1½-inch ring, and be leveled off to 1¼ inches below the finished surface. The broken stone is then thoroughly rolled with a 10-ton steam roller, and dry trap rock screenings that will pass through a 1-inch ring are rolled into it until the interstices are well filled. The top course is then applied of the same material, but broken to pass through a 1½-inch ring. For this course the binding material is mixed with trap screenings and clean sharp gravel, not exceeding half an inch in any dimension, which is well worked in with a 10-ton steam roller water being applied at the same time to facilitate the packing, and when the desired level is reached enough water is used to flush to the surface and sufficient gravel added to form a wave before the roller. Each layer is thoroughly amalgamated with that below, which, if necessary, is roughened for the purpose.

The centre of the pavement is crowded up 18 inches in the middle, and the unpaved portions of either side are well rolled, so as to form a good shoulder to support the edge of the paving. Suitable gutters are formed on either side of the road and under-drains constructed where necessary.

The contractor is paid for labor and material for macadamizing pavement at so much per square yard, and for grading, opening gutters, etc., at so much per cubic yard. He is required to keep the work in repair for one year, and 10 per cent. is deducted from all payments due him to cover the expense of making such repairs should he fail to do so.

The cost of repairs and maintenance is estimated at not over $250 a mile per year.

Some roads constructed as above have been in use in the country for the last fifteen or twenty years, and with proper maintenance have given very good satisfaction.

CAN MACADAM BE ACCURATELY MEASURED WHEN LAID ON THE ROADWAY.*

LEXINGTON, Mo.

SIR: Please inform me either through the columns of your paper, or by letter, whether you can get an accurate answer for the measurement of macadam, if you have the length, breadth and thickness, if said rock is properly spread. Also, please inform me whether you can ascertain the

---

* xx, 220.

amount of macadam on a block after it is spread. By answering these questions you will greatly oblige.

W. S. CLAGETT.

The answer to your first question seems so obvious that we must conclude that you wish to estimate the shrinkage that takes place when the road metal or macadam is compacted by rolling to a proper surface, or, in other words, that you wish to know how much loose stone it will take to cover a given surface of macadam road to a given thickness, or that you may have contracted for a given amount of broken stone to be used in making a macadam road, and having neglected to measure it in advance, now wish to estimate it from the finished work. The only satisfactory way to make an estimate would be to experiment under similar conditions with a measured quantity of loose stone, and having found the proportion of shrinkage, the original volume of the stone can be easily calculated from the finished work.

If not convenient to make such an experiment, or the importance of the work is not sufficient to warrant it, perhaps the following facts may enable you to arrive at a reasonably satisfactory conclusion.

One cubic yard of solid rock will, when broken and loosely piled, occupy about 1.9 cubic yards, and when carefully piled, about 1.6 cubic yards. If the voids in well rolled macadam bear the same proportion that they do in carefully piled broken stone, which would seem to be a reasonable supposition, then if you multiply the cubic contents of your finished macadam work by, say, 1.2, you will get the amount of loose stone that was used to make it, or if you multiply the amount of finished macadam by $\frac{5}{8}$, you will get the amount of solid rock that had to be quarried for it.

The only way to get at the amount of macadam after laying is to measure the surface, and multiply by the average thickness, as ascertained by making a sufficient number of holes.

MACADAM IN CITIES.*

In a report on pavements made to the Common Council of Topeka, Kan., the following brief summary is given of the experience of a number of cities with macadam.

The authorities of Toronto say: "Our experience is enough to utterly condemn it. We have spent more than $10,000,000 in macadamizing streets, which became seas of mud after a few hours' rain. It will actually cost the residents on Yonge Street thousands of dollars to get rid of the macadam and put themselves in the favorable position of having only a dirt road to be dealt with."

---

*xv, 375.

In Fall River the stone used is harder, but "macadam roads
will not stand the wear due to heavy teaming."

In Sandusky, O., limestone is used for the macadam, and "it
wears away so rapidly that it requires almost constant attention" to
keep it in order.

The Board of Public Works of Washington consider it as only
suitable for country roads.

Cincinnati reports "the chief thing permanent about the limestone macadam of that city has been the permanent *expenditure* for
repairs."

St. Louis recommends the abandonment of limestone macadam
on account of rapid wear, mud and dust.

Mr. Hill, City Engineer of New Haven, reports that though
cheapest in first cost, even the best macadam is more expensive to
maintain.

In Kansas City, St. Joseph, Omaha, Atchinson and Leavenworth
it is being taken up, or is only used on streets where pleasure driving is the rule.

The committee therefore decide they "would prefer the present
dirt roads."

PAVEMENT PROTECTOR FOR HYDRANT FLUSHING.[*]

This device is intended to prevent the injurious washing of
macadam pavements by the flushing of fire hydrants. It is simply
a box about 3 feet long by two 2 feet wide and high, with a

PAVEMENT PROTECTOR FOR FLUSHING HYDRANTS

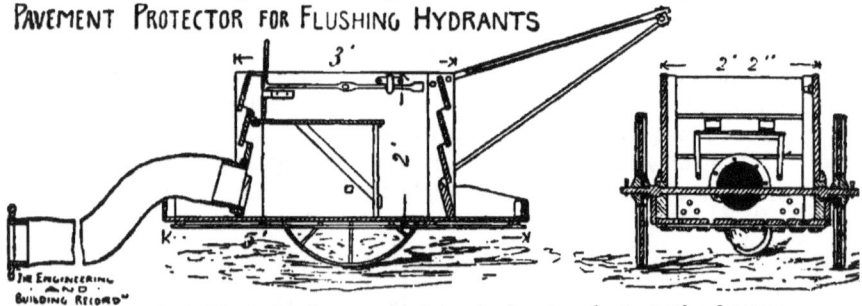

sort of shelf projecting at each end about a foot at the bottom.
The bottom and ends of this box are loosely made of slats, so that
water can escape freely in all directions but in no large streams in
any direction. It is connected to the hydrant to be flushed by
about 8 feet of large canvas hose, and the force of the escaping
water being dissipated through the numerous openings prevents
injury to the pavement. It is mounted on wheels for convenience
of transportation and has proved very useful. The arrangement
was devised by R. C. P. Coggeshall, Superintendent and Engineer
of the New Bedford (Mass.) Water-Works.

---

[*] xix, 133.

A FIRST PRIZE TREATISE ON THE SCIENCE OF ROAD-MAKING.*

Introduction.—This treatise was written in answer to the printed circular of a Committee of the Board of Agriculture, calling for "treatises upon the science of road making, and the best methods of superintending the construction and repair of public roads in this Commonwealth." [Massachusetts.]

This circular was issued about the middle of December, and as the time for writing and sending in the called for essays was limited to January 28, the writer has thought it best, no specific character being prescribed for the treatises, to attempt to write one suitable to be so called from the stand-point of the public, rather than from that of the civil engineer, and, giving results rather than the methods of arriving at them, to be as concise as possible.

The Science of Road Making.—Starting, then, with the first of the two subjects mentioned in the circular—the science of road making—we can divide this into three periods : (1) laying out a road ; (2) making the road-bed, which includes all earthworks, cutting and filling, culverts, drains, bridges, even tunnels, etc.; and (3) the making of the road surface ; to which, not improperly, might be added, (4) keeping the road in repair.

Laying Out a Road.—The considerations which determine the best location of a road, are those arising from the nature of the travel it is proposed to accommodate ; that is, from the admissible grades, radii of curves, etc. Given two points it is desired to connect, with no intermediate point where the road is to touch, that route is the best which will cost least to build and maintain, the grades and curves being kept within bounds ; and to find this location constitutes the whole problem of the engineer.

In older countries, where trade and manufactures are more settled and unchanging than in the United States, the probable future travel upon a road about to be laid out and built, forms a material element in the data that govern its alignment and grades. A very able and clever article upon this subject may be found in the Journal of the Society of Civil Engineers and Architects at Hanover, for the year 1869, and also in pamphlet form. It is in the German language, written by Launhardt, Superintendent of Highways (and a civil engineer) in the Hanoverian provinces.

The Romans built all of their roads in perfectly straight lines, up hill and down, at a very great expense, as being absolutely the

---

* By Clemens Herschel, Civil Engineer, of Boston, now of Montclair, N. J. xix, 187. This treatise was written in 1870 and reprinted in 1877. We now give it, by request, on account of its intrinsic merit, and because the principles on which it is based are unalterable, and the methods, with such primitive materials, can change but little.

shortest distance between two points.   At a later period in history, it was argued that a road must be winding to be agreeable, and many were so built only for this reason.   The modern road builder or engineer in general, ignores any such considerations, and has for his aim only to achieve the most, at the least present and future expense.

As regards curves in roads in a hilly and mountainous district, we have then the rules never to make a smaller radius than 20 feet, and that only in extraordinary cases.   On roads where long logging or other wagons may be expected, the smallest radius ought to be 50 or 60 feet ; and, in general, 40 to 45 feet is none too much.

A rule sometimes followed in constructing mountain roads, is, where the inclination is 1 or 2 in a hundred, * heavy teams require 40 feet and light ones 30 feet radius ; with a grade of 2 or 3 in a hundred, heavy teams require 65 feet and light ones 50 feet radius. Where a reverse curve (shaped like the letter S) occurs, there should be a straight piece connecting the two curves (Fig. 1).   On the

FIG. 1   FIG. 2

contrary, where the two curves to be connected are concave in the same direction, the connecting link should be curved also, and not straight (Fig. 2).   On the length of the curves the grade should be made easier than on the parts of the road immediately adjoining.

As regards grades, to start with mountainous paths, we find pedestrians able to walk up an inclination of 100 in 120 ; mules ponies, etc., 100 in 173.   For roads, Telford's rule was, that for horses attached to ordinary vehicles, to trot up a hill rising 3 to one hundred, was equal to walking up one of a 5 in a hundred grade.

Experiments have shown that—

1. On a road falling 2 in a hundred, vehicles would run down themselves.

---

* In describing grades, the first figure gives the vertical height which is ascended in a horizontal distance given by the second figure.   Both figures must of course be taken to refer to the same unit of length, thus : 100 feet in 120 feet, 100 inches in 120 inches, or 100 miles in 120 miles, all express the same inclination to a level plane, and are more general in their application than the ways of expressing grades in so many inches to the foot, or feet in one mile, etc., etc.

2. On the same kind of road, but having an inclination of 4 in a hundred, light vehicles had to be held back lightly, loaded ones with considerable force.

3. On a road having a fall of 5½ in a hundred, light vehicles had to be held back with considerable force, or if a brake was applied they had to be pulled, whereas heavy or loaded vehicles had to be braked to keep the horses from being speedily exhausted.

On inclinations steeper than 5 in a hundred, the rain-water running down the road is apt to do some damage to the road surface.

The regulations of different countries having a long experience in road building, such as France, Prussia, Baden, etc., vary somewhat, but the following is the general result:

In treating of roads, it often renders the subject much clearer to divide them into three classes : First, second and third class roads, or, as we might also say, state, county and town roads. Accepting this nomenclature, we have this : For first class or state roads, the greatest inclination should not exceed 3 to 5 in a hundred; second class or county roads, 5 to 7 in a hundred ; third class or town roads, 7 to 10 in a hundred. A road rising 10 in a hundred is not supposed ever to have any heavy teams upon it. In ascending a hill it is well and proper to decrease the grade as the top is reached. and in the same measure as the horses get tired. Thus, if a first class road starts up hill with a grade of 4½ per hundred, it should gradually diminish to 4 and 3½ in a hundred, and end near the top with a grade of 3 in a hundred. Launhardt, the Superintendent of Highways, and engineer, mentioned in the previous note, has a valuable article on the subject of the best grades for highways, in the engineering journal there mentioned, for the year 1867 ; reprinted also in pamphlet form. He shows in this article that, according to the received formula that expresses the relations between the tractive force, the velocity in feet per second, and the daily working hours that go to produce the maximum amount of work that can be got out of a draught-horse, a uniform grade between any two points, except perhaps in curves, and, if desired, for resting places, is the grade that tends to enable the horse, or other draught-animal, *to produce the most work per diem.* If a grade of 4 or 5 in a hundred must needs be kept up for some distance, then it is well to have resting places 40 or 50 feet long, having a grade of only 1½ or 2 in a hundred, in the line of the road at proper intervals. An expedient adopted by Telford, the eminent English engineer, in order to avoid making a piece of road a mile long, on a less grade than 5 in a hundred, on account of the increased cost this would have occasioned, and yet not have this part of the road too much more tiresome for the horses than the rest, was to make the road

surface on this mile of a much better quality than on the remainder ;
the additional cost required for the improved road-bed amounting to
only about one-half of what it would have cost to reduce the grade to
say 4 in a hundred, as will be again referred to under the head of
track-ways. In sharp curves the grade should be only 1 or 2 in a
hundred or level.

Table I. gives the effects of various grades on the amount a
horse can pull.

TABLE I.

Calling the load a horse will pull on a level, one :

Then, on a grade of 1:100, a horse can pull....... ....0.90
"   "   1: 50,   "   " 0.81
"   "   1: 44,   "   " 0.75
"   "   1: 40,   "   " 0,72
"   "   1: 30,   "   " 0.64
"   "   1, 26,   "   " 0.54
"   "   1: 24,   "   " 0.50
"   "   1: 20,   "   " 0.40
"   "   1: 10,   "   " 0.25

To determine whether it is most advisable to go over or around
a hill, all other considerations being equal, we have this rule :   Call
the difference between the distance around on a level and that over
the hill d, the distance around being taken as the greatest, and call
h the height of the hill.

Then, in case of a first class road, we go around when d is less
than 16 h.

And in case of a second class road we go around when d is less
than 10 h.

When the height of a necessary embankment gets to be more
than 60 or 65 feet, a bridge or viaduct will be found cheaper, and
the same measure, 60 feet, applies in case of tunnels, they being
cheaper at that depth than open cuttings.

Under the head of laying out roads something should be said
of their width. Speaking only of such roads as are not apt to turn
into streets from their proximity to towns and cities, it is well not
to make them too broad, for the less the width the less the cost of
construction and maintenance, and a good 23-foot road is much
better than a poor one 40 or more feet wide. Each rod (16¼ feet)
in width adds two acres per mile to the road. An agreeable form
of road is to have on each, or on one side of the same a strip 5 or
6 feet wide, sodded, and then a sidewalk equal in width to one-
eighth the width of the roadway. The intervening strip above
mentioned is planted with trees, and at intervals of 200 to 250 feet
furnishes storage places, 30 or 40 long, for the materials used in

the road repairs. The width of first, second and third class road-ways may be given as 26, 18½ and 13 feet, with a tendency during the last ten years to have none, except in the vicinity of cities, wider than 24 feet, and the rest correspondingly narrower. In view of the changes constantly going on in this country in the value and settlement of land, it would probably be well always to lay out a road 50 or 60 feet wide, but to build the road proper of the width above indicated.

TABLE II.

Number.	KINDS OF EARTH.	Cost per cubic yard to loosen.	Cost per cubic yard to load in wheel-barrows.	Cost per cubic yard to load in carts or wagons.	Amount to be added for keeping and repairing tools.
		IN PARTS OF A LABORER'S DAY'S WAGES.			
1	Loose earths, which are loam, sand, etc., inclusive of loading.	⅛-⅛	..	..	⅟₁₆
2	Heavier earths, such as sticky clay, which does not readily leave the shovel, etc.	⅛-¼	⅟₁₆	⅟₁₂	⅟₁₆
3	Earths which must be loosened with a pick before they may be shoveled.	⅟₁₂-⅛	⅟₁₂	⅟₁₂	⅟₁₆
4	Solid banks of gravel or clay, earths containing boulders, etc., in which one man only loosens as much as another man shovels...	⅛ ⅟₁₂	⅟₁₀	⅟₁₂	⅟₁₆
5	Same material, worst kind, brick and mortar heaps, earth full of roots, etc., in which it takes two men to loosen what one man shovels.	⅛-⅟₁₂	⅟₁₀	⅛	⅟₁₀
6	To break up stone which is in layers or seams, requiring the use of the crowbar only, but no blasting.	⅟₁₂-⅛	⅟₁₂	⅛	⅟₁₀
7	Blasting rocks in an open cut, according to the hardness of the rock, to the position of the seams etc.*	⅟₁₂-¼·⅟₁₂-¼·⅟₁₂-⅜·⅟₁₂	⅟₁₂	⅛	⅟₁₆
8	In forming and shaping embankments.	⅟₁₂	.	..	⅟₂₀

* To excavate rock to a given line and level—that is, to trim a cutting may cost double these figures per yard.

With all these rules and data in mind, the real work of actually laying out the road on the ground and on a map is next in order, and this comes so entirely within the province of the civil engineer, and is a matter requiring so much explanation and study, that it cannot well be introduced within the limits of this treatise. It is in this part of the work that a little skill and labor well spent may be productive of very great saving in the cost of the whole work, and it should not be left to the inexperienced or unskillful.*

* Gillispie, i   s treatise on "Roads and Railroads," gives two forcible instances of the amount those roads which might properly be called chance roads, can be improved by a road-maker of skill and understanding. An old road in Anglesea, England, rose and fell between its two extremities, 24 miles apart, a total perpendicular amount of 3,540 feet; while a new

Making the Road-bed.—Under this head are included earth-works, drains, culverts, bridges, stay walls, etc., etc., all matters requiring a special kind of skill to construct properly. The writer believes it impracticable to write a book which shall at once be interesting to and therefore valued by the public, and of value to the professional man, and thinks an attempt so to do results always in a failure in both directions. True to the determination expressed in the introduction, he proposes, therefore, to treat under this head mainly with those parts of the subject in which the public at large is most interested, for example, the data for the cost of earthworks, general information relating to drainage, bridges, etc.

Earthworks.—The basis of all values is the daily wages of a common unskilled laborer, and in the data given below, this figure, whatever it is from time to time and in various places, must be taken as unity, or the standard measure.

The cost of earthworks may be divided into three parts: (1) cost of loosening the earth, (2) cost of transport, and (3) cost of forming the transported earth into the desired shape. The cost of the first part depends materially on the kind of earth to be handled. The cost of the second, mainly on the distance the earth is to be moved.

We find by experience that in digging and loading or throwing $\frac{10}{16}$ feet horizontally with a shovel, we obtain for different materials the results of Table II.

TRANSPORT OF EARTH.

Throwing with a Shovel.—This is to be done only from $\frac{10}{16}$ feet in distance or from $\frac{6}{8}$ feet vertically. To throw 5 feet vertically

---

road, laid out by Telford between the same points, rose and fell only 2,257 feet, so that 1,283 feet of perpendicular height is now done away with, which every horse passing over the road had previously been obliged to ascend and descend with its load. The new road is besides two miles shorter. The other case is that of a plank-road built in the State of New York, between the villages of Cazenovia and Chittenango. Both these villages are situated on Chittenango Creek, the former being 800 feet higher than the latter. The most level common road between these villages rose, however, more than 1,200 feet in going from Chittenango to Cazenovia, and rises more than 400 feet in going from Cazenovia to Chittenango, in spite of this latter place being 800 feet lower. That is, it rises 400 feet where there should be a continual descent. The line of the plank-road laid ou by George Geddes, civil engineer, ascends only the necessary 800 feet in one direction, and has no ascents in the other, with two or three trifling exceptions of a few feet in all, admitted in order to save expense. The scenes of similar possible improvements are scattered all over this and the rest of the States ; and these facts are still more or equally to be borne in mind in laying out new roads, where the ounce of prevention may take the place of the pound of cure.

costs as much as 12 feet horizontally, that is to say, if 30 feet horizontally cost per cubic yard one day's wages, at 8.4, the same distance vertically will cost about 2½ times as much, or more exactly, one day's wages divided by 3.5, whence is seen the economy of using windlasses, etc., instead of "stages,"* in shoveling earth vertically. Table III. gives the cost of shoveling earth certain distances, expressed in the number of cubic yards a laborer's day's wages will pay for.

TABLE III.

Distance of throw, in feet.	Vertical or horizontal.	Whether done at one operation or by means of so-called "stages."	Number of cubic yards which can be transported at the cost of one laborer's day's wages.	Remarks.
0-10,	Horizontally,	No " stages."	23.5	
10-20,	"	1 stage.	12.6	} Wheelbar-
20-30,	"	2 stages.	8.4	} row cheaper
0-5,	Vertically,	No stages.	14.1	
5-10,	"	1 stage.	8.8	

Wheelbarrows.—The usual distance of transport suitable for the use of wheelbarrows is $\frac{100}{200}$ feet. In exceptional cases it may be more, but perhaps never above 500 feet, and then only for moderate quantities. In going up hill, the greatest inclination is to be not more than 1 in 10, and a man can push only ⅔ as much on this inclination as on a level. Three feet vertical transport costs as much as $\frac{90}{100}$ feet horizontally. Whenever possible, planks should be laid for the wheelbarrows to run on. The best timber for this purpose is beech-wood, and the cost of keeping such planks is only about $\frac{1}{20}$ or $\frac{1}{30}$ per cent. of the cost of transport per cubic yard. See Table IV.

Patent Portable Railroad and Hand Cars.—These have lately been introduced in this country and appear to be coming into general use and favor. The company owning this improvement, as it seems to have a right to be called, claim that by means of their track and cars, which can be used everywhere that a wheelbarrow or a horse-cart can go, and in a great many places where these vehicles cannot go, they effect a very large saving, as much in some cases as five-sixths of the cost by the other means of transport. There are no data published as yet to make tables from similar to the foregoing; from the company's pamphlet, however, one given case which occurred on Staten Island in 1867, may be analyzed as in Table V.

* By a "stage" is meant the operation of one shoveler lifting and throwing what another has thrown in front of him.

TABLE IV.

DISTANCE OF TRANSPORT, IN FEET.	Number of trips per day of ten hours, made with one man at barrow, and one to load.	Contents of wheel-barrow load in cubic feet.	Number of cubic yards which can be transported at the cost of one laborer's day's wages.
10-20............	120	$2\frac{1}{4}$	23.5
20-50............	110	$2\frac{1}{4}$	16.9
50-70............	100	$2\frac{1}{4}$	14.4
70-100............	98	$2\frac{1}{4}$	13.8
100-150............	96	$2\frac{1}{2}$	13.3
150-200............	94	$2\frac{1}{2}$	12.8
200-250............	92	$2\frac{1}{2}$	12.4
250-300............	90	$2\frac{1}{2}$	12.0
300-350............	88	$2\frac{1}{2}$	11.6
350-400............	86	$2\frac{1}{2}$	11.2
400-450............	84	$2\frac{1}{2}$	10.9
450-500............	82	$2\frac{1}{2}$	10 5
500-550............	80	$2\frac{1}{2}$	10.2

TABLE V.

Distance of transport, in feet...................... 550

Number of trips per day of ten hours, with one man
at two cars, and two to load 150

Contents of car in cubic feet............. .... 11.34

Number of cubic yards which can be transported at
the cost of one laborer's day's wages........... 60

One-Horse Carts.—The table for this kind of transport may be stated about as follows: 1 foot vertical costs as much as 14 horizontal. See Table VI.

TABLE VI.

DISTANCE OF TRANSPORT IN FEET.	Number of trips made per day of ten hours, assuming only four minutes to load, dump, etc., per trip.	Contents of cart load in cubic feet.	Number of cubic yards which can be transported at the cost of a laborer's day's wages.
300............ ...	86	8	17.1
500...............	67	8	13.6
1,000...............	43	8	8.6
1,500...............	31	8	6.4
2,000......	25	8	5.0
2,500...............	21	8	4.3
3,000......	18	8	3.6

Ox-cart transport is 10 or 12 per cent. cheaper than the above, but takes more time.

Other methods of transport, such as horses or engines on temporary tracts would hardly ever be applied to road-building, but belong more properly under the head of railroad construction.

Shrinkage.—In calculating the cost of earth-works, the so-called shrinkage of earth must not be overlooked. Earth occupies on the average one-tenth less space in embankment than it did in its natural state, 100 cubic yards shrinking into 90. Rock, on the contrary, occupies more space when broken, its bulk increasing by about one-half. The shrinkage of gravelly earth and sand may be taken at 8, of clay 10, loam 12, surface soil 15, and of "puddled" clay 25 per cent. The increase of bulk of rock is 40 to 60 per cent.

To make use of all these data in calculating the probable cost of a piece of road, there is of course still wanting the equally essential factor which gives the number of cubic yards to be dug and moved and the distance of transport. These are got from the plan, profile and cross-sections of the proposed work, an engineer's knowledge being requisite to make the necessary drawings and calculations.

Drains and Culverts.—The drainage of roads is of two kinds, surface and sub-drainage. The first provides for a speedy removal of the rain-fall on the surface of the road and the cutting and embankments on which it is carried ; the second, for the removal of that part of the rain-fall which nevertheless does penetrate into the body of the road covering. With a perfect sub-drainage the winter's frost, having no water to act upon within the body of the road, is robbed of its great power to destroy the same, and it also prevents the road surface from becoming soaked and thence destroyed in the summer. The need of surface drainage is self-evident. This last named is to be provided for at this stage of the building of the road, the sub-drainage being more properly a part of the building of the road covering or top. For this purpose ditches, one on each side generally, are absolutely necessary, both when the road is on a level with the surrounding country and when it is in a cutting. They may become necessary also in the case of embankments ; for example, when an embankment is built across low ground. Where these side ditches cross under the embankment we have a culvert ; also whenever any small valley, having a constant or intermittent stream of water, is crossed by such an embankment. It is very bad policy to make such culverts of wood, unless indeed they are so situated as to be constantly under water ; the cost of replacing them after the embankment and road has been built over them is disproportionately great. They should be made of stone or brick ; lately, cement drain-pipe, oval or egg-shaped, has been used to advantage in their construction.

All ditches, drains and culverts should have a fall throughout their entire length. Their size will depend upon the amount of water they may be expected to carry, and this again on the rain-fall

that may occur on the area which they drain. Extraordinary showers have occurred of two inches in half an hour, but only over a
very limited area, and two inches in an hour may be taken as a
large allowance. This is the basis of the Central Park drainage
calculation, and is larger than usually taken—none too large, however, for safety.

The determination of the proper width and height of culverts,
that will enable them to pass the requisite quantity of water without
damming it up, is a question in practical hydraulics, easily enough
settled, in cases of doubt, by the proper gaugings and observations
made upon the spot, but which is answered only in a very crude
and imperfect manner by any general rules that may be given. And
yet it may prove a very important question at times. There is now
(1877) pending in Massachusetts, a suit for damages that may involve claims to the amount of about half a million dollars, in which
one great center of attraction is nothing but a simple railroad culvert, and the question: Was it as large as it ought to have been?
and the writer passes every day, when home, by a culvert which for
some 150 or 200 years has dammed the waters of a brook back
about three miles, from one foot to say 20 inches at the culvert,
vertically, and done this right along two or three times per annum;
and at the present time it contributes in this manner, more than its
proper share towards the flooding of about 500 cellars. These two
cases may serve to call attention to the great damage that may
accrue from making culverts too small, and show whence comes the
rule : in cases of doubt, make the culvert plenty large enough. The
following rough and approximate rules for determining the quantity
of water that a culvert will be called upon to pass through it, are
taken from a German pocket-book for road engineers. Compute
the cubic feet per second from the drainage area that lies above the
culvert, and for the different lenghts of valley from the corresponding rain-falls per hour (the rain-fall is given in inches per hour, instead of in decimals of a foot per second, only for the purpose of
avoiding the printing of long decimals).

Length of Valley in Miles.	Inches Per Hour.
2.5 or less	1.2
2.5 to 5	0.75
5. " 7.5	0.45
7.5 " 10	0.30
10. or more	0.15

As culverts grow larger and wider, with the amount of water
they are to pass under the road, they develop finally into

Bridges.—Bridge-building is a life's study, taken by itself, and in
some of its parts it is not half appreciated and known as yet among

the public. Prominent among these is beauty of design and *appropriateness to the situation.* There is, perhaps, nothing else that will so much improve the appearance and attractiveness of a road as a beautiful bridge. So also in cities we find that a street will of its own accord, seemingly, improve in appearance when a good and handsome bridge has been erected on its line, the owners and builders of the adjoining buildings taking the bridge for their pattern and model. Nor must it be supposed that a handsome bridge must necessarily cost more than an inappropriate or homely, uncouth structure ; it need never be the case. Very often the chief beauty of a structure lies in the fact of its carrying the most with the least expenditure of material. No one bridge is proper in every situation, and herein many mistakes are made. The correct way to build a good bridge is the same or a similar way to that followed in first-class buildings—namely, to have plans drawn for the same, and receive estimates and offers to build according to these plans. It is not well to allow the offices of designer, superintendent and contractor to be united in one person or firm, and is expecting too much from human nature.

Making the Road Surface.—There are two subordinate kinds of road surface, if the term road can properly be applied to them— namely, that of foot and riding paths ; these may be disposed of first, before proceeding to the more important consideration of the road surfaces proper, those used by vehicles of all descriptions.

Footpaths.—For the surface of a footpath little solidity is necessary, except in city sidewalks, which are not supposed to be treated of here ; but we do need a material that shall become and stay compact soon after it is laid. Coarse sand, screened gravel, stone chips and dust, make good paths ; should these materials be too free from any earth or clay, a little of the same may often be added to advantage to act as a binding material. Wherever the ground underneath the surfacing is not porous or likely to remain porous enough to let all the water that may soak through drain away, a layer of such porous material must be filled up before the top surface is put on. Oyster shells, or large stone chips, gravel stones or pebbles, etc., make a good foundation of this sort. The top covering should have a slope, best in both directions from the center of the path towards each side of about 1 in 16 ; the thickness of the foundation course to be 3 to 5, and that of the top 3 to 4 inches. No gravel path, or sidewalk, will afford good walking at the season of the year when the frost is coming out of the ground. Carting on more gravel is in vain; it often is no better than mere foolishness. If village communities will get this idea firmly into their minds, and, instead of a fruitless struggle against the laws of nature and of gravel, will build stone screening sidewalks, with a *good foundation course* underneath, as above described, or else some sort of hard sidewalk covering, they will save themselves much expense, many muddy feet, and no small amount of consequent and annual discontent, not to say profanity and ill feeling. *Heavy* rolling will save much time in finishing the whole process; the roller should be used unsparingly and throughout the whole construction of the path, on the foundation as well as on the top.

FIG. 3

Riding-Paths.—From the nature of the travel these are intended to accommodate, their surface is of a peculiar nature. Inasmuch as a horse, in galloping, tends to throw the soil he treads on backwards with his hind feet, the surface must be kept somewhat loose and soft to make riding on it easy and agreeable. This requirement makes it impossible to have any slope on the surface (the loose material would wash away if there were any), and hence we must rely here wholly on sub-drainage, and not attempt any surface drainage. The top is made of coarse sand, *free from clay* or other binding material, laid on two and one-half to three and one-half inches thick, and spread out level. Under this is a solid foundation, about four inches thick, made of coarse gravel and clay, and having a slope of about 1.20, so that the water will run off along its top surface to either side, where it must further be disposed of by drains or ditches. In case of riding-paths too wide to be so simply built, Figure 3 shows the method to be used. The foundation is made in several slopes, at the lowest parts of which are placed drains, running in the direction of the path, but communicating from time to time with the side ditches or drains. Should, however, the ground underneath be porous enough, the drains may be dispensed with; and if in their stead holes be dug along the lowest lines, marked *a a*, and these filled with large stone, the water will, through them, drain away into the ground.

Roads.—To make a good road surface is a very simple operation after it is only once understood, and the fundamental principles thereof once comprehended they can hardly be forgotten. Everything connected with the construction, the use and maintenance of roads, was, in times past, before the invention of railways, the subject of exact observations and experiments, many and varied in their character. Old engineering works that treat of road making are excellent reading upon this subject at the present day ; upon road construction, and, no less, upon the need of better road legislation. Some, perhaps the most, of the evils we suffer in the shape of bad common roads, are merely the result, the necessary consequence, of our bad systems of common road management, which are derived from our antiquated legislation upon that subject. Legislation of this kind has changed but little in a hundred years, and is producing the same evils to-day that it did a hundred years ago. Hence it is explainable, that the complaints concerning bad roads, and bad road management which we read in books, of fifty and sixty years ago, sound to our astonished ears as though they had been written but yesterday. On this subject may be consulted : The life of Telford, the great English road builder, who died some fifty years ago ; (also among " The Lives of the Engineers," by Samuel Smiles). Besides this, we have the results of a great number of years of experience in older countries, and there would seem to be little to invent, but much to learn, in this branch of construction. Though less progressive than other branches, there are, nevertheless, improvements in road-making, especially in road-making machinery and tools; and no treatise on this or any other living subject can be considered complete a very few years after it is written.

Ancient roads were made with a surface as nearly resembling the solid rock as possible. So, in China, roads were made of huge granite blocks laid on immovable foundations. In time these became worn with ruts, especially in the joints or seams of the stones, and the surface generally so smooth that animals could hardly stand, far less trot on it. They are now for the most part deserted, and left to be covered up by land-slides, etc., to one side of the new roads of travel.

The invention of McAdam consisted in having no large stone at all on the roadway, but having it all pounded into small fragments and spread over the road-bed. This has, without fear of efficient contradiction or shadow of doubt, been proved by trial to be a worthless proceeding, though at one time popular, and even now only too often done, either from ignorance or laziness. The separate fragments of stone, having no bond among themselves, are liable to sink into the underlying ground or road-bed, evenly or unevenly as it may chance, more in one place than in another, and thus never come to rest or to an even top surface. Between these two extremes of an ancient Chinese solid rock road and that of McAdam, lies the true principle of road-making, which consists in giving every road two component parts ; one—the foundation,—to be solid, unyielding, porous, and of large material ; the other—the top surface,—to be made up of lighter material, and to be made to bind compactly and evenly over the rough foundation. This constitutes the whole principle to be followed; and, let it be repeated, that to dump the road material directly on the ground, without first preparing a foundation for it, as is so frequently done, is a waste of time, labor and materials, by no possibility resulting in a good road. On this one fundamental idea, which is never abandoned, however, there are a number of variations. Besides these roads, whose characteristics is the foundation they are all built on, we have paved roads, or pavements, of a great many kinds, and roads with track-ways, also of various kinds.

Foundation Roads.—The roads of this kind, with macadam for the top surface, are called Telford roads by English writers, from Telford, who first built them in England. The Central Park " gravel roads " belong under this head, gravel taking the place of the macadam of the Telford roads. These foundation roads are of far greater importance than any other kind for State, county or town roads, also for parks and driveways. The top surface of all these roads must have a certain inclination to cause efficient surface drainage. Various authorities give various rules for the amount of this inclination or side-slope. It would seem just that it should depend on the nature of the top covering, being less for more solid than for looser or softer materials, and also on the grade of the road.

In Baden, one of the smaller German States, but which is worthy to be taken as a model in matters of road building, and in France, the rise at the center is given as $\frac{1}{40}$—$\frac{1}{60}$ of the width of the road, according to the nature of the material ; that is, inclinations of 1 in 20, and 1 in 30. The rules in Prussia prescribe inclinations of 1 in 24 for roads falling more than 4 in a hundred ; 1 in 18 for roads on a grade of between 2 and 4 in a hundred ; and 1 in 12 for those on a grade of less than 2 in a hundred. When first built the center should be made some four inches too high to allow for after settling.

FIG. 4

Macadam Top.—The cross section of such a road is shown in Fig. 4 ; the thickness of the foundation $b = a$, the thickness of the top covering at the center, and is six, four or five and three and one-half inches in thickness for first, second and third class roads. If the stone for the foundation—for which most anything will do, and that kind should be taken which is cheapest to procure—happens to be got out cheapest in larger pieces than the above dimensions, it will do no harm. This foundation course is sometimes set so as to present an inclination on top, and the cover then put on of a uniform thickness over the whole breadth. This is perhaps best, but is somewhat more expensive. It will do, in nearly all cases, to set the foundation course on a level, or as near so as the stones will allow, and then make the top crowning, by making the covering say three-quarters of an inch or an inch less thick at the edges than in the center. The stones forming the foundation should not be set in rows, nor ever laid on their flat sides, but set up on edge and made to break joints as much as possible ; that is, set up irregularly. After they are set up, the points that project above the general level may be broken off, and the interstices generally filled with small stone. More or less care and work are necessary in this part of the operation, according to the importance of the road and the depth and character of the material used for the top covering. To roll the road at this stage is to be recommended ; afterwards it becomes a requisite. The point never to be lost sight of, is that this foundation course must remain porous, must be *pervious* to water, so that all rain-water that shall soak through the top covering will find, through it, means of escape to the ground underneath ; thence, according to the nature of the subsoil, it is left either to soak into the ground or must be further led away by appropriate drains.

Road Covering.—Of very great importance is the *material* used for the top or road covering. In the order of their value for macadam, we have: I. Basalt. II. Syenite and Granite. III. Limestones. IV. Sandstones.

It will be evident that a very much greater quantity of the soft stones would be required to repair a certain road, than of a harder kind, and on a road lying out of the way of a hard-stone quarry or deposit, the question will arise which is cheapest, to pay more for the raw material and get good stock, or pay less and use the worse. There have been some interesting results in places where this matter has been the subject of experiment, continued for a number of years. Thus, on a road in Baden which was formerly macadamized with rock costing only fifty cents per cubic yard, it was finally found cheaper to take harder rock from a distance, costing $1.78 per cubic yard, the saving being both in less quantity of material used and less labor required in repairs. Just where the limit is must be found in each case by long continued experiment, which it is well worth the trouble to make, both to save expense and also to have the best possible road, the harder material making a road better at all times, at the same or less cost. After the right kind has been determined, none other should be mixed with it, and should any inferior piece accidentally or designedly get into the stock to be broken up, it should be picked out and thrown aside. The stone is broken up into macadam, either by hand or machinery. Wherever any considerable quantity of macadam is in present or future demand, a stone-breaker is certainly a saving over hand labor, though it is difficult to draw the line exactly, where hand-labor or machine-labor is cheapest.

Probably no town that pretends to keep thirty or forty miles of road in good repair ought to be without one of these labor-saving machines. Those most in use are made by Blake Bros., of New Haven, Conn.

When broken by hand and for country roads, the stones should be broken on the storage places already mentioned, which are to be established along the side of the road every 200 to 250 feet. The laborer is not to pound the stones on a heap of such, but to use one large stone as a sort of anvil to break the others on. He is to use a light hammer, except for pieces containing more than four or five cubic feet, and may use a ring with a handle attached to hold the stone he desires to break.

In order that the road shall get an even surface the macadam must all be of one size, and the proper size for the macadam depends on the degree of hardness of the rock. If too small it turns to dust, if too large the top will not pack even. The size is regulated by the use of a ring as a guage—every stone being obliged to be capable of falling through this ring in any direction it may be dropped. Hard stones should be 1 to $1\frac{1}{4}$, softer ones $1\frac{1}{2}$, and the softest 2 inches in diameter. Larger sizes give less perfect roads. In loading and otherwise handling macadam, a many and close-pronged pitchfork should be used instead of a shovel, so as not to mix in any earth or sand and to sift out the stone dust and chips.

The macadam being properly prepared and loaded up, it is spread over the foundation in two or three successive layers. Each layer should be rolled, but the top and last one must be rolled to make a good road. Nor will rolling alone do the work. Two other helps are needed: the use of a binding material, to act as a cement between the broken stone, and sprinkling. It is difficult to prescribe in words just what to use as binding material, and just how much to sprinkle and roll; common sense will in most cases be a safe enough guide. In the macadamized streets of Paris the rule is to roll till a single piece of macadam, placed under the roller, will be crushed without being pressed into the road surface. Gravel somewhat mixed with clay by nature, but not too much, is probably best as a binding material. Clean coarse sand is very good. Other substances will do, where it would cost too much to procure either of the above.

A late writer in the *Journal of the Society of Civil Engineers and Architects*, at Hanover, calls attention to the practice in Bohemia of making foundation roads by setting first the foundation course, spreading a thin layer of the binding material on that, and the broken stone on top of this again. The subsequent rolling has the effect of forcing the binding material, slowly and gradually, from beneath, upwards into and through the broken stone. The writer states that he himself has tried a system of road construction that consists of a combination of the two methods hitherto used, with good results—namely: First, a foundation course, a thin layer of binding material on this, then the broken stone, another thin layer of binding material, and then wet down and roll.

Rollers.—The subject of rollers is one demanding some attention. In general, people are apt to over-estimate the value of a roller with respect to its weight. It will be evident, on reflection, that a roller should be as heavy per inch in length of roller, as a loaded wagon wheel is per inch of tire; or, in other words, if we have a wagon with tires 2½ inches wide, and on each wheel a load of say one ton, the roller should weigh two-fifths ton for every inch in length, or a roller three feet long should weigh about 14½ tons, or else a wagon as above described would exercise more pressure on the road-bed per square inch than the roller and consequently would cut into the roller surface and produce ruts.

The proper width of tire, or proper load upon any vehicle for a given width of tire, is a question that occasionally attracts attention. Bokelberg, a good German authority on the subject, in an article in the *Journal of the Society of Civil Engineers and Architects*, at Hanover, 1858, comes to the conclusion that for four-wheeled vehicles, upon a broken stone road, the loads should vary with the widths of tires, as follows :

Width of tires. Inches.	Load in pounds.
2 to 3	5,000 to 6,600
3 to 4	6,600 to 8,800
4 to 5	11,000
5 to 6	13,000
6 to 7	15,000
7 and over	16,500

Further conclusions are that the best width of tire, measured when they are new, for the transportation of freight, is from four to seven inches; this width being best for the easy traction of the load no less than for a minimum wear of the road surface. To make the tires wider than seven inches does not diminish the force required to move the load, and unnecessarily increases the dead weight of the wagons.

Road-rollers are of two principal kinds : those pulled by horses and those propelled by steam. The latter are, for many reasons, the best. In the first place they can be made as heavy as desired, without proportionally increasing the cost of propelling them, and being self-propelling, the only track they make is that of the roller, whereas with horse rollers, the hoof-marks of the horses are a great objection. Then, again, in the amount of work they will do at a certain cost, they excel horse rollers. They may be briefly described as a sort of locomotive mounted on three or four very broad and heavy wheels, these latter being the road-rollers.

An excellent pamphlet on the subject of steam road-rollers is the "Report on the Economy of Maintenance and Horse Draught Through Steam Road-Rolling," by Frederick A. Paget, E. & F. N. Spon, 1870. Readable articles on the same subject are : "Steam Road-Roller," *Engineering*, October 4, 1867; "Paris Kind of Steam Road-Roller," *Engineering*, May 7, 1869; "Cost of Operating Steam Road-Rollers, *Engineering*, June 18, 1869; "Good Steam Road-Roller," *Engineering*, January 14, 1870; "Economy of Steam Road-Rolling," *Engineer*, April 1, 1870; "How to Use the Road-Roller During Alternate Thawing and Freezing," Annales des P. & C., 1877, p. 125. In the spring and fall on finished roads, and occasionally during the first construction or reconstruction of roads, the surface becomes *sticky* mud, and to roll the road at those times, or to travel on it, tears up the covering and spoils the whole. If at such

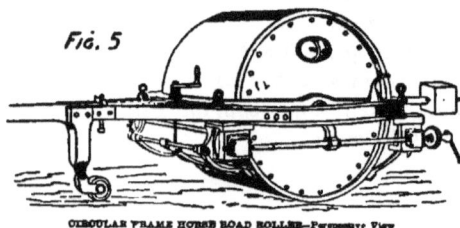

FIG. 5

CIRCULAR FRAME HORSE ROAD ROLLER—Perspective View

times the roller be constantly sprinkled and kept wet while it is being used, it will shed the mud or road covering, instead of tearing it up, and will consolidate the road in a very superior manner. And this method requires less water (only about one gallon to one and a half gallons per 100 feet of travel) than the method formerly used under these circumstances, of converting the *sticky* mud into liquid mud, by copiously wetting down the whole road.

There are several varieties in use in France and England, and two, at least, of the English kind have been imported into this country, one for the New York Central Park, the other for the Arsenal Grounds in Philadelphia. The cost of the Central Park steam road-roller, made by Aveling & Porter, of Rochester, Kent, England, was about $5,000 set up in New York, and the amount of work it will do in one day, at a running expense of $10, has been given as equal to that of a seven-ton, eight-horse road-roller in two days at $20 per day, or, in other words, it will do the same work at one-fourth the running cost and in one-half the time, of a first-class horse road-roller.

Since 1870, many other steam-rollers have been bought by various parties in the United States. Thus, there is one owned by Daniel Brennan, a road contractor, in Orange, N. J.; the city of New Haven, Conn., has run one with great success for several years;

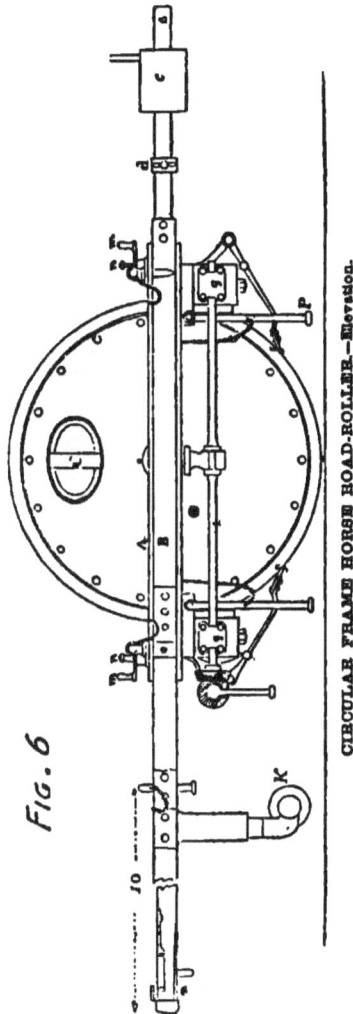

FIG. 6

CIRCULAR FRAME HORSE ROAD-ROLLER.—Elevation.

after many years of agitation on the subject, the city of Boston now owns and operates a steam road-roller, and so on.

The best horse road-roller of which the writer has any cognizance is the one shown by the annexed drawings in plan, elevation and in perspective. (See Figs. 5, 6 and 7.) It originated in Chemnitz, Germany, but can, of course, be easily made by any machine-shop or foundry. The hollow roller is made of cast iron, and is so arranged that it may be filled with water when it is to be used in

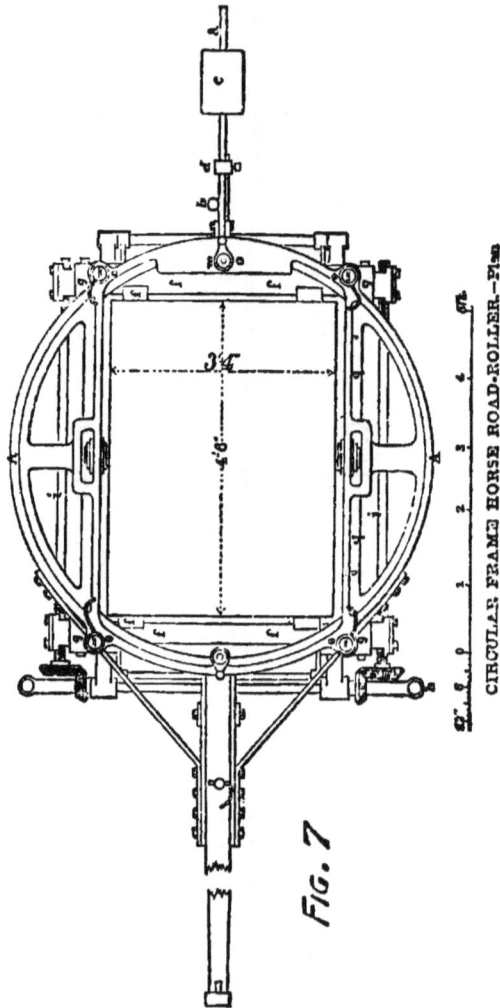

FIG. 7

CIRCULAR FRAME HORSE ROAD-ROLLER.—Plan.

heavy rolling; when not in use and about to be moved from place to place, the water is allowed to run out, thus materially lessening the load. A circular cast-iron frame A, surrounds the roller, and carries the axle bearings of the same. The outside of this frame is turned to form a groove in which a strong wrought-iron ring is fitted in such a manner that it will turn easily around the former. This wrought-iron ring consists of two semicircular parts, at whose junction the pole is attached on one side, and on the other an extension bar, carrying the balance weight c, which may be shifted by means of the set clamp d, or turned up by means of the hinge b. Pins going through the holes at e, fasten this ring or allow it to be turned for the purpose of pulling the roller in the contrary direction, when desired. The brake is shown at ff, and consists of four wooden brake-blocks, attached by iron shoes to a bar behind them and having rubber packing between the shoes. The screws shown and the handles h are used to operate these brakes. The cranks m, working the screws n, operate the scrapers l, which are used to keep the roller clean in muddy weather. The frame A, is made heavier at o, so as to have increased weight there to balance the whole frame-work in turning around. The support p, and the guide-wheel k, might be dispensed with. A great saving in time and in movements hurtful to the road is effected by making the frame circular as described, this allowing the roller to be turned with the greatest ease. The dimensions are figured on the drawing. A roller of this kind, 4½ feet in diameter and 3½ feet long, and weighing some four tons when empty, would cost perhaps $560 to $600; one 5'x3'x8", weighing about 5¼ tons (empty), some $700 to $750. Leaving off the brake would diminish the cost about $50.

Before leaving the subject of macadam top roads, it ought to be mentioned that a bed of rubble stone, 10 or 12 inches deep, merely spread uniformly over the road-bed as a foundation, is better than nothing at all, but can never make the same quality of road as the rough paving described above.

The following data are to be used in estimating the cost of the kind of road just described. Rough foundation paving, pieces five or six inches long, filling up crevices and ramming the whole with hand rammers, costs, after the material has been brought to the spot, one day's work of a common laborer for every four square yards, this assuming that the paver gets 1⅔ common laborer's wages. Same kind of paving if set in sand will cost one day's work of a common laborer for every 2¼ square yards. These figures for the cost of setting rough pavement for a foundation course have been objected to by an American road contractor as entirely too high, he claiming to set 20, and even 50, square yards to a man per day. An explanation of these different figures probably lies in the phrase "ramming the whole with hand rammers"; in the general quality of the work done, etc., the writer's own opinion is, that no very fine work is necessary in the construction of the foundation course. Its duties are to remain pervious and not to settle unevenly. The same contractor above mentioned wrote, in 1870 : " I put down and keep in perfect order for a year from the time of completion, a 12-inch road (6 to 7 inch foundation, 5 to 6 inch surfacing) at a distance from the quarry of three miles (materials exclusively quarried trap rock) for $1.50 per square yard ; wages of men average $2.25 per day, and of transportation, $1.25 per cubic yard.   This includes my profit."

To make macadam by hand costs, for sizes from 1¼ to 1½ inches of very hard rock, one day's work for every 0.6 to 0.44 cubic yards, for less hard rock, one day's wages will make 0.7 to 0.6 cubic yards, and of soft rocks 1.76 to 1.17 cubic yards.   In 1872 the estimated cost of crushing stone by the Blake machine ranged from 30 to 60 cents per cubic yard ; to crush the same stone by hand it was estimated would cost from $1.20 to $3 per cubic yard.   To spread 14 to 12 cubic yards of macadam is also about a day's work.

Gravel Top.—Instead of the macadam top described in the preceding articles, screened gravel may be used.   These roads are the favorite ones in Central Park, New York, and are probably the best roads there are for pleasure drives.   It is a matter of some doubt yet whether they do as well for heavy trucking as they do for light vehicles.   The foundation for these gravel roads should be the same as the rough paving for the macadam road ; some pieces were built in the Central Park having a rubble stone foundation but they are not recommended by their builders.   The gravel to be used for the top must be selected with some care ; it should be of a hard kind of stone, clean, that is, free from clay, etc., of the right color, etc. It is put on in two layers, each rolled, and the top one made compact and firm, by spreading and mixing in some good binding material, sprinkling and rolling.   There need be no fear of making a poor road by using the smoothest, most water-worn pebbles, free from all sand, etc., in making a road-top.   The upper portions of the river Rhine are remarkable for the clean, smooth pebbles that form its bed to a very great depth.   These pebbles are dredged up and used in road-building, making an excellent road-covering at a small expense.   In gravelly soil all the materials that are needed for a good road are frequently found on the spot ; they only need sorting out and re-laying.   For this reason a common gravel sieve often constitutes the principal instrument whose judicious use will make a good road out of a miserable string of ruts and cobbly elevations.   It would be only necessary to sift out and separate the soil under the road to a sufficient depth, into cobbles, coarse gravel, fine gravel, and sand ; then replace them in the order named, and with the proper thickness of layers of each ; wet down and roll, and the result will be a good road.   As regards the advisability of well constructing roads, the following from the Bath (Me.), *Times*, of May 11, 1870, is not without instruction (the Waltham roads therein spoken of are also mentioned in the extracts from a report, which is printed in the appendix): " I will here submit a comparison of the cost of our roads with those of the town of Waltham, noted for its *good* roads. Waltham has 51 miles of roads, the expense, including everything, of maintaining their highways, except sidewalks, for seven years previous to 1868, was $3,357 per year, or $66 per mile. In 1868, with 60 miles of road, including probably the building of nine miles, the cost was $6,000 or $100 per mile. The city of Bath has not over 32 miles of roads. The average cost of repairs on our roads for the past three years is $10,153; not including the expense of sidewalk, $317 per mile. At his rate, if we reduce the cost of repairs of our roads to $100 per mile we could afford to hire money at 7 per cent. and expend $100,000 upon their *permanent* improvement, and it would be vastly cheaper to do so than to continue our present system." There are many miles of such roads in Baden and in the Bavarian Rhine provinces.

Keeping Roads in Repair.—This subject properly finds its place here, being a matter of skill and a thing of debate only in the case of what we have called foundation roads ; pavements and trackway roads, to be considered after this, need no special directions as regards their repair or maintenance.

After a road has been properly rolled, and the surface made compact and smooth, it should always be maintained in that condition, no matter how great is the amount of travel on it. " A stitch in time saves nine," here as well as elsewhere. The tendency is to produce ruts ; these gather water; this soaks into the road-bed and spoils the whole. The problem can be put in this way : To have a good road it is necessary that there be no dust or mud on the same, and that there be no ruts ; therefore remove the dust and mud as fast as they are formed, and fill up the ruts as fast as they are made. The whole matter is here in a nutshell. It may be thought, at the first view, that this is too expensive a system. Its principal beauty lies, however, in the fact that it costs less per mile of road kept one year than the pernicious system of annual or semi-annual repairs, as will be shown and proved. The above two rules—sweep off the mud and dust as fast as they are formed, and fill up the ruts and bad places with new material as fast as they appear—are all that is necessary to be carried out in order that there be continually a good road. Without continual repairs, there can be no such thing as a constantly good road—a proposition that cannot too often be repeated. By repairing a road annually, or twice a year, it matters not which, the result is, strictly speaking, a good road at no time during the whole year. The road is wretched just after repairs ; it becomes passable after a while, and deteriorates from that day forward, until it is again made wretched ; and so on, *ad infinitum*, according to the present only too commonly followed system. By the other method is offered us a road as smooth as a floor, year in year out, and, let it not be forgotten, at a less expense.

A French engineer, named Trèsaguet, was the first in 1775, to call attention to this proper method of making road repairs. His system—the above described one—was adopted in Baden in the year 1845, and has been long in universal use in all the active European countries. Table VII. gives the actual average quantity

TABLE VII.

YEAR.	Cubic yards used per mile in one year to repair roads.
1832...	218.6
1839...	198.7
1851...	127.2
1855...	91.4
1856...	89.4
1860...	93.4

of road macadam used per mile of road in Baden to make the repairs in one year, and shows the decrease after 1845. Table VIII gives, in the first column, the cost of materials and labor required to repair one league for one year according to the old way—this column being calculated for the years following 1845 from the cost of the preceding years, and allowing for the increased value of labor and materials—while in the second column we have the actual cost as it was with the system followed at the time.

These figures are taken as given by the Chief Engineer of the Baden Public Works, Mr. Keller. He quaintly adds : " These tables give clear evidence in favor of the reduced cost by the adopted system. That roads are better now than they formerly were, everybody knows." Another German engineer expresses himself to the same effect in a little different way : " It costs no more," says he, " to keep the roads in repair now (1864) than it did twenty years ago, when this method (of continual repairs) was not in use, although labor is now three times and materials are twice as dear as they then were." There seems to be no doubt of the superiority of the continual repair system in every respect, producing very much better roads, and at the same time costing less. It need only be tried with us to be thenceforth adopted.

TABLE VIII.

YEAR.	Cost of Repairs of one league of Road.	
	By old way of so doing, in florins.	By system of continual repairs, in florins.
1835	1,002	1,002
1840	1,086	1,086
1845	1,170	$975\frac{28}{60}$
1850	1,254	$965\frac{40}{60}$
1855	1,339	$835\frac{6}{60}$
1860	1,423	$978\frac{54}{60}$

How to Repair Roads on the Continuous System.—We suppose the material for road-covering to lie in regular measured heaps, all ready to be used, at the storage places, once or twice above mentioned, as being 200 to 250 feet apart alongside of the road, but not encroaching upon it. Then, for every two or three miles of road, a so-called road-keeper is employed to do the necessary work and repairs. An enumeration of his duties will comprise at the same time an essay on the art of road repairing.

1. The road keeper is to remove the dust formed in dry weather by sweeping with a brush broom. This is done to greatest advantage just after a slight shower. In muddy weather, it is essential that the mud be removed by means of brooms or hoes. A little mud on the surface causes ruts, and much mud softens up the whole road-surface. The mud is to be raked up in heaps alongside of the road, there left to dry and then carted off. To hinder as much as possible the formation of any mud, the surface drainage must remain unimpaired; should it be out of order, the water standing on the road is to be swept off. To diminish the wear of the road in dry times, the road should be sprinkled.*

2. Inasmuch as the covering gradually wears off, notwithstanding all precautions, it must be renewed, and should be so renewed gradually in the same measure as it wears off. The best time to put on new road metalling is during continuous wet weather.

* Bowles, in his book, "Our New West," mentions the case of the stage road from Sacramento to Virginia City *via* Placerville, 150 miles long, and having an annual traffic of 7,000 or 8,000 heavy teams, and whose proprietors found that the simplest and cheapest way of keeping it in repair during dry weather was to sprinkle the whole of it—150 miles of mountain road.

3. In filling up holes, the bottom of the same is to be swept clean of mud, then filled up level with the remainder of the road, not in a heap so high above it as to obstruct travel.

Every care should be taken to have the new material join as speedily as possible with the old portion of the road, and it should be so well laid that it will give the least possible hindrance to vehicles, which will then not avoid the patched places.

4. When many ruts occur in a short distance, the deepest only are to be filled at first. After the patching in these has become solid, then the rest are to be attended to. Long ruts or wheel tracks are not to be filled up the whole length at once, but only short pieces at a time. If this precaution is neglected, vehicles avoid such places, and new ruts are formed elsewhere.

5. Inasmuch as more material is worn off in a dry season than can be put on, there are then when wet weather comes, large places to be repaired. These must be mended by degrees, never filling up a piece larger than 8 to 10 feet by 4 to 7 feet at a time, and not having these pieces too near together; when these have become solid, then some more may be filled in and so on till the whole is done.

Should it, however, become absolutely necessary to repair a piece of road in dry weather, the place where the new macadam is to be deposited must be loosened up with a pick, then the new material put on and a solid top formed by the judicious use of stone-dust or other binding material and sprinkling with water and pounding down with the shovel, or by what may be called "puddling" until the whole be solid. Should a frost or very dry weather occur immediately after macadam has been put on the road in wet weather, so that the same will not join on to the rest of the road surface, the whole must be removed, cleaned and returned to the storage heaps for future use. A layer of macadam over the whole road should never be put on without treating it immediately afterwards in the manner described above for building new roads, that is, mixing in binding material with the top course and rolling it in wet weather or after sprinkling.

The road-keeper is naturally also the person to see to the proper delivery on the part of the contractors, if such there be, of the road material in the prescribed places and to attend to the measuring of the same.

In short, and to sum up, it is his business to keep the road in good order, and with proper men and surveillance the desired result is achieved easily, and at a less cost than by any other system. The quantity of macadam required to keep a certain length of road in repair varies very much ; it depends, as we have seen, on the care with which the repairs are made, naturally also on the kind of stone used and on the amount of travel over the road. For a width of road = 20 feet, the average quantities required per year to keep a length of ten feet in repair, on the system of continuous repairs, has been given, as in Table IX.

TABLE IX.

		Cub. ft.	Cub. yds.
1.	Good material and heavy travel............	15–20 —	.55–.74
2.	Good material and medium amount of travel	10–15 —	.37–.55
3.	Good material and light travel......... ...	5–10 —	.18–.37
4.	Medium material and heavy travel.........	20–25 —	.74–.92
5.	Medium material and medium amount of travel...	15–20 —	.55–.74
6.	Medium material and light travel....	10–15 —	.37–.55
7.	Third rate material and heavy travel........	25–30 —	.92–1.01
8.	Third rate material and medium amount of travel	20–25 —	.74–.92
9.	Third rate material and light travel..... ...	15–20 —	.55–.74

These are the quantities as given by one authority, but from a comparison with the amounts actually used during a period of ten years on 39 roads, having various amounts of travel upon them and being repaired with all kinds of road metal, it would seem that the above figures are very ample. ,

The exact relation between the quantity of road material that is necessary to keep a road in repairs, and the amount of travel over it is still a matter of intelligent observation and discussion. The quantity required does not seem to be proportional solely to the amount of travel, even with one and the same kind of stone used on the same road ; as will appear also when it is considered that were there no travel over the road at all, the surfacing would, nevertheless, wear out by the action of the frost, the rain, etc. As recent an article as the "Annales des P. and C.," 1877, p. 226, is devoted to this subject, and does not arrive at any definite general conclusion.

Repairs of Macadamized and Much Frequented Streets in Cities.— In this case, where the amount of travel in one day is often greater than that of a month or more on the town road, the system of continuous repairs ceases to be the best available, on account of the incessant throng of vehicles not giving any repaired place a chance to become solid before it is again plowed up and scattered. Thus in the city of Paris, on the Boulevards, etc., the continuous system has been abandoned, and the practice now is to let the street gradually wear down three to four inches, then close half of it (divided "fore and aft") to travel, loosen it all up with picks, and put on a layer three or four inches (best not to put on more than that), spread a thin layer of sand over this, sprinkle and roll heavily. It often happens that the men put too much of the sand on; in that case, the road, after it is all done, is finally well watered and the roller is again passed over it a number of times.   This operation causes the superfluous binding material to come to the surface in the shape of thin mud and leaves the road covering as hard and smooth as mosaic, making a most excellent driveway.   It emits a sonorous, ringing sound on being driven over, and remains clean and without mud throughout the heaviest rain-storms.   The rolling of the streets in Paris is done by a company owning a large number of steam-rollers; in paying them for work done the city was obliged to go back to first principles for a measure of such work, it being found impossible to estimate correctly by the square measure of surface rolled to such and such a degree of hardness.   The measure adopted is that of weight multiplied into the distance it has been moved, or "feet-pounds," as we should say.   It has been found from many years' experience that to roll one cubic meter of macadam requires 4 to 5 "kilometer-tonnes," and this is true whether the layer of macadam be $3\frac{1}{4}$ or 10 inches thick.   Expressed in our measures, this is 11,020 – 13,775 feet tons to 2,000 pounds = 2.09 – 2.61 mile tons per cubic yard of macadam.

The advocates of the steam road-roller claim that by means of that machine they are enabled to make a road that will wear out evenly and uniformly for four or five inches, so that the operation of patching need never be resorted to.   The steam road-roller can also be used for "picking" up a road, for which purpose the roller is armed with sharp spikes, and is then driven over the surface to be "picked" up.

Pavements and Trackways.—No essay on roads would be complete without some mention of these two species of road surface, though the use of the former is confined principally to streets, and that of the latter is out of date.

Pavements are either of stone, wood, iron, various concretes, asphalt, and may be of still other substances.

Stone Pavements.—The modern sizes of paving stones may be seen from the following cases: The Boston size is $4\frac{1}{2}''$x$3\frac{1}{2}''$x$7''$ deep; New York Belgian, $6''$ to $8''$ x $5''$ to $6''$ x $6''$ to $7''$ deep; new Broadway pavement, also called Guidet pavement, $3\frac{1}{2}''$ to $4\frac{1}{2}''$ x $10''$ to $14''$ x $7\frac{1}{2}''$ to $8\frac{1}{2}''$ deep.   This last is laid with the long sides of the stones across the street; and, as far as the author's judgment goes, is the best size for stone pavement there is.   The Boston size is too small, and allows of no bond between the separate paving stones.   Further, the weakest part of each stone being its edge, it follows that the more edges there are in a given surface of pavement, the speedier will it wear out, each stone becoming rounded and slippery.   It is only the excellent workmanship and great care displayed in setting these stones in Boston that prevents these facts from being at once apparent to all. When it is added that in setting pavements, the natural soil, except it be sand or fine gravel, is in all cases to be excavated 12 to 19 inches, and then filled up 5 to 12 inches, according to the solidity of the subsoil, with clean, coarse sand or fine, clean gravel, and the paving stone set in this and well rammed down with hand-rammers; about as much is said on this topic as can be said without going into long details.

From 4½ to 6 cubic feet of sand are required for every square yard of paving. In setting two different pavements, the same written rules may be exactly followed in either case, yet one be much better than the other, so much depends here upon good, careful, conscientious workmanship.

Wooden Pavements.—There are so many kinds of these that it would be out of place to enumerate and describe them all here. Their advantages are less wear on tires and horses, less noise and smooth traction; a disadvantage is their slipperiness in the winter. There seems to be a sort of notion that wood pavements and coal tar must go hand-in-hand; but there certainly is no necessity for this. Coal tar is applied as a preservative to the wood; but it must be acknowledged that many better ones are known, and indeed are used, to the utter exclusion of coal tar, in all cases where it is desired to preserve wood, except in this of wood pavements.

No wood should be used in paving that has not been first subjected to some approved method of preservation, or impregnation, as it is frequently called. · The best manner of setting the same is still a mooted point, which it would be presumptuous at present to decide.

A valuable contribution to the subject of wooden pavements is the report of the Commission appointed by the City of Boston to consider this subject, in 1872, City Document No. 100, 1873. The Commission comes to the conclusion that the best way to preserve the wood that is put down is by the method called Burnettizing, after its inventor, Sir H. Burnett, of England, in 1838. It consists of treating the wood to be preserved with chloride of zinc. The Commissioners wisely add: ` "Your Commissioners are of the opinion that if the city adopts any method of preserving blocks to be used for pavements, some additional security should be had that the treatment of the wood shall be thorough and complete." As regards the construction of the pavement, the Commissioners recommend spruce blocks (for this section of the country), lay stress on the necessity of a solid, uniformly constituted and rolled gravel foundation, and then say: " The rows or blocks should be set square across the street, and should be about four inches thick at top, with spaces of about one-half inch between the rows. This may be done with blocks of uniform thickness set apart, or with tapering blocks half an inch thicker at bottom than at top. The latter arrangement is the more costly, but it is believed by some that it will stand better, by reason of its covering the whole surface of the foundation. Longer trial is necessary to settle this point beyond dispute. Blocks of only a short chamfer at the top leave the inter-space too narrow, as the blocks wear down." The Commission named consisted of "two chemists, two practical mechanics, and one civil engineer."

Cast-iron pavements are out of favor on account of their great cost, and concrete pavements are a matter of experiment as yet.

Asphalt pavements are chiefly used in Paris. They are slippery in wet weather, and produce a very disagreeable, penetrating dust in dry weather. It is necessary to prepare a bed of macadam to lay them on, and they are not used in Paris except in streets where the gas-pipes are carried either in the sewers or under the sidewalks, as any leak of gas would destroy them. Their use is a matter of doubtful economy.

Trackways are, as has been mentioned, out of date. Where a common road does not suffice nowadays, a railroad is built; but time was when trackways were of considerable importance. They consist, if of stone, of large, flat stones, say 12 inches deep and 4 to 6 feet long by 14 to 16 inches wide, solidly bedded in two parallel rows, at such distance apart as to make of each row a track for the wheels. The space between is paved. They are, of course, very expensive, but cost little to repair, and enable a horse to pull a very great load. As has been mentioned, Telford made use of such a stone trackway to avoid cutting down a hill on his Holyhead road. There were two hills, each a mile in length, with an inclination of 5 in 100. It would have cost $100,000 to reduce this grade to 4½ in 100, but nearly the same advantage, in diminishing the tractive force required, was obtained by keeping the 5 in 100 grade, with moderate cuttings and embankments, and making stone trackways, at a total expense of less than half the former amount.

"Plank roads," once so much in vogue in the United States, may not improperly be classed among roads with trackways, and, with them, also among the things that were. From their perishable nature, they can never advantageously do more than help the development of a new country, and in this, as well as other States, are yearly becoming more and more impracticable on account of the constantly increasing price of lumber.

On the Resistance to Motion or the Force Required to Move Vehicles on Different Kinds of Roads.—Before, as well as since the introduction of railways, engineers in England, Germany and France made many experiments on the force necessary to pull different vehicles, at various speeds, over various surfaces. To enumerate the details of all these experiments would be perhaps useless; a few general results only are here given.

Experiments, as above indicated, were made by Edgeworth, Count Rumford, Bevan, MacNeill, Minard, Navier, Perdonnet, Poncelet, Flachat, Morin, Kossak, Umpfenbach, Gerstner, and no doubt others, a list of authorities that proves the subject to have been well nigh exhausted. The experiments of Morin, made in 1838–41, appear to have been made with a degree of care and accuracy, leaving nothing more to be desired. Table X. is an extract from his results,* and gives that fraction of the weight of the vehicle and load which is required to move them on a level road :

TABLE X.

CHARACTER OF THE ROAD.	2-wheeled carts.	Trucks, 4-wh 3 and 4 horse.	4-horse stage coaches, on springs.		2-horse carriages, body on springs.	
Firm soil, covered with gravel 4 to 6 inches deep....................	$\frac{1}{12}$	$\frac{1}{8}$	$\frac{1}{8}$		$\frac{1}{8}$	
Firm embankment, covered with gravel 1¼ to 1½ inches deep......	$\frac{1}{16}$	$\frac{1}{11}$	$\frac{1}{10}$		$\frac{1}{10}$	
Earth embankment, in very good condition	$\frac{1}{11}$	$\frac{1}{15}$	$\frac{1}{16}$		$\frac{1}{16}$	
Bridge flooring of thick oak plank...	$\frac{1}{16}$	$\frac{1}{26}$	$\frac{1}{14}$		$\frac{1}{42}$	
BROKEN STONE ROAD.			Walk.	Trot.	Walk.	Trot.
In very good condition, very dry, compact and even..	$\frac{1}{18}$	$\frac{1}{54}$	$\frac{1}{18}$	$\frac{1}{41}$	$\frac{1}{40}$	$\frac{1}{42}$
A little moist or a little dusty.... ..	$\frac{1}{38}$	$\frac{1}{18}$	$\frac{1}{14}$	$\frac{1}{27}$	$\frac{1}{31}$	$\frac{1}{28}$
Firm, but with ruts and mud........	$\frac{1}{23}$	$\frac{1}{14}$	$\frac{1}{21}$	$\frac{1}{18}$	$\frac{1}{28}$	$\frac{1}{18}$
Very bad, ruts 4 to 4½ inches deep, thick mud...............	$\frac{1}{13}$	$\frac{1}{14}$	$\frac{1}{12}$	$\frac{1}{10}$	$\frac{1}{12}$	$\frac{1}{16}$
Good pavement { Dry.............	$\frac{1}{60}$	$\frac{1}{56}$	$\frac{1}{57}$	$\frac{1}{56}$	$\frac{1}{57}$	$\frac{1}{57}$
Good pavement { Covered with mud	$\frac{1}{29}$	$\frac{1}{50}$	$\frac{1}{21}$	$\frac{1}{12}$	$\frac{1}{28}$	$\frac{1}{11}$

* A full account of Morin's experiments on the resistance to motion of vehicles, on the wear caused by different vehicles on roads, and on the loads different vehicles should carry so as to produce the same wear, may be found in Morin, " Expèrience sur le tirage des Voitures," Paris, 1842.

To take an example, suppose we have a truck weighing with its load 9,000 pounds. How many pounds traction will be required to move the same?

Ans.—On firm soil, gravel 4 to 6 inches deep—that is, a newly repaired road, as we often find it ($\frac{1}{9}$ by table), 1,000 pounds; on best kind of embankment ($\frac{1}{29}$ by table), 310.3 pounds; on broken stone road in good condition ($\frac{1}{54}$ by table), 166.6 pounds; on broken stone road, deep ruts and mud ($\frac{1}{14}$ by table), 643. pounds; on a good pavement ($\frac{1}{65}$ by table), 138.5 pounds. Or, since the tractive force of a médium horse when working all day is said to be about 125 pounds, we need in the first case, 8 horses; in the second case, $2\frac{1}{2}$ horses; in the third case, about $1\frac{1}{4}$ horses; in the fourth case, about 5 horses; and in the fifth case, only one good horse to move the same entire load all day.

These facts, expressed in Table X. in striking yet perhaps dry figures, can be nearly as well given in popular language.

Says a correspondent (Dr. Holland), of the Springfield *Republican*, writing from England, after describing the kind of horses in use there:

"Now, with all these horses the rule follows that every pound of muscle does just as much work on the road as two pounds do in America. The cab and omnibus horse does twice as much as the same horse does in America. The draft horse does as much in the dray as two ordinary dray horses in America, and the little horses, which are driven mainly in butchers' carts and grocers' carts, will tire a cab horse to follow them with no load at all.

"In connection with these statements it should be recorded that the speed of all vehicles in the streets of London, whether the localities be crowded or not, is at least a third faster than it is in corresponding streets in American cities. The ordinary speed of vehicles in London, in which passengers or light loads are transported, is one which is considered not entirely safe in Main Street, Springfield, Mass., and one which, in some streets of Boston or New York, would be at once checked by the police. A man who sits in a ' Hansom' finds himself driven at an unprecedented pace through crowded thoroughfares, and, Yankee though he may be, he will often wonder whether he is going to bring up at last without a broken neck.

"I mention this matter of speed, particularly, because it shows that even more work is done by one horse in London than by two in New York. He not only draws as large a load, but he travels with greater rapidity. The streets of London present such a spectacle of headlong activity as no American city can show, in consequence of the rapid passage of all sorts of vehicles through the streets. I might add to this statement, touching the superior speed of the London horses, a word about the greater weight of the carriages which they are obliged to draw behind them. All carriages are built more heavily in Great Britain than in America. They are built to last, and many of them seem to me to be superfluously heavy.

"The point which I wish to impress upon my American reader is simply this: that the English horse, employed in the streets of a city, or on the roads of the country, does twice as much work as the American horse similarly employed in America. This is the patent, undeniable fact. No man can fail to see it who has his eyes about him. How does he do it? Why does he do it? These are most important questions to an American. Is the English horse better than the American? Not at all. Is he over-worked? I have seen no evidence that he is. I have seen but one lame horse in London. The simple explanation is that the Englishman has invested in perfect and permanent roads what the American expends in perishable horses that require to be fed. We are using to-day, in the little town of Springfield, just twice as many horses as would be necessary to do its business if the roads all over the town were as good as Main Street is from Ferry to Central. We are supporting hundreds of horses to drag loads through holes that ought to be filled, over sand that should be hard-ened, through mud that ought not to be permitted to exist. We have the misery of bad roads, and are actually or practically called upon to pay a premium for them. It would be demonstrably cheaper to have good roads than poor ones. It is so here. A road well built is easily kept in repair. A mile of good macadamized road is more easily supported than a poor horse."

Other results of Morin's experiments are as follows :

1. The force required to draw a vehicle, is directly proportional to the load and inversely so to the diameter of the wheels ; in other, more common, words, the tractive force increases in the same ratio that the load increases, and the diameters of the wheels decrease.

2. On a paved or well built macadam road, the tractive force is independent of the width of the tires, provided the same is more than three or four inches. On compressible roads, such as new tgravel, on a meadow, etc., the tractive force diminishes with an increase in the width of the tires.

3. Other circumstances being equal, the tractive force is the same for vehicles with and without springs, as long as the horses are not moving faster than a walk.

4. On paved and well macadamized roads the tractive force increases with the velocity, according to the law, that beyond a velocity of $2\frac{1}{4}$ miles per hour ($3\frac{7}{16}$ feet per second) the increase of the tractive force is in direct proportion to the increase in velocity ; his increment is, however, less, the softer the track or road and according as the vehicle is best provided with springs.

5. On soft earth embankments, or on sand or sods, or on streets newly covered with gravel, the tractive force is independent of the velocity.

6. On a well made pavement of regular shaped stone, the trac-tive force, horses on a walk, is about three-fourths of that on a good macadam road, but with horses on a trot, the two are about equal.

7. The wear on the road is greater the smaller the diameter of the wheels, and greater in the case of vehicles without than for those with springs. Most road-rollers, as now in use, have too small a

diameter, besides being too light, and consequently do not properly compress the road surface.

8. The tractive force, as well as the wear on the road, is greater in the case of vehicles that have their wheels placed at an angle with the vertical by reason of the ends of the axle-trees being bent down, than for those that have their wheels set plumb and the center line of the axle-trees level.

Part II.

On the " Best Methods of Superintending the Construction and Repair of Public Roads in this Commonwealth."

In looking for a solution of this question the people of the Commonwealth may turn as they choose, either to the West or to the East, to see a guiding star : to the city of Chicago, or to the city of London, both under a republican form of government, alike or similar to that under which we live. It lies in the establishment of a Board of Public Works, composed of a number of able men, well paid for their services, gradually changing in their membership in the board who shall have this and only this as their occupation and who can therefore be held responsible for their acts. This is the system that has been adopted both in London and in Chicago and with remarkable success and resultant benefits. There are many other systems in use in foreign countries, all of which, however, seem to be inapplicable here, placed as we are under so different forms of government ; hence, though well acquainted with the systems adopted in France and in Germany, the writer has not described them here.

Experience in London.—The history of the " Metropolitan Board of Public Works of the City of London " is about as follows :

What is known as the city of London consists in reality of a great number of what we should call towns, there called parishes, and of which the *city of London* is only one single member. Each one of these parishes had, and still has in most respects, its own local government and in consequence took care of its drainage, its streets, etc., etc., as seemed best and as it liked, some better, some worse and some not at all. This state of things in the matter of drains and sewers finally led to a most deplorable condition of affairs ; there was not, nor could there under these conditions be, such a thing as a *system* of sewers and consequently a proper and adequate drainage ; the death-rate increased to an alarming extent and matters came to be universally regarded as past all endurance. What could be the remedy ? No well-grounded complaint could be made against the majority of the men composing the various local governments, since they were good and honest citizens, and hence no change in the separate governments could ever bring relief. The fault lay not in the men, but in the system of ruling they were called upon to fulfil—that is, in the incompetent and faulty tread-mill of government they were annually called upon to keep in its usual operation. It was then seen that by having an elected power to supervise and regulate the sewage affairs of the whole metropolis, a complete *system* of drainage could be carried out, and thus only. Such a regulating power is exercised by the Metropolitan Board of Public Works, chartered by Act of Parliament and composed of members elected from all parts of London. It is, perhaps, in place here to explain what is meant by a *system* of sewers, as the same definition will hold good in other matters ; as for a *system* of roads, of drainage and irrigation of lands, etc. Perhaps the best illustration would be to refer one to the veins and arteries in the human body, or to the body of a tree, from its trunk through the branches growing smaller and smaller down to the smallest twig that may be on it. It will be at once seen how different any arrangement, in which may be detected the wisdom to contrive, the strength to uphold and the beauty to adorn, like this, is from a miserable patchwork such as cannot but arise where the separate parts of one whole are each left to guide themselves without any unity of action or design, as to their final resultant. The London Board of Public Works had some extraordinary powers conferred upon it, such as the right to levy assessments on real estate benefited by their improvements, and others. Originally constituted merely to plan and execute a system of sewerage for the metropolis, this Board of Public Works soon showed itself so useful and beneficial in its actions that other matters were placed in its charge, such as the laying out of new streets the building of the Thames embankment—a work of exceeding great magnitude and importance—and there seems to be no doubt that in all public works London will find it advantageous to employ its Metropolitan Board of Public Works.

Experience in Chicago.—In the city of Chicago there has been a Board of Public Works almost from the very start. It arose there from the union of the water-supply and the sewerage commissioners, and has existed since May, 1861. No less than in London, it has proved to be of great benefit to the community ; and it would have been impossible, under any other system, to have executed in so satisfactory a manner the many and useful public works for which Chicago is famed. At the risk of introducing in this place some very dry reading, a general synopsis of those parts of the city charter which relate to the Chicago Board of Public Works is here given. The whole may be found in a copy of "Laws and Ordinances, Chicago, 1866."

SECTION 1. Establishes a body known as the *Chicago Board of Public Works*, to consist of (3) three members, chosen by the people, one from each division of city.

The first *three* chosen for one, two and three years; after that, one each year for three years,

SEC. 2. Each member of Board shall receive annual salary of $3,000 (by Act of February, 1866); give bonds for faithful discharge of duties; pay over all moneys, papers, etc., at expiration of his term, or when ordered by City Council.

SEC. 3. Board to elect president and treasurer, and make by-laws.

SEC. 4. Majority constitutes quorum; records to be kept of proceedings; copies of all plans, estimates, etc., to be kept; report (annual), to be rendered on or before ——————— each year, or when required by City Council. Each member authorized to administer legal oaths.

SEC. 5. Board shall have special *charge* and *superintendence*, subject to the laws and ordinances of City Council, of all streets, lanes, alleys, etc., in the city of Chicago, and of all walks and crossings in the same, and of all bridges, docks, wharves, public places, landings, grounds and parks in said city, and of all halls, engine-houses, and other public buildings in the city belonging to city, except school-houses, and of the *erection* of *all* public buildings; of lamps and lights in streets, etc., and in public buildings, and repairs of same; of the harbor works and improvements; of the city sewers and drains and of the water-works; of the fire-alarm telegraph, and all public works and improvements hereafter to be commenced by the city, as well as such other duties as may be prescribed by the City Council by ordinance.

SEC. 6. All *applications or propositions for improvements*, or new works of kind specified in section five, shall hereafter be first made to Board of Public Works, or if first made to City Council, shall be by them referred to Board. Upon receiving application, Board shall investigate the same, and if they find such work necessary and proper, shall thus report to City Council, with an estimate of the expense thereof. If they do not approve of such application they shall report the reasons for their disapproval, and the City Council may then, in either case, reject said application or order the doing of work or making of public improvement, after having first obtained plans and estimates thereof. The Board may also in like manner recommend, whenever they think proper, any improvement of the nature above specified, though no application has been made therefor.

SEC. 7. Shall be duty of Board to procure for city full *plans and estimates* of contemplated improvements, when so ordered by Council.

SEC. 8. Whenever any public improvement shall be ordered by City Council, and money appropriated, Board shall *advertise for proposals* for doing work; plans and specifications of same first placed on file in office of Board, which plans and specifications shall be open to public inspection; advertisement to state work to be done, and to be published ten days at least. The bids shall *be sealed bids*, directed to Board, and accompanied by bond to city, signed by bidder and two responsible sureties, in sum of $200, conditioned he shall do work if awarded to him; in case of his default to do so, etc. Bids to be opened at time and place mentioned in advertisement.

SEC. 9. All *contracts* shall be *awarded* to lowest reliable bidder, and who sufficiently guarantees to do work under superintendence and to satisfaction of Board; *provided*, that the contract price does not exceed the estimate, or such other sum as shall be satisfactory to Board. Copies of contracts to be filed with City Comptroller.

SEC. 10. Board reserves right, in contracts, to decide questions as to proper performance of work and meaning of contracts; in case of *improper construction* may suspend work and re-let same, or order entire reconstruction; or may re-let to other contractors and settle for work done, etc.

In cases where contractor properly does work, Board may, in their discretion, as work progresses, grant to said contractor estimate of amount already earned, reserving 15 per cent. therefrom, which shall entitle holder to receive amount, all other conditions being satisfied.

SEC. 11. In case prosecution of any public work be suspended, or bid be deemed excessive, or bidders be not responsible, *Board may*, with written approval of treasurer, where urgency of case and interests of city require it, *employ workmen* to perform or complete any improvement ordered by Council; *provided*, that the cost and expense shall in no case exceed the amount appropriated for the same.

SEC. 12. All *supplies of materials*, etc., when costing over $500, to be purchased by contract, subject to same conditions as letting out work.

SEC. 13. Whenever Board think necessary for interests of city, to protect same from damage or loss, shall report thus to aldermen, and reasons for same, asking power to give *contracts without notice* required above, and aldermen may grant request; *provided*, three-fourths vote for it.

SEC. 14. Whenever Board is of opinion *work may be* better *done without contract*, shall so report to Council, and same may authorize Board to procure machinery, materials, etc., hire workmen, etc.; *provided*, a three-fourths vote be in favor of granting authority.

SEC. 15. All contracts and bonds by Board to be in name of city.

SEC. 16. *No member to be interested in any contract;* all contracts made with any member interested, city may declare void; any member so interested shall forfeit his office and be removed therefrom; the duty of every member of Board and of every officer of city to report delinquency, if discovered.

SEC. 17. All existing contracts executed by city, by water or sewerage department, etc., to be carried out by Board.

SEC. 18. Board shall nominate each year the various *officers*, now provided for by ordinance, which serve in the departments under their special charge. the city engineer, superintendents sewers, streets, etc. Shall be empowered to employ from time to time such other superintendents, clerks, etc., as they may deem necessary, subject to ordinance as regards pay, etc.

SEC. 19. Board to have charge and superintendence of works made for city, and paid for by private individuals or by State. Plans for same to be approved by Board.

SEC. 20. Board shall, on or before ——————— every year, submit to auditor, by him to be presented to Council with annual estimate, *statement of the repairs and improvements* necessary to be undertaken for current year, and of the sums required by Board therefor; report to be in detail; report, having been revised by Council, sums required shall be provided for in annual tax-levy. All moneys to be paid to any person out of moneys so raised, shall be certified by president of Board to auditor, who shall draw warrant on treasurer therefor, stating to whom payable and to what fund chargeable; such warrant to be countersigned by president of Board.

SEC. 21. Board to keep *accounts*, showing moneys received and spent, clearly and distinctly, and for what purpose. Accounts to be always open for inspection of auditor or any committee appointed by City Council.

The object of introducing this synopsis here has been to give a complete picture of just what such a Board of Public Works is. It will be seen upon a little examination how entirely different a thing it is from the usual and only too customary "committee." Perhaps the greatest fault of a committee is its entire lack of what might be called body and soul. If corporations, as has been said, have no souls, a committee may be said to have neither body nor soul. It is alive to-day, wields great power, decides vital and important questions, and yet is nowhere to-morrow, and seemingly even its component atoms have vanished from the face of the earth. It is amusing and yet sad, when the action of some such committee has caused trouble to read some time after, that it all "is exceedingly discreditable to whoever is responsible for it." How much better to have a conservative, expert and reliable body, the members of which have no other business than to attend to their duties as such, who are well paid for it, and consequently can at any time be held strictly responsible for their actions. With such a power, wisely governing and regulating the roads of this Commonwealth, it would be an easy matter to make thorough improvements in the legislation concerning roads and in the roads themselves.

These are two changes the need of which is generally felt at present and has found expression in various ways.

It may be well to quote one at least, notable for saying very much in little compass,—of these calls for improvement, in this connection, and adding some more as belonging to their subject in the form of an interesting appendix. Says Governor Claflin in his Inaugural: "Few things are of greater importance to a community, or a surer test of civilization, than good roads. Those of our citizens who have visited Europe are unanimous in the opinion that our public roads are far inferior to those of other countries, where the means of easy and safe communication are better appreciated. The science of road-making is apparently now well understood; or, if it is, the present modes of superintending the construction and repair of roads are so defective that the public suffers to an extent of which few of us are aware. It may be found upon investigating the cause of our miserably poor and ill-constructed roads, that the laws relating to this subject need revision, so as to give more uniformity in their construction and the repair of our highways. It is evident, also, that the science of road-making should have a prominent place in the course of applied mathematics at the Massachusetts Agricultural College."

We stand in this matter of roads at precisely the same point that the good people of London did ten or a dozen years ago in the matter of their drainage, and our remedy is the same. The fault lies in the machinery of Government ; originally built up to cater to the wants and needs of a newly settled country—a colony breaking a path through the wilderness—it has long since ceased to satisfy the demands of the present *State* in no matter so essentially as in that of its government and laws relating to common roads and highways. This is a subject requiring special knowledge, to be acquired only by long experience or the shorter method of imbibing the experience of others, which on analyzing it, is all that any *study* amounts to ; formerly it was not so, and most any one sufficed to make improvements on Indian paths. We need then an *expert* government on this point.

There should be a distinction made between first, second and third class, or between, as they might be called, State, County and Town roads; the first two should not be left to be dealt with as it is the pleasure of each town. A chain cannot be perfect unless every link in it is so ; no more can a road. The State must attend to the State and County roads and set a proper example at least to be followed by the towns in the case of their roads. We need then a higher power than that of the towns.

It has been previously shown how we need a power that can be held responsible and is somewhat permanent, and to put it all together, we need, to order and maintain our highways, a Massachusetts Board of Public Works. For some years, it would have its hands full in improving the existing main roads and laying out some new ones, but in course of time, as in the older countries of Europe, its principal business would be the *maintenance* of the roads. It must be remembered that the Board of Public Works is merely the intelligent servant and adviser of the legislative and executive ; whatever sums the Legislature appropriates for certain objects, that is taken by the Board and made to yield its most in the shape of work accomplished. Beyond this and keeping its accounts, it has nothing to do with money or taxation.

The small State of Baden, a part of Germany, has been heretofore mentioned as a model in road construction and the care of the same. From the brief history of the roads of that country and their repsent management, we may take some useful notes. The account is that of the Chief Engineer of the department of "Roads and Hy-draulic Engineering," which has this matter in charge and is there-fore reliable.

"In Baden the condition of the roads has been the subject of great care. Within the last forty-five years many millions have been spent upon them and experience has shown this expenditure to be one of those most advantageously spent. As most of the roads are well laid out and as there are plenty of them, there remains now (1863) mainly the keeping in repair of the roads to be attended to and not to build any new ones. Our endeavor now is to do this at the minimum of cost. Statistics gathered on this subject show good results and point out to us the means of arriving at still better ones. The present road law was made in 1810. That part of the old law which relates to the maintenance of roads is still in force, but that part requiring labor as a road-tax was abolished in 1831, and likewise most of the road police regulations. The appropriation for roads had to be increased 250,000 florins to pay for the abolished road-tax labor and to make up 170,000 florins previously received from tolls, which were also abolished in 1831. The system now is as follows : All town roads are taken care of by the towns. The State merely appoints and pays a roadmaster, so called, who superintends fifteen or twenty roadkeepers and reports on the state of the roads, the reasons for their bad condition, if that be the case, what is needed, etc. The law for second-class or county roads was formerly, that when they were of importance to several towns, they had all to help maintain the same. As this gave rise to continual bickering and quarrelling, in which the road suffered most, it was changed in 1856. They are now taken care of under the direction of the State and paid for partly by the State and partly by the towns in which they are situated. Most of the roads under this head are those which have risen in importance since the building of railroads, and are generally those that lie perpendicular to the direction of the railroad they are influenced by. The towns not having the means very often to properly improve and repair such, it was found necessary and ex-pedient to give them the aid of the State, and in order to procure the necessary funds, all roads that run parallel to railroads and all those that had lost their importance by the construction of railroads, were in 1855 stricken from the list as State roads. These latter, as the name implies, are wholly under the care and kept up at the ex-pense of the State. Table XI. gives the total lengths of the roads at various periods.

TABLE XI.

In 1835, the total length of the State roads was..... 1,430.8 English miles.
In 1855,   "   "   "   "   "   ..... 1,500.8   "   "
In 1855, by excluding several State roads, this last
      length was reduced to............. .... 1,142.4   "   "
In 1861, it had increased to...................... 1,190.0   "   "

Second-class roads (keeping partly paid for by State).

In 1835, the length of these was......... .... 467.6 English miles.
In 1861, the length of these was.... .... 630.0   "   "

The areas, population, and population per square mile of Baden, Prussia, France, Hanover and Massachusetts, according to recent census, are as follows :

Country.	Year.	Area, Sq. Miles.	Population.	Population per Sq. Mile.
Baden....................	1871	5,891	1,461,562	248
Prussia.......	1871	134,045	14,643,698	184
France	1872	204,088	36,102,921	177
Hanover............	1871	14,857	1,963,080	132
Massachusetts	1875	7,800	1,651,912	212

Baden did have, at the time when her population per square mile was less than it is now, and Prussia, France, Hanover and many other countries that could be named, have now got, and for the past 40 or 50 years have had, a system of common road management and resultant common roads, of the character above described; while Massachusetts with a population of 212 per square mile, and corresponding wealth, and others of the States of the Union, have a species of highway management, and its resultant and corresponding sort of highways, which, in thinking of the roads of the countries named, are but as evidences of a partial civilization.

" So that the State had, in 1861, in all, 1,820 English miles of road to maintain, the towns helping to pay on 630 miles thereof. *

---

* According to "Chambers' Encyclopædia," Baden has an area of about 5,900 square miles, and had a population in 1858 of 1,335,952. It is probably this, or a little less, at the present time. Massachusetts has an area of about 7,800 square miles, and, according to the average of the computed populations in the supplementary tables of the census of 1865, it is, in 1870, 1,343,604.

Or, population per square mile in Baden......... .... ─ 226.43
    "   "   "   " in Massachusetts.... ..... ─ 172.26

By the same tables, accompanying the State census of 1865, we find that the population per square mile in Massachusetts will equal that of Baden, above given, somewhere between 1890 and 1891. It exceeds that of Prussia, and probably equals that of France at the present day, both of which countries have systems of roads and road-repairing but little, if any, inferior to those of Baden.

"The statistics of the road repairs are kept in the following manner: The roadkeepers are required to keep a record of all draught animals that pass in either direction. Horses that are being ridden, animals not before a vehicle, and teams going to and from the fields, are not counted. These records are kept only during the working hours. Likewise, not during the whole year, but only four months in each year, so selected as to give an average amount of travel. The travel on the road on Sundays and out of working hours is taken from a few observations; it is a very small percentage of the whole. At the end of the year these records and observations are collected and graphically represented on a map of the whole State. The different roads are drawn of a different thickness of line, according as the amount of travel on them is greater or less. The quantity of road metal used per yard of road, and the kind of metal used, give the data for another such map, in which the different colors of the roads represent the different materials used in their repair, and the figures on them and their thickness show the number of cubic yards per mile required to keep the road in order. Finally, we have a third map, which indicates, by the thickness of the several lines representing the roads and by the figures on them, the total cost per mile of repairing the road one year."

With this picture of a country happy and prosperous, in the possession of good and well kept roads, it may be well to leave the subject.

Massachusetts wants, for her proper development, much better roads than she now has; and reckoning for a period of, say fifty years, she can have these good roads, and have them kept in order at a less cost than that of keeping up the present poor ones for the same time. Besides this, we should see in the one case a healthy state of internal communications and trade; in the other an absence of both. Let each citizen so act and do his part, that these benefits may accrue to the Commonwealth.

NOTE BY THE SECRETARY OF THE MASSACHUSETTS BOARD OF AGRICULTURE.

For the sake of arriving at some practical end, I have requested the gentlemen to whom the prizes for essays were awarded, to suggest what form of legislation would be desirable as a change, from our present inefficient system of road management, to one which would promise better, more economical, and more satisfactory results. The large and varied experience and observation of these gentlemen, all of whom are competent engineers, entitle their opinions and judgment to favorable consideration; and the following, submitted by them, may serve as a basis or outline for future legislation:

AN ACT for the More Perfect Construction and Maintenance of the Common Roads or Highways Throughout this Commonwealth (Massachusetts).

SECTION 1. Establishes a body to be known as the State Board of Highways and Bridges, to consist of three skillful civil engineers, or persons practically expert in the science of road-making, to be appointed by the Governor, with the advice and consent of the Council, and to have their office in the State House.

SEC. 2. It shall be the duty of the Attorney-General, personally or by his deputy, to give his counsel and opinion on such matters as he may be called upon by the Board, for which service his compensation shall be

SEC. 3. The first appointment of members of the Board of Highways and Bridges shall be made on or before   , and there shall be appointed one member each for the terms of one, two, and three years; after that there shall on or before   each year be appointed one member for the term of three years.

SEC. 4. Each member of the Board shall receive an annual salary of   dollars ; give bonds for the faithful discharge of his duties ; pay over all moneys, papers, etc., at the expiration of his term, or when ordered by the Governor and Council.

SEC. 5. Board are to elect a president and treasurer, and make their own by-laws.

SEC. 6. A majority of the Board constitutes a quorum ; records to be kept of all the proceedings ; copies of all plans, estimates, etc., to be kept; report to be rendered on or before   each year, or when required by the Governor and Council. Each member authorized to administer legal oaths.

SEC. 7. Said Board shall prepare and submit to the   Legislature a plan for the systematic classification of all the highways and townways in this Commonwealth into two or more of the following three classes:

Class 1. State roads, to be controlled and maintained wholly by the State.

Class 2. District roads, to be controlled and maintained by the State, but the expense thereof to be borne by the towns and cities of the districts in which said road shall lie, and the State, in such proportions as said Board shall apportion.

Class 3. Town roads to be controlled and maintained as now provided by law.

The construction of new road roads, of the three classes above specified, to be done as follows:

Class 1. State roads, to be laid out and built by the State, through the Board of Highways and Bridges.

Class 2. District roads, to be laid out, etc., by the County Commissioners, as now provided, but the Board to have the final approval or disapproval of the proposed plans and profiles for said road, and also to have the charge and superintendence of their construction.

Parties aggrieved by the refusal or neglect of County Commissioners to lay out a road, to have the right to appeal to the Board of Highways.

Class 3. Town roads, to be laid out and constructed as now provided by law.

Sec. 8. The paying of road taxes by labor is hereby abolished, and all road taxes are hereafter to be paid in cash.

Sec. 9. Board shall have the special charge and superintendence, subject to the laws and resolves of this Commonwealth, of all the highways and bridges and the public works appertaining thereto, which are or shall be executed or maintained wholly or in part by this Commonwealth. They shall also perform such other duties as may be required of them by the General Court or the Governor and Council.

Sec. 10. Whenever any highway or bridge, or public work appertaining to these two, shall come partly within the province of this Board and partly within that of any other State Board already constituted, then such subject shall be discussed and decided upon in a joint convention or conventions, composed of equal numbers of this and the said other State Board, and some member by them chosen as presiding officer.

Sec. 11. All applications or propositions for improvements or new works of the kind specified in section nine as coming within the province of this Board of Highways and Bridges, and intended to be laid before the Legislature, shall hereafter be first made to this Board. Upon receiving such application, Board shall investigate same, and if they find such work necessary and proper, shall thus report to the Legislature, with an estimate of the expense thereof ; if they do not approve of such application, they shall report their reasons for their disapproval.

The Board may also, in like manner recommend, whenever they think proper, any improvements of the kind above specified, though no application has been made therefor.

Sec. 12. It shall be the duty of the Board to procure for the Legislature full plans and estimates of contemplated works or improvements when so ordered by the Legislature.

Sec. 13. Whenever any work shall have been authorized or ordered by the General Court and the money appropriated therefor, Board shall advertise for proposals for doing said work; plans and specifications of the same first to be placed on file in office of Board, which plans and specifications shall be open to public inspection; advertisement to state work to be done, and to be published ten (10) days at least. The bids shall be sealed bids, directed to Board and accompanied by bonds to the Commonwealth, signed by bidder and two responsible sureties, in sum of two hundred (200) dollars, conditioned he shall do the work if awarded to him, in case of his default to do so, forfeits, etc. Bids to be opened at time and place mentioned in advertisement.

Sec. 14. All contracts shall be awarded to the lowest responsible bidder, and who sufficiently guarantees to do work under superintendence and to satisfaction of Board ; provided that the contract price does not exceed the estimate or such other sum as shall be satisfactory to Board. Copies of contracts to be filed with State Auditor.

Sec. 15. Board reserves right in contracts to decide questions as to proper performance of work and meaning of contracts; in case of improper construction may suspend work and relet the same; or order entire re-construction; or may relet to other contractors and settle for work done, etc. In cases where contractor properly does work, Board may in their discretion as work progresses, grant to said contractors estimates of amount already earned, reserving 15 per cent. therefrom, which shall entitle holder to receive amount, all other conditions being satisfied.

SEC. 16. In case prosecution of any public work be suspended, or bid be deemed excessive, or bidders be not responsible, Board may, with written approval of Governor, where the urgency of the case, or interests of the Commonwealth require it, employ workmen to perform or complete any work ordered by the Legislature, provided that the cost and expense shall in no case exceed the amount appropriated for the same.

SEC. 17. All supplies of materials, etc., when costing over five hundred (500) dollars, to be purchased by contract, subject to same conditions as letting out work.

SEC. 18. Whenever Board thinks necessary, for interests of the Commonwealth, to protect same from damage or loss, shall report thus to Governor and Council and reasons for same, asking power to give contracts without notice required above, and Governor and Council may grant request, provided three-fourths vote for it.

SEC. 19. Whenever Board is of opinion a work may be done better without a contract, shall so report to Legislature, and they shall procure machinery, materials, etc., hire workmen, etc., to do said work, whenever so authorized by the Legislature.

SEC. 20. All contracts and bonds by Board to be in the name of the Commonwealth.

SEC. 21. No member of the Board to be interested in any contract; all contracts made with any member interested, Governor may declare void. and shall remove such member so interested from office. It is the duty of every member of the Board and every officer of the Commonwealth to report any such delinquency, if discovered.

SEC. 22. Board shall be empowered to employ such engineers, clerks or other assistants, as shall be provided for by the Legislature.

SEC. 23. Board shall, on or before   every year, submit to the Auditor, by him to be presented to the Legislature with his annual estimate, a statement of the repairs and new work needed for the current year, and of the sums required by the Board therefor; report to be in detail; all sums appropriated therefor to be included in the annual tax-levy.

SEC. 24. All moneys to be paid to any person out of moneys so raised, shall be certified by President of Board to Auditor, who shall draw warrant on Treasurer therefor, stating to whom payable, and to what fund chargeable; such warrant to be countersigned by President of Board.

SEC. 25. Board to keep accounts, showing moneys received and spent clearly and distinctly, and for what purpose. Accounts to be always open for inspection of Auditor or any committee appointed by the Legislature.

NOTE.—In the year following the printing of these essays, Massachusetts abolished the payment of road taxes in labor. (See Sec. 8 above.) Towns have also been authorized to appoint "superintendents of highways." Other plans for the improvement of common roads are still (1870) a matter of public discussion and agitation. The politic thing to have done in 1870, would have been, plainly, to have recommended the appointment of an expert commission to revise the road laws of the State, and to submit such revision to the Legislature. And such a course remains a desirable one to pursue at the present day in Massachusetts and in other of the States of the Union.

PART III.

———

THE ENGINEERING AND BUILDING RECORD'S

PRIZE ESSAYS ON

ROAD CONSTRUCTION AND MAINTENANCE.

Introduction to Part III.

The following from *The Engineering and Building Record* of March 29, 1890, refers to the matter found on the following pages :

The Competition for Essays on Road Making and Maintenance instituted by *The Engineering and Building Record* has, like the Competition for Water Tower and Pumping Station designs, proved very successful. The competition closed March 1, and the 21 essays received have in the intervening time received the careful consideration of the Committee of Award.

At a time when there is such a general awakening to the importance of having good roads this valuable contribution to the literature of the subject is most opportune. In addition to the prize essays, which will be published in successive issues of this journal, we are happy to announce a discussion of certain features in the several essays by the Committee of Award, which, in view of the experience of these gentlemen, will add materially to the practical results of this competition. The following is the report of the Committee :

To the Editor of The Engineering and Building Record :

The Committee invited by you to award prizes amounting to $150 for Essays on Road Construction and Maintenance have examined the 21 papers submitted to them, and make the award as follows :

"CESA." *First Prize,* $75.
    S. C. Thompson, 832 East One Hundred and Sixty-first Street, New York City.
"BONA VIA, BONUM OPUS." *Second Prize,* $50.
    I. F. Pope, 905 Brazos Street, Austin, Tex.
"A SIMPLE SCRATCH." *Third Prize,* $25.
    John P. Pritchard, Quincy, Mass.

Also as worthy of Honorable Mention, abstracts of each to be published, we would name the following :

"GRANITE STATE."
    Prof. John V. Hazen, Chandler Scientific School, Hanover, N. H.
"PALISADES."
    Samuel L. Cooper, Engineer Finance Department, City of New York.
"ROY."
    Frank B. Sanborn, Brookline, Mass.
"ZAMORA."
    A. T. Byrne, Civil and Mining Engineer, 361 Fulton Street, Brooklyn, N. Y.

As a brief declaration of principles, "TO THE POINT," A. L. Phillips, Pencoyd, Pa., is also worthy of publication.

The result of this competition is a matter for congratulation, and the essays are, as a whole, a contribution to the literature on this important subject both valuable and practical. We do not wish, however, to be understood as indorsing every proposition made in any one of the papers.

At a future time we will submit a brief criticism that we think should go on record with the essays herewith recommended for publication.

                    F. Collingwood,
                    Edward P. North,
                    James Owen.

ROAD CONSTRUCTION AND MAINTENANCE.*

BY S. C. THOMPSON (" CESA "), NEW YORK

In writing upon this subject, it is the intention of the author to treat the different matters under consideration, so as to present the method of doing the work, and the reasons for it, in a manner to be easily grasped by the popular mind, rather than to collate and arrange information for the professional reader.

Commencing with Road Construction it can be divided into—first, locating, or laying the road ; second, making the road-bed, which includes all earth-work, drainage, and everything necessary to get the foundation in proper shape ; and third, surfacing, or making the top of the road.

Location.—First let us see what are the elements which effect the location of a road, and upon what is the location of a road dependent.

The problem in its simplest form is, of course, given two or more points to be connected by a road-way, without any limitations whatever, but the usual considerations which determine where a road shall be built are often varied and troublesome.

Elements that Determine Location.—Taking them in the order which they are likely to occur, we have, first, the amount of money that can be expended, and as a necessary part of this, and largely determined by it; second, the allowable grades, and maximum curves permissible.

Select Best Possible Route.—When it has been determined to connect two places with a road, the best possible route over which to build the road should be determined, and then the construction should conform as nearly as the contingent circumstances will permit to the location laid down.

Make the line as nearly straight as practicable, and when changes of line occur connect them by regular curves of proper radius.

When the line is intended to be straight make it so absolutely.

Curves.—Where curves are required make them as flat as possible, and on a first-class road where long trucks or loads are liable to come, only in extreme cases should it exceed a curve of 50 feet radius.

---

* *The Engineering and Building Record* Competition.—First Prize Essay.

Minimum Radius to Curve.—On mountainous roads it sometimes becomes necessary to increase the curve to a radius of 20 feet.

Heavy Traffic vs. Passenger Service.—In discussing the question
of permissible grades it will, perhaps, be best to consider the roadways as being used for two specific purposes, one, where heavy
traffic has to be provided for, and the other where passenger service
is of the greatest importance, or, as it might be designated, loads
vs. time. In a general way it may be said that when the conditions
necessary for the first are complied with, the second are met equally
well.

To Determine Grades.—To determine what grades can be
allowed, it becomes necessary to see what difference in results are
caused by a difference in inclination.

Reduction of Draught by Increase of Inclination.—A man can
walk up a slope of 100 in 120, and a horse or mule can ascend an
incline of 100 in 175, and it has been found by experiment that a
horse pulling his maximum load on a level, can pull but ⅔ as much
if the slope is made 1 in 50, and this gradually lessens until with a
slope of 1 in 10 he can draw but ¼ as much as his level load, or to
make it more complete, see the following table made from experiments by Gayffier* and Parnell† upon French and English roads,
the results of the latter being the average of three velocities, calling
the load which a horse can draw on a level 100.

Experiments of Gayffier and Parnell.

On a rise of 1 in 100 a horse can draw only.................. .90*
    "   " 1 " 50 "   "   " .... .81*
    "   " 1 " 40 "   "   " .... .72†
    "   " 1 " 30 "   "   " .... .64†
    "   " 1 " 26 "   "   " .... .54†
    "   " 1 " 20 "   "   " .... .40†
    "   " 1 " 10 "   "   " .... .25*

Experiments of Sir Henry Parnell to find Draught Required.—
And as the capacity for draught is limited by the steepest grade, no
slope should be allowed up which a fair load cannot be drawn. The
experiments of Sir Henry Parnell show that to draw a stage coach
over the same road, having different degrees of inclination, the following results were obtained :

FORCE OF DRAUGHT REQUIRED.—LEVEL ROAD CALLED 100.

Inclination.	At 6 miles per hour.	At 8 miles per hour.	At 10 miles per hour.
1 in 20	268	296	318
1 " 26	213	219	225
1 " 30	165	196	200
1 " 40	100	166	172
1 " 600	111	120	128

No Counter Grades to be Allowed.—In laying out a road, with regard to grades, have a continuous inclination in one direction, and do not allow any counter grades, for in ascending, each foot descended on a counter grade, means just so much more rise to be overcome.

Divide Roads in Classes According to Grades.—Different writers have proposed to divide the roads into classes, and limit the grade for each class. In some cases the grades are fixed by law.

The maximum grade established by the French Government Board of Engineers is 1 in 20.* The Holyhead Road in Wales uses 1 in 30 as a maximum, except in two cases. The road over the Simplon Pass averages 1.22 on Italian side, and 1 to 17 on Swiss side, with one case of 1 to 13, and in this State several turnpike roads are limited by law 1 in 11.

Herschel's Classification of Roads.—C. Herschel, M. A. S. C. E., in his prize Essay on the Science of Road-Making,† proposes to divide in three classes—viz., State, county and town roads, and limit the inclinations, first, 3 to 5 in 100; second, 5 to 7 in 100; third, 7 to 10 in 100.

Prof. S. F. Miller ‡ suggests making them of two classes, the first not to have a gradient of more than 1 in 20, and the second not to exceed 1 in 10.

Vertical Rise Equivalent to Lengthening Roads—A vertical rise is equivalent to an increase in the length of the road proportional to the angle of inclination.

Equivalent Length of Level Road; Experiments of Sir John Mac-Neil.—The force expended in carrying a given load from any point to another at a higher elevation, is equal to the force of traction together with the force necessary to lift the load to the elevation attained. Experiments made by Sir John MacNeil show that a wagon loaded with six tons, drawn three miles per hour, upon a slope of 1 to 30, one mile is equivalent to 2.7 level miles, and for a stage coach weighing three tons, drawn six miles per hour, the equivalent level road for one mile is 1.62 miles.

On Long Incline Have First Portion Steepest.—Where a long incline becomes necessary it will be found economical to make the first portion the steepest and decrease it as it ascends, and if the slope can be varied by occasional level stretches the efficiency of the road will be greatly increased.

The road-bed on steep slopes is subject to greater wear from the feet of horses in ascending, and is much more subject to serious erosion by heavy rains.

---

* Gillespie's Roads and Railroads, p. 42.
† Report to Massachusetts State Board of Agriculture, 1869–1870.
‡ Massachusetts State Report, 1869–1870.

Minimum Grade.—Where the inclination of the road is not greater than 1 in 33 it will not pay to materially increase the length to reduce nearer to a level.   So far, in treating grades they have been considered as obstacles to ascent, and we will further consider how they may be obstacles to descent.

Experiments of Sir Henry Parnell and Dr. Lardner on Angle of Repose.—On a descent the inclination should not exceed the angle of repose for a vehicle, which varies according to the road-bed, from 1 in 40 to 1 in 10.   On a road with a slope just sufficient so that the force of gravity would overcome the friction, a loaded vehicle would descend with unaccelerated velocity and a considerable speed could safely be maintained in descending.

If the inclination is materially steeper than as shown above, so that a speed of only one-half or one-third as much can be safely maintained, it would be equally cheap to have a level road that was twice or three times as long.

To Ascertain Angle of Repose.—The angle of repose, for any given road-bed, can be easily ascertained from the draught upon a level, with the same character of surface; that is, if the force necessary, on a level, to overcome friction is $\frac{1}{30}$ of the load, then the same fraction expresses the angle of repose, for that surface.

WIDTH OF ROADS.

The proper width of roads and road-beds are matters which ordinarily receive little attention.   A road-bed is frequently constructed so wide as to add so much to the cost of maintenance as to render it impracticable to keep it in proper repair.

Revised Statutes of New York Fixing Width of Public Roads.—The revised Statutes of New York State require all public roads laid out by the Commissioners of Highways to be not less than three rods wide, and require " all roads that have been used as public highways for 20 years to be opened to *two rods at least.*"

English Roads.—In England turnpike roads are obliged by statute to be laid out four rods wide, and 22 feet wide to be bedded with stone.

Roman Military Roads.—The Roman Military Roads were established 12 feet in width when straight, and 16 feet when crooked.

Classification by French Engineers.—French engineers make four classes of roads :   First, 66, 22 stoned ; second, 52, 20 stoned ; third, 33, 16 stoned ; fourth, 26, 16 stoned.

In England turnpike roads near towns are 60 feet wide, by-roads 20 feet wide.

Holyhead Road.—The width of the Holyhead Road varies from 37 feet where level to 22 feet in precipitous places.

Effect of too Wide Roadway.—Where the roadway is too wide it usually results in no part being kept in good repair, while if it was narrowed the whole could be kept in first-class condition at less expense, and a well kept road of even 20-feet width is far preferable to a road but half maintained of double the width.

In laying out it may be advisable to take a strip considerably wider than the intended road-bed, so as to provide for possible contingencies in the future when the land becomes more valuable. *Lay out* sufficiently wide, but build only so much as can be kept in thorough repair.

These general rules about what should be done are all well enough in their way, but it will be found much more satisfactory in results to obtain the services of a competent engineer, and let him enter into the details involved in each particular case and decide from the conditions that arise.

The judgment and skill introduced in this way will frequently save costly and troublesome mistakes, and leaves nothing to chance or ignorance.   (See Appendix 1.)

ROAD-BED.

Requirements of a Good Road-bed.—The essential requirements of a good road-bed are, that it shall be practically unyielding— smooth on the surface, and impervious to water ; and without these requirements there can be but little durability.

General Requisite for Road-beds.—Before calling attention to particular kinds of road-bed there are certain requisites apply to all kinds.

The first and most important of these is *thorough drainage.*

No matter what the material may be a proper attention given to drainage will be found to be a good investment, both as to first cost and future maintenance.

Advantages of Drainage.—Some of the advantages of thorough drainage will readily present themselves.

In this climate the worst enemies to building, or properly maintaining a road-bed, are water and frost, and if the first is kept out the second will have little or no effect, as the surface will not be effected, and heaving will be reduced to a minimum.   Again, if a road-bed is thoroughly drained, it dries much more promptly, and has less mud and less dust.

The construction of road-beds may be properly divided into two parts—viz.: the foundation and surface.   Under the head of foundations is included all necessary sub-drainage works, culverts, ditches, etc.

Earth Roads.—For earth roads, as commonly built, there is but little to be said, and they should only be tolerated in a new country or where there is absolutely nothing but earth, of which to make a road.

Yet with earth alone a very passable road can be made and maintained, if sufficient care is taken to have it thoroughly drained, and the surface of proper shape.

The persistent care with which some of the so-called Road-Surveyors, in the country, excavate the material which has washed into the gutters and replace it upon the center of the road, seems to indicate a belief that the powers of man surpass and are superior to those of nature.

Cost of Handling Earth, etc.—Regarding the cost of handling earthwork of various kinds, many elaborate tables have been compiled of the cost, the distance to which it can be profitably cast with shovels, how far it can economically be carried in barrows, etc., all of which are of great value to an engineer in the determination of methods of conducting work, and the economy of management, but to the average individual who may be called upon to make or supervise the making of a road, they are but little better than so much Latin or Greek.

Handling Earth with Shovels, etc.—The generally accepted rules in handling earth, are: first, with shovels it can be cast horizontally from 5 to 10 feet, and vertically it can be lifted or thrown from 5 to 7 feet, and the cost of moving to the maximum distance either horizontally or vertically will be found to be nearly the same.

Wheelbarrows.—Where wheelbarrows are used, it will be found that they can be economically used up to a distance of perhaps 200 feet, but when it becomes necessary to carry to a greater distance than this, it will be better to use carts. Any increase in the inclination of the barrow road will have the effect to decrease the amount of material that can be handled, and an inclination of 1 in 10 will reduce the load that can be wheeled to two-thirds of what can be handled on a level road.

To overcome a vertical rise of 3 feet, will cost as much as wheeling from 80 to 100 feet horizontally.

Hauling with Carts.—With one-horse carts, it is estimated that one foot vertical costs as much as 14 feet horizontal.

Time Required for Dumping.—For a long haul it will be found that two-horse carts are much more economical than single carts; and dumping carts in every way preferable to the much used pole-bottomed carts or Studebaeker wagons employed in the vicinity of New York. A series of observations were made by the writer upon the relative time required to dump a load and get away, with the two kinds of carts.

The time was taken when the team was stopped on the dump, and when the driver started to drive away ; and it was found that with the dump cart the average time for dumping and getting away was three-quarters of a minute, while with the other kind it averaged about three and one-half minutes.

Shrinkage of Earth.—In determining the amount necessary to make a specified embankment, the shrinkage of the material used should be carefully considered, as it will be found to vary from 8 to 15 per cent., according to the nature of the material, gravel and sand shrinking the least and clay the most, the latter sometimes reaching 25 per cent.

*Gillespie gives an easy approximate rule for determining the cost of moving, with single carts, one cubic yard of earth any distance, on a level, deduced from the formulas of Ellwood Morris, C. E. (See *Journal of the Franklin Institute*, Sept., 1841.)

Gillespie's Rule for Cost of Moving Earth.

For   300 feet + wages of cart and driver by.............. 24
 "   500 "   "   "   "   19
 " 1,000 "   "   "   "   12
 " 1,500 "   "   "   "   9
 " 2,500 "   "   "   "   6
 " 3,000 "   "   "   "   5

The proportional expense of carrying long distances will be found to be less than short distances, as less time is lost is filling and dumping.

Drainage.—Drainage for a road should be of two kinds, surface and sub-drainage. The first provides for the speedy removal of surface water or rain-fall on the surface, and the second for the removal of water which penetrates beneath the surface, and into the body of the road covering.

Effect of Perfect Sub-drainage.—As a result of perfect sub-drainage, there is little or no effect from frost, and the surface and body of the road does not become softened and destroyed in summer.

Surface Drainage.—For surface drainage, ditches should be provided along each side of the road having sufficient fall to promptly carry away any water that reaches them. Where it becomes necessary to carry the water across the roadway culverts should be provided.

All drains should have a continuous fall throughout their entire length, and the size will depend upon the inclination and the amount of water they are expected to carry.

---

* Roads and Railroads, p. 131.

The amount of rain-fall over the area to be provided for is an important element in deciding upon the sizes required. For Central Park provision was made for a rain-fall of two inches per hour over the entire surface, but this is largely in excess of the usual requirements.

ROAD SURFACES.

Kinds of Road Surfaces.—These may be divided into Gravel, McAdam, Telford, and the pavements of various kinds, such as granite, wood, asphalt, brick, etc.

Gravel Roads.—In portions of the country where gravel is easily obtainable, a very satisfactory road can be made by making the surface for a greater or less depth of gravel.

Manner of Construction.—Prepare the foundation so as to allow for prompt drainage, and shape as the finished road is intended to be; make the sides of the road planes, and not curves, and then roll thoroughly to get a solid foundation.

Put on a layer of gravel from 6 inches to 8 inches in thickness, sprinkle thoroughly and roll till very compact and firm. Next spread another layer from 4 inches to 6 inches of gravel, and sprinkle and roll till the desired hardness and smoothness are obtained.

Use Binding Material if Necessary.—If the gravel has no binding material in it, a sufficient amount may be incorporated in the last layer to cause it to take a good bond.

Hard Pan and Bank Gravel.—Where it is possible to get blue gravel or hard pan and clean bank gravel, the two can be so mixed as to give a surface almost like concrete in hardness. When the two are used together a two-horse grooved roller for the first layer will be found very effective, and the material should be quite wet while rolling. The surface can then be finished with a steam roller, or with a smooth roller, sufficiently loaded to give the requisite weight.

Weight of Roller.—In completing the surface of a gravel, or other road, where rolling is required, the weight of the roller should be as much per inch as the weight per inch on the tire of the heaviest vehicle likely to pass over it. For ordinary traffic a very durable and economical surface can be produced in this way.

GRAVEL SURFACE WITH RUBBLE FOUNDATION.

Gravel and Rubble.—There are in the vicinity of New York City many very good examples of the above named road—viz.: in Central Park, also Avenue St. Nicholas from One Hundred and Tenth Street northward, and others.

Avenue St. Nicholas.—The method of constructing this latter road* will perhaps furnish as good an illustration as can be readily found.

On the prepared road-bed, which was made of the same shape as the finished surface, and well rolled, was placed rubble stone not exceeding 9 inches in the greatest dimension, and most of them less. These stones were placed as uniform as possible and closely covered the surface. On top of this was placed a light layer of quarry chips which was then rolled with a 6½-ton roller till the spawls were well compacted and wedged into the interstices of the rubble foundation. On this was spread a light layer of gravelly earth or hard pan, moistened by sprinkling carts, and then rolled to fill all spaces.

The gravel was applied in two layers, all of the large stone raked out, and the whole kept moistened while rolling. Care was taken not to have it too wet.

The first layer was thoroughly rolled before applying the second.

Grant on Gravel Roads.—Before the roller was put on the top layer a light coat of gravelly loam or hard pan about one-fifth the bulk of the gravel was thoroughly mixed in, by raking, until well incorporated with the gravel. A thorough rolling completed the process, and the roller could be used occasionally to good advantage after the road was open to travel. W. H. Grant, M. A. S. C. E., who had an extensive experience in building gravel and McAdam roads, says, "It is believed that gravel roads, constructed with either rubble or Telford foundations, are better suited for light and pleasure travel, are more agreeable for carriages and horses, less difficult and expensive to maintain, require less attention in watering, and raise less dust than roads finished with McAdam surface."

Cross Section of Roadway.—The top surface of all roads should have, in addition to the gradient, a side slope from the center, varying according to the hardness of the surface.

In a general way, whenever it is practicable, the side slope of the road should equal or exceed the longitudinal gradient, so that the water passing over the surface may take a course to reach the gutter not sharper than 45 degrees with the center line. In some of the smaller German States, where road-making and maintenance has been reduced to more of a science than with us, they allow the rise at the center from $\frac{1}{20}$ to $\frac{1}{40}$ of the width of the road.

Rise of Road-bed in Germany and Prussia.—In Prussia the side slope is prescribed and bears a certain relation to the gradients of the roads, varying from $\frac{1}{14}$ in roads falling more than 4 in 100, to 1 in 12 for roads falling less than 2 in 100.

---

*See first Report of the Commissioners of Public Works.

MC ADAM SURFACE.

McAdam, from whom the road so called takes its name, maintained that the foundation of the road should not be of large stone at all, nor should it of necessity be of an unyielding nature, and he made the comparative cost of maintaining a road with a road-bed of solid rock and one having a soft soil or morass for foundation, as 7 to 5.

Size of Stone as Fixed by McAdam.—McAdam maintained strenuously that a stone over one inch in diameter was determined in a road, as it had a tendency to tip when a wheel came on it, and thus moved the adjacent material; he afterward substituted weight for size and made 6 ounces the maximum stone that was to be used. He has a considerable following among the French and English engineers, but in this country the Telford foundation is generally preferred.

Parnell's Specifications.—The specifications for Telford, as given by Parnell's Treatise, London, 1833, give the following directions: Prepare the road-bed of the required shape, and on this set stones by hand, to form a close pavement. Set the stones carefully on their broadest edge, lengthwise across the road, the upper faces not to be more than four inches in breadth. Break off all irregularities with a hammer, and fill all interstices with stone chips, well rammed in between the larger pieces, to make a compact mass.

On this place a layer of stone as nearly cubical in form as practicable and about 2½ inches in diameter to a depth of 4 inches, and then 2 inches more to be added as a second layer, and finally the whole to be covered with 1½ inches of gravel, free from clay or earth.

Binding Material Detrimental.—Parnell says that *the presence of binding material* on a new road is a positive detriment, as it prevents solidity by getting between the stones. The custom to-day is much as he prescribes, excepting some improvements or modifications. Neither McAdam nor Telford used rollers.

The Use of Rollers.—The use of rollers in road making, which is now considered a necessity, was brought prominently before English readers by a paper written by Sir John Burgoyne in 1843. He recommended as the greatest attainable weight 261 pounds per inch of roller. (See Appendix.)

There are three types of McAdam road, as given by him—traffic made, and horse and steam-rolled.

With a horse roller it is extremely difficult to compact hard stone without binding, and no steam roller is heavy enough to compact trap or granite without binding.

Horse vs. Steam Roller.—The horse roller will make a better road than can be made by traffic, but this is very much inferior to a steam roller with the maximum weight per inch of roller.

Where the surface is rolled with a sufficiently heavy horse roller it is cut up badly by horses' feet, and then the admixture of manure and dirt prevents a thorough bonding and compacting of the surface.

The steam roller, in addition to being heavier, does not cut up the surface and allows a harder binding to be used, and the action of the roll has a tendency to arrange the stone more compactly than would be done by weight alone.

Southern Boulevard.—About New York City are many good examples of Telford and McAdam pavement, and considering the surface rendered, none perhaps more worthy of mention than the Southern Boulevard constructed by W. E. Worthen, M. A. S. C. E., commencing at Third Avenue and running easterly and northerly from the north side of Harlem River through the annexed district.

Manner of Building.—In constructing it, the earth was first rolled thoroughly, and then two layers of $2\frac{1}{2}$-inch trap rock, the first 6 inches in thickness and the last 4 inches in thickness, placed upon it; these were thoroughly rolled with a two-horse roller, and then $2\frac{1}{2}$ inches of screening were spread over the surface and well rolled.

The screenings were applied very wet. The road had a very steady, and much of it heavy, traffic, yet it remained in very good condition for quite a number of years with trifling repairs.

As long as the surface of a roadway of this kind remains unbroken it stands very well; but, as is the case at present with the Southern Boulevard, it has been partially excavated to make place for a railroad, and the material has never been gotten back as compactly as before, and the road has been badly used up.

Road Built by E. P. North, M. A. S. C. E.—A road built by E. P. North, M. A. S. C. E., was composed of 2-inch stones in layers 6 inches in thickness, rolled with a roller 6 feet in length, weighing two tons when unloaded and three and a half tons with a load, and drawn by four or six horses.

Special attention was paid to get the stone as free from dirt, etc., as possible, and for this purpose it was handled with ten-tined forks, which allowed most of the finer portion to drop through, thus leaving the stone clean. The layers were thoroughly rolled, and then the dust and small stones were spread and rolled in, and finally about $\frac{3}{4}$-inch of clayey earth was spread on top and the whole thoroughly rolled, the roller passing 144 times over the surface. The roller used weighed from 75 to 100 pounds per inch of roll, but he considers them too light to be economical. (See Appendix, 2.)

The Glasgow and Carlisle Road.—The Glasgow and Carlisle Road was built with the roadway 18 feet in width and a crown of 4 inches in the center.

Metaling was put on the road in two layers. The bottom course was 7 inches in depth, placed with the broad end down and the top surface not over 3 inches in diameter; the bottom stones were carefully placed, breaking joints as far as possible, then the top spaces were filled with smaller stones, packed in by hand and securely wedged.

The top layer was composed of stone broken to pass through a $2\frac{1}{4}$-inch ring, spread and thoroughly compacted, and then covered with one inch of gravel. There was a drain from the bottom course every hundred yards. The drains were carried clear across the road and into water courses at the outside.

After this road was completed and in use for a while, it was found that the repairs necessary required from 80 to 120 cubic yards of material per annum, per mile.

McAdam Center on Earth Roads.—In constructing roadways having a portion only macadamized it will be found of great benefit in rolling to lap the roller on to the edge of the earth portion and thus compact the earth so as to give a suitable shoulder to help hold the metaling, then continue the rolling toward the center.

Thickness of Roads.—The thickness necessary for a good road-bed is given differently by different authorities. McAdam advocated 7 to 10 inches and considered the last ample for any service, but the custom in this country for heavy traffic is to make the thickness not less than 12 inches, though some roads have been built in Bridge-port, Conn., which have been made considerably less in thickness, and so far have reflected great credit upon the engineer in charge, by their thorough construction.

Where the traffic is so heavy that McAdam or Telford are not economical, it becomes necessary to use some kind of pavement.

The more common kinds are granite or trap-block, wood, brick and asphalt.

GRANITE OR TRAP-BLOCK PAVEMENT.

These blocks, with various modifications for foundations, constitute the principal pavement in this country for exceedingly heavy traffic, and are durable, fairly smooth, and easily laid.

Size of Stones.—The general requirements for the laying of pavement in New York City are—the stones may vary from 7 to 9 inches in depth, 8 to 12 inches in length, and from $3\frac{1}{2}$ to $4\frac{1}{2}$ inches in thickness, and nearly rectangular in shape.

The road-bed is prepared by excavating to a depth sufficient to give a good sand bed, usually from 3 to 5 inches in thickness.

The sand must be clean and sharp, without too large an admixture of large stone, that is one inch or over.

The preparation of the road-bed must be made without plowing, and it is thoroughly rolled if required with a steam roller weighing not less than ten tons.

After the road-bed is prepared, the sand is placed thereon, and bond stones set at the proper places, giving a crown of 6 to 8 inches to the roadway, according to the width. The blocks are laid at right angles to the street, allowing one inch joint. The whole surface is then covered with sand which is swept into the joints as the work progresses.

Ramming.—The rammers follow after, and with pinch bars adjust the stones to give fair and even joints, and then ram thoroughly, after which the sand is again strewn over the surface, and brushed into the joints. (See Appendix.)

Various Foundations.—Different foundations are sometimes used, such as concrete, etc., and the joints run full of cement or mastic of some kind ; the latter method makes a surface practically impervious to water, and wears very satisfactorily.

Where a sufficiently rigid system of inspection, over opening the streets for connections with water, gas or steam pipes, and sewers is maintained, this block pavement stands the heaviest traffic as well as anything in use.

WOODEN PAVEMENT.

Wooden pavement has been tried quite extensively in the different cities of this country, but so far have not proved an unqualified success.

Advantages.—The advantages of a wooden pavement are slight resistance to draught, noiselessness, easily cleaned—they cause less wear and tear on vehicles—are pleasant to ride over, and the first cost is less than stone blocks.

Objections.—The objections are slipperiness in wet weather, non-durability, both from wear and decay, and there have been found serious objections to them on sanitary grounds. Still where the foundations are properly made and drained, and the block treated so as to prevent rapid decay by any of the various processes, and especially where stone blocks are not easily available, they have proven a very satisfactory improvement in the character of the roadway.

Manner of Laying.—The various methods of laying will only be outlined, and a reference to the requirements for laying on the Strand, in London, will perhaps give the most modern and satisfactory practice.

Requirements for the Strand Pavement in London.—The excavation for the roadway is carried to a depth of 15 inches and a layer of Portland cement concrete 9 inches in thickness placed thereon and shaped to conform to the finished surface of the road. Concrete is composed of seven parts of stone, etc., to one barrel of cement. The top is floated on to the required shape and then the blocks of best Baltic redwood, or equally good timber, creosoted and placed endwise upon the concrete.

The joints between the blocks are then run with cement grout or hot bituminous mastic. Cement grout is composed of one part cement and three parts sand. The top of the pavement is then covered with a layer of sand or gravel and opened to traffic. (See Appendix.)

BRICK PAVEMENT.

In certain portions of this country brick has been used as a pavement and found to answer the requirements fairly well.

Brick Pavement in Des Moines, Iowa.—In Des Moines, Iowa, there are about ten miles of streets paved with this material, and they are proving quite satisfactory.

The method of laying is to surface the road according to plan, then to roll with a roller weighing 150 pounds to the inch, or ram with rammers weighing 81 pounds.

In embankment the layers are made 9 inches in thickness and thoroughly rolled or rammed. On the foundation as prepared is spread a layer of sand 3 inches in thickness, and on this the bricks, which are specially burned for the purpose, are laid flat, breaking joints in the laying. The joints are completely filled with sand and on top a course of sand 1 inch thick is spread.

The next course of brick is laid on edge, and the joints thoroughly filled with sand, then one inch of sand is spread over the top and it is ready for traffic. Brick used for the same purpose have been used in West Virginia and have proven satisfactory.

ASPHALT PAVEMENT.

Until quite recently asphalt has been used but little as a pavement for roadways in this country, and the climatic conditions are so different from France and England that at present it is a little problematical just how far, and under what conditions, it will prove successful when used alone.

Many of the pavements laid with it in Paris and London have proven very satisfactory, and some work has been done with the pavement in this country, but with the extremes of temperature from 10 degrees to 110 degrees in this climate, it will give the pavement a very severe trial, unlike anything it is called upon to undergo in either England or France.

Some of the pavement already laid in this country has seemed to prove fairly satisfactory, as in Boston and in Washington, but in many cases, under a severe traffic, it has been found wanting.

When used in combination with other material it has proven a very valuable adjunct, but with the limited information available and the comparatively short time it has been in use, the writer is not prepared to give an unqualified indorsement to its use in this country, or at least in this portion of it.

Constant Attention Required to Keep Surface in Good Condition.— Having obtained a smooth, properly compacted surface on a road, it should always thereafter be maintained in first-class condition; but it will be found that this can only be done by constant vigilance and attention.

Repairing roads is not unlike a fire just starting—a very little properly directed effort will stop it; and with great force it applies to a road when the surface begins to get in bad condition.

In speaking of repairing roads we shall confine our attention to roads having an earth, gravel or macadam surface, as repairs on pavements need no description here.

On earth or gravel roads a very effective and economical instrument to help keep the surface in condition, is what is described by E. P. North, M. A. S. C. E. (see Trans. A. S. C. E., May, 1879), as a hone.

The Hone or Scraper.—This is made of various shapes and sizes, but in a general way consists of a timber, or heavy plank. faced on one side with a steel plate to prevent rapid wear, and provided with handles on the back to guide it, when being drawn along the road surface. It is attached to the team in such a manner as to draw at an angle with the line of the road, and is very effective in filling ruts, and removing small stones from the surface of the roadway.

This is best used while the road is somewhat softened by rain.

To Keep a Road in Good Condition.—To keep a road in good condition there should be no ruts; and to prevent ruts, it becomes necessary to prevent any accumulation of mud or dust.

If ruts are formed water collects, and the adjoining material becomes softened and more easily cut up and the drainage much increased.

Secret of Successful Road Repairs.—Here we have the secret of successful road repairs—viz.: *continual* repairs. Keep the dust swept from the road, the ruts filled up with suitable material, and the surface of the road will keep in good order.

If repairs are not continually made there can be no constant good road.

Sprinkle During Dry Weather.—When a long continued dry spell occurs it will be found economical as well as effective, in preventing excessive wear on a road, to sprinkle frequently.

Trésaquet, a French engineer, was the first to call attention to the continuous system of repairs, and it was adopted in Baden in 1845, and it has extended over nearly all European countries.

Amount of Material and Cost in Baden, 1845–1860.—In tables giving the amount of material used in making repairs previous to 1845, and subsequent to that time, it was shown that in 1832, 218.6 cubic yards of material were used per mile for repairs, while from 1855 to 1860 there was only about 90 cubic yards per mile required, and the cost of maintenance was reduced during the same period about 40 per cent.

Statements of French Engineers.—The statements of Gen. John F. Burgoyne, taken from the reports of French engineers, show equally good results from the continuous method of repairing, giving better roads at a reduction in the cost of maintenance of from 12 to 40 per cent. At the time these statements were made French engineers calculated that by maintaining the roads in the best possible condition, the cost of draught of merchandise over the roads in France might be reduced one-third, or about $30,000,000 saved to the public in one year.

Repairing on the Continuous System.—The method of repairing roads on the continuous system can be briefly outlined as follows:

Have the material for road covering deposited at convenient intervals along the side of the road, from 200 to 300 feet apart. Have a man detailed to look after the repairs of the road for every two or three miles, and let his sole duty be to keep his section in repair. The tools necessary for him to have will be a broom, hoe, shovel, rammer, wheelbarrow and water pot.

Duties.—He should keep all dust swept from the road in dry weather, remove all mud in wet weather, see that the surface drainage is maintained, and if there is any standing water on the road sweep it off and fill up the depressions. The covering which wears off should be replaced gradually, and the best time to replace metaling is during wet weather. Where there are ruts they should be thoroughly swept out to free from dust or mud, and then fill up with the prepared material.

Repairing.—If large breaks take place they should be brought up gradually, especial care being taken not to produce surfaces which will be avoided by the vehicles passing over, and proper attention should be given to make the new work bind to the old as promptly and completely as possible. The quantity of macadam necessary to keep a road in proper condition varies according to the traffic and the quality of the material used, but the following table gives perhaps as fair an estimate as any.

Amount of Material Needed for Repairs.—The table is calculated for a roadway 20 feet wide and 100 feet in length, and gives the amount necessary to keep in repair for one year :

    (1) Good material and heavy travel, 5.5 to 7.5 cubic yards.
    (2) Medium material and heavy travel, 7.5 to 9.2 cubic yards.
    (3) Poor material and heavy travel, 9.2 to 10.2 cubic yards.
    (4) Good material and average travel, 3.7 to 5.5 cubic yards.
    (5) Medium material and average travel, 5.5 to 7.5 cubic yards.
    (6) Poor material and average travel, 7.5 to 9.2 cubic yards.
    (7) Good material and light travel, 1.8 to 3.7 cubic yards.
    (8) Medium material and light travel, 3.7 to 5.5 cubic yards.
    (9) Poor material and light travel, 5.5 to 7.6 cubic yards.

Where it is not convenient to divide the care of the road in short sections, it will be found that a road roller, in connection with the maintenance, will be both profitable and satisfactory.

Use Roller After Frost Comes Out of Ground.—In the spring, after the frost is well out of the ground, the whole surface of the roadway should be rolled with a heavy roller, preferably with a steam roller of maximum weight.

Use of Roller Considered Profitable.—In the opinion of the writer, there is nothing that gives more satisfactory returns for the amount invested, either in making or maintaining a road, than the free application of a heavy roller to the surface.

Uniform System of Construction and Maintenance Desirable.—For the most efficient and economical maintenance of the public roads, it would seem to be essential that a uniform system of construction and maintenance be carried out through the whole State or country. Create a State or National department to have general charge of all public roads and bridges.

Executive Officers and Division of the Work.—The head of the department should be a man of large experience, good executive ability, and thoroughly conversant with the art of road building. His duties should be to arrange and direct the carrying out of all details relative to the construction and repairs.

For convenience, the State should be divided into districts and sub-districts, with a competent civil engineer in charge of each district, and assistants in charge of each sub-district, who, in their turn, have control of a certain gang or working force, and who must supervise the details of all construction and repairs for their respective districts.

Expensive Machinery Made Available.—Under a general system, stone crushers, steam rollers, and other expensive machinery might be employed such as towns would not be able to command.

Suitable material could in this way be prepared wherever found without regard to town lines.

Two Methods of Doing the Work.—Either of two methods could be adopted for doing the work: by men stationed singly, as previously mentioned, or by gangs working together, assigned to districts of such extent as they can thoroughly keep in repair.

Advantages of the System.—By substituting this or some similar method of managing the repairs on the roads, in place of the present unsatisfactory and expensive custom, we should undoubtedly be able, in a short time, to produce a permanent and noticeable improvement in the character of our roads, with the most economical expenditure of time and money.

It would also establish a system of individual accountability which would be a great help to keep up the morale of the force employed, for each one would be directly responsible to some one in authority over him, and any dereliction in duty could be easily detected and punished.

With the presentation of the foregoing we will leave the subject of maintenance of roads, firmly believing that if there could once be a system in successful operation, to produce good roads, it would matter but little after that what cost was involved in making of the roads, or what skill was necessary to maintain them, a grateful public would gladly pay the necessary taxes to do both.

From this we will proceed to suggest some of the advantages of good roads.

ADVANTAGES OF GOOD ROADS.

Let us look at some of the advantages to a community resulting from good roads.

Economic Considerations.—First let us look at it from an economic standpoint, as that is the position which strikes most people promptly and forcibly.

It has been determined by careful experiments that two or three times as great a load can be drawn on a broken-stone road as on gravel, and four or five times as much on a good pavement of rectangular stones.

Experiments of MacNeil.—Very many experiments have been made to determine the actual force required to draw a given load over different surfaces or pavements, and Sir John MacNeil, in his experiments, determined that the traction for a load of one ton for different surfaces was as follows :

1. On good pavement the force necessary to move the load was 33 pounds.

2. Broken stone on old flint road, 65 pounds.

3. Gravel road, 147 pounds.

4. Broken stone on rough pavement bottom, 46 pounds.

5. Broken stone on bottom of beton, 46 pounds.

Many experiments have been made on the force required, mostly by English and French engineers, and the experiments have left but little to be desired.

The most concise and complete compilation of these different results, with which the writer is acquainted, is that made by Rudolph Hering, M. A. S. C. E., and read before the Engineers' Club in Philadelphia, in March, 1882, and recently printed in *The Engineering and Building Record.* (See Appendix.)

Roads being the only means of communication with railroads and markets, it follows that if a heavier load can be drawn, and better time made on a good road than on a poor one, that the area benefited by railroads and made tributary to markets is increased, in direct proportion to the goodness of common roads.

Good Roads Shorten Distances and Save Wear and Tear, etc.— It will readily appear that a farm four or five miles from market, on a good road, is virtually nearer to the market than one located but two or three miles away, but located on a poor road. It is estimated that a saving of as much as 25 per cent. in animal power alone can be saved by the improvement of the roadway, besides the saving in time, and the wear and tear on vehicles.

An eminent writer says : " The road is that physical sign or symbol by which you will best understand any age or people. If they have no roads they are savages, for the road is the creation of man and the type of civilized society."

The Best Roads.—The best roads in the world to-day are those of England, France and Germany, and their excellence is largely due to the fact that in each country they came under *national* supervision, and the resulting highways, the finest in existence, are the cheapest to maintain, and in every way the most satisfactory to those who use them.

American Roads.—" The American roads are far below the average; they are among the worst in the civilized world and always have been, largely as a result of allowing local circumstances to determine the location. No general system of building or maintaining, and haste, waste and ignorance in building." To summarize the advantages of good roads :

They attract population, increase the value of property, decrease the cost of transportation, and thus encourage the greater exchange of products between one section and another; and being feeders for the railroads, they directly bring distant places more nearly together, and promote intercourse and the development of commercial life.

The writer has been compelled in looking for the results of elaborate experiments to gather them from foreign sources—that is, he has not been able to find any records of any extensive or continuous investigations upon traction, friction and kindred subjects made in this country.

He has availed himself liberally of the investigations and writings of such men as Gillespie, Gillmore, Herschel, North, Owens and others.

APPENDIX.

(1) Two forcible examples of the improvement that can be made, by careful engineering skill, in the chance location of roads are given in Gillespie's Roads and Railroads. (See pages 36 and 233.)

An old road in Anglesea, laid out in violation of this rule (to have no counter grades), rose and fell between its extremities, 24 miles apart, a total perpendicular amount of 3,540 feet; while a new road laid out by Telford between the same points, rose and fell only 2,257 feet; so that 1,283 feet of perpendicular height is now done away with, which every horse passing over the road had previously been obliged to ascend and descend with its load. The new road is besides more than two miles shorter.

(2) A more modern exemplification of the true principles of road making are furnished in the construction of a plank road between Chittenango and Cazenovia, N. Y., under the supervision of Mr. Geddes.

Both villages are situated on Chittenango Creek, the latter being 800 feet higher than the former.

The most level common road between these villages rises more than 1,200 feet in going from Chittenango to Cazenovia, and rises 400 feet in going from Cazenovia to Chittenango, in spite of the latter place being 800 feet lower.

The line of the plank road, however, by following the creek, ascends only the necessary 800 feet in one direction and has no ascents in the other, with two or three trifling exceptions, allowed in order to save expense.

Rollers.—It is perhaps superfluous to give any description of the different kinds of rollers in use, but being desirous of calling attention to one mentioned by C. Herschel in Report of Massachusetts State Board of Agriculture, 1869–1870, I will briefly suggest rather than describe them.

The roller described and also delineated, is hollow, of cast iron, and so constructed as to be filled with water when additional weight is required, which can be emptied when going from place to place.

A circular cast-iron frame surrounds the roll and carries the axle bearings. The outside of this frame is turned to form a groove, in which a strong wrought-iron band is fitted in a manner to turn easily. The wrought-iron band is made in two parts, and when put together incloses the pole on one side of the roll and the balance weight on the other side on an extension bar. This ring is arranged so as to turn on the frame, when it is desired to go in the opposite direction. Allowing the ring to turn instead of turning the roll saves much wear and tear on certain parts of the road. A roller of this kind, 4½ diameter and 3½ feet long, weighing 4 tons empty, would cost $560–$600, and a larger size from $700–$750. It originated in Chemnitz, Germany.

The ordinary, smooth cast-iron roller for use on the farm is of but little use on the road on account of its lightness.

The grooved iron roller, formed of two sets of rolls, one about 8 inches less in diameter than the other, is a very effective implement in constructing banks, or rolling the lower layer for a gravel road, but is unsuitable for use on the surface.

The steam roller is the most thorough and satisfactory machine, and is much to be preferred over any other kinds. They are made of various weights, some of those in France weighing as much as 440 pounds to the inch of roll. They cost from $4,000 to $7,000, and cost about $10 per day to run them.

Appended find the general form of specifications used in New York City for block pavement.

EXCAVATION FOR FOUNDATION.

Excavation.—(*a*) The roadway shall be carefully excavated where necessary and brought to the required pavement sub-grade, and all superfluous and extraneous matter removed.

Rock and Masonry Taken out Below Sub-grade.—(*b*) Any rock or masonry which is above the pavement sub-grade shall be removed to a depth of one foot below the same.

Sub-grade.—(*c*) Pavement sub-grade shall be uniformly one foot and one-tenth below the intended pavement surface. The foundation shall nowhere rise above said sub-grade. If it should fall below, owing to shrinkage or on account of removal of rock or masonry, or other cause, no new filling shall be added other than sand of the same quality as that used for bedding the paving stones.

No Plowing Allowed.—(*d*) In the excavation below grade and preparation of foundation no plowing will be allowed.

Preparation of Foundation.—(*e*) The foundation shall be carefully brought to an even surface, conforming to the required sub-grade, and should there be any spongy material or vegetable matter in the bed thus prepared, it shall be removed and the space filled with clean sand carefully rammed so as to make such filling compact and solid, and when required by the engineer, the entire road-bed, after having been brought to the required sub-grade, shall be rolled with a steam roller weighing not less than ten tons, until the surface is firm and compact, to the satisfaction of the engineer.

When the steam roller cannot reach every part of the road-bed, the bottom at those portions shall be tamped or rolled with a small roller and sprinkled with water if required; and should the material of the bottom not admit of satisfactory rolling, such material must be removed and proper material substituted by the contractor at his own expense.

Price for Paving to Cover Excavation, etc.—(*f*) The cost of all excavation of whatever character required in executing the work shall be included in the price paid for the pavement.

BLOCK PAVEMENT.

Quality and Dimensions of Stone Block.—(*a*) The stone blocks are to be of a durable, sound and uniform quality of trap-block, each measuring on the face or upper surface not less than 8 inches nor more than 1 foot in length, and not less than 3 inches nor more than 4 inches in width, and not less than 7 inches nor more than 9 inches in depth; to be split and dressed so as to form, when laid, close end joints, and side joints not exceeding 1 inch wide top and bottom, with fair and true surfaces on top, bottom and ends; and are in all respects to be equal to the specimen blocks deposited at the office of the engineer, as hereafter provided.

Sample Blocks.—(*b*) Before any paving blocks are placed upon any part of the work, the engineer shall approve of the quality and finish of samples of the same, which shall be furnished at his office by the contractor.

When Stone Blocks May be Hauled on Line of Work.—(*c*) After any portion or portions of the street or avenue, not less, each, than the distance between two intersecting streets, shall have been brought to the required grade and the curb-stones set and crosswalks laid, and not until then (unless permission in writing is given by the engineer), the contractor shall haul upon the line of the work at each of said points a sufficient quantity of stone blocks to pave such portions.

To be Carefully Culled.—(*d*) The stones will be carefully inspected after they are brought on the line of the work, and all blocks which, in quality and dimensions, do not conform strictly to these specifications, will be rejected, and must be immediately removed from the line of the work. The contractor will be required to furnish such laborers as may be necessary to aid the inspector or inspectors in the examination and culling of the blocks; and in case the contractor shall neglect or refuse so to do, such laborers as in the opinion of the Commissioners of the Department of Public Parks as may be necessary will be employed by said Commissioners, and the expense incurred by them will be deducted and paid out of any money then due or which may thereafter grow due to the said contractor under this agreement.

Rejected Materials to be Forthwith Removed.—(*e*) After inspection, as provided above, all rejected stones shall be immediately removed by the contractor from the line of the work. The contractor will then be required to pile such stones as may have been approved, neatly on the front of the sidewalk, and not within three feet of any fire-hydrant, and in such manner as will preserve sufficient passage-way on the line of the sidewalks, and also permit of free access from the roadway to each house on the line of the street. *After this inspection has been made, and after all the rejected stones shall have been removed entirely from the line of the work, and the accepted stones piled in the manner aforesaid, and not until each of these conditions shall have been faithfully fulfilled will the contractor be permitted to proceed with the preparation of the road-bed for the new pavement.*

(*f*) It being expressly understood that the work is to be prosecuted in sections of not less than the space between any two intersecting streets, and when the blocks are over 300 feet in length, then in sections not less than 150 feet nor more than 300 feet, and that these provisions relative to the hauling, inspection and removal and piling of stones shall apply to the work on each of said sections on the whole line of work.

Over Rail Ties and Other Places.—(*g*) Between, in and one foot outside of railroad tracks, over vaults, around sewer, man-holes, frames, and in such other places as the engineer may deem proper, the contractor shall use for the pavement stone blocks of such lesser dimensions as the said engineer shall direct, but the same general dimensions on the top surface shall be maintained.

Laying the Pavements, etc.—(*h*) The stone blocks are to be laid at right angles with the line of the avenue, except at intersecting streets and in other special cases, when they shall be laid at such angles as directed, with such crown and at such grade as the engineer may direct; each course of blocks shall be of uniform width and depth, and so laid that all longitudinal joints shall be broken by a lap of at least two inches, and that all such joints shall be as close as possible.

Ramming.—As the blocks are laid they shall be covered with clean, fine sand, which shall be raked until all the joints become filled therewith; the blocks shall then be thoroughly rammed to a firm, unyielding bed, with a uniform surface, to conform to the grade and crown of the street. No ramming shall be done within 25 feet of the face of the work that is being laid, and in doing all ramming the contractor shall employ one rammer to every two pavers.

Covering Sand.—Whenever the pavement for not less than 200 feet, and not exceeding 260 feet, shall have been constructed as above described, it shall be covered with a good and sufficient second coat of clean, sharp sand, and shall immediately thereafter be thoroughly rammed until the work is made solid and secure, and so on until the whole of the work embraced in this agreement shall have been well and faithfully completed, in accordance with these specifications.

Wooden Pavement on The Strand, London, England.—For complete specifications from which the abstract in the body of the article was taken, see *Engineering and Building Record* for December 9, 1889.

Resistance to Traction on Roads.—Resistance to traction is very nearly proportional to the weight.

It increases on paved roads with the velocity, owing to the increase of concussions, but it remains nearly constant on roads with a compressible surface, as earth, sand or turf (Rumford & Morin). The smoother the pavement and the less rigid the carriage—*i. e.*, the better it is hung, the less does the velocity affect the resistance.

Whenever the velocity is not above a pace, springs do not affect the resistance ; when it is greater, the resistance decreases as the carriage is better hung. On soft roads springs may slightly increase the resistance at low speeds. The destruction of the road and carriage is greater when the latter is without than with springs (Morin).

The advantages of a paved over a macadam road, in point of traction are considerable for heavy loads, less for stages, and nearly nothing for light carriages (Debaune).

According to Morin, the resistance to traction increases slightly with the number of wheels, but is inversely proportional to their diameter. According to Dupuit it is inversely as the square root, according to Clark, as the cube root of the diameter. The destruction of the road and the carriage is always greater as the diameter of the wheel is less (Morin).

On a paved or hard macadam road the resistance is independent of the width of the tire, when this quantity exceeds about 4 inches (Dupuit and Morin). When it is between 2 and 4 inches for heavy loads and low speeds, the broader tire will give proportionately less resistance. For light loads the width has no influence on the resistance. On earth and other compressible roads the resistance increases inversely as the width of the tire (Rumford and Morin). In Europe a common rule is to make the width of a tire one-half an inch for every 125–175 pounds upon the wheel, according to the character of the roads. The least width is usually about 2 inches.

From the above we may gather that the most economical conditions for traction are :

1. A hard and smooth surface, both when speed is to be attained and where heavy loads are to be hauled, and

2. Large wheels, especially for heavy loads.

3. For light loads the width of tire is immaterial, and is therefore better as narrow as the character of the road will permit. For heavy loads the resistance increases inversely as the width of tire. While on hard roads this effect is less, it is very decided on soft roads. The width should not exceed 4 inches at least on hard roads, and it is rarely less than 2 inches, except for light pleasure carriages.

4. Easy springs not only act preservingly on roads and carriages, but materially decrease the resistance of traction at trotting or running speeds.

The following table presents the results of experiments on the resistance of vehicles to traction or different classes of roads The fractions which are generally rounded off, indicate the part of the whole weight, which is equivalent to the resistance of drawing it on a level road. An examination of this table will clearly show the great economy in horse-power by using the hardest and smoothest roads. For instance, if 1 horse can just draw a load on a level road on iron rails, it will take $1\frac{1}{4}$ horses to draw it on asphalt, $3\frac{1}{4}$ on the best Belgian block pavement, 5 on the ordinary Belgian pavement, 7 on good cobble-stones, 13 on bad cobble-stones, 20 on an ordinary earth road, and 40 on a sandy road.

Character of Road.	Resistance in Terms of Load.	Velocity.	Authority.
Sand................................	$\frac{1}{4}$	Pace	Bevan
Sandy Road....................	$\frac{1}{14}$	3', 12' per sec	Morin
Loose Gravel............	$\frac{1}{9}$	Pace	Bevan
Gravel 4 inches thick........	$\frac{1}{10}$	Pace	Morin
Common Gravel Road..........	$\frac{1}{14}$	Pace	MacNeil
Gravel Road....................	$\frac{1}{23}$	3' per second	Rumford
Gravel Road...	$\frac{1}{25}$	12' per second	Rumford
Hard Rolled Gravel.	$\frac{1}{20}$	Pace	{ Bevan { Minard
Wet Turf......................	$\frac{1}{4}$	Pace	Morin
Hard and Dry Turf.............	$\frac{1}{13}$	Pace	Morin
Hard and Dry Turf	$\frac{1}{25}$	....	Bevan
Ordinary Earth Road..	$\frac{1}{10}$	Pace	Bevan
Hard Clay.....................	$\frac{1}{20}$	....	Bevan
Hard and Dry Earth Road.... ..	$\frac{1}{25}-\frac{1}{30}$	Pace	Morin
Ordinary Cobble-stone.....	$\frac{1}{8}$	Trot	....
Ordinary Cobble-stone..........	$\frac{1}{14}$	Pace	....
Good Cobble-stone (3½-inch).....	$\frac{1}{17}$	Trot	Kossack
Good Cobble-stone (3½-inch).....	$\frac{1}{30}$	Pace	Kossack
Good Carriage with Springs	$\frac{1}{35}$	Pace	Kossack
Macadam little used.............	$\frac{1}{12}-\frac{1}{13}$	....	Morin
Bad Macadam...................	$\frac{1}{17}$	Pace	Gordon
Old Macadam..................	$\frac{1}{25}$	....	Navier
Ordinary Macadam............	$\frac{1}{25}$	Trot	{ MacNeil
Ordinary Macadam.............	$\frac{1}{18}$	Pace	{ Perdon't { Kossack
Good Macadam (wet and slightly muddy)........................	$\frac{1}{10}-\frac{1}{24}$	....	Morin
Best French Macadam.	$\frac{1}{30}$	....	Navier
Very hard and smooth Macadam	$\frac{1}{30}$	....	MacNeil
Best Macadam..................	$\frac{1}{35}$	Trot	Rumford
Best Macadam	$\frac{1}{75}$	Pace	Rumford
Best Macadam................	$\frac{1}{54}-\frac{1}{60}$	....	Gordon
Best Macadam...........	$\frac{1}{54}-\frac{1}{70}$	....	Morin
Ordinary Stone Block......... ..	$\frac{1}{57}$	....	{ Perdon't { Poncelet { Minard
Ordinary Belgian Block.........	$\frac{1}{50}$	Pace	MacNeil
Good Stone Block	$\frac{1}{17}$	Trot	Rumford
Good Stone Block	$\frac{1}{50}$	Pace	Rumford
Belgian Block, Boulevard, Paris..	$\frac{1}{55}-\frac{1}{65}$	....	Navier
Good Belgian Block	$\frac{1}{30}$	Trot	Rumford
Good Belgian Block	$\frac{1}{60}$	Pace	Rumford
Good London Block.......... .	$\frac{1}{49}$	....	Gordon
Well laid Belgian Block..........	$\frac{1}{65}$	....	MacNeil
Good Belgian Block	$\frac{1}{45}-\frac{1}{57}$	....	Morin
Planked Roadway (bridge).......	$\frac{1}{30}-\frac{1}{57}$	....	Morin
Asphalt	$\frac{1}{111}$	....	Gordon
Granite Tramway...............	$\frac{1}{144}$	....	Gordon
Iron Tramway...................	$\frac{1}{200}$	....	Gordon
Sleighs on snow, 3 inches thick, ¾-inch runners, temperature 26°	$\frac{1}{20}$	....	....

ROAD CONSTRUCTION.*

BY I. F. POPE ("BONA VIA, BONUM OPUS"), AUSTIN, TEXAS.

General Remarks.—The construction of country roads may be divided into three parts.   1. The alignment.   2. The earthwork, or formation of the road-bed.   3. The metaling, or formation of the hard road surface.

Width of Road.—Before, however, any of these can be undertaken, the width of the proposed road must be determined.   Any excess width beyond the requirements of traffic is a waste of money. As two vehicles can conveniently pass one another in a width of 18 feet, the metaled or hard central surface should be made that breadth.   Provision must also be made for foot passengers, and for a softer footing for horses during dry weather than the hard central surface.   An earthen surface eight feet in width on each side of the central metal surface will answer both these purposes.   The use of this less expensively constructed surface during dry weather will also be economical, as it will save the wear of the more costly metaled surface.   For all these reasons combined, the width of a country road should be as follows :

A central macadamized or metaled width of 18 feet, with earthen sides 8 feet broad on each side; or a total width for traffic of 34 feet, raised, for drainage purposes (except where in cutting) not less than six inches above the level of the area it passes over.   There should also be an earthen berm 8 feet broad on each side of the raised road for trees and for stacking the metal required for repairing the road. The total width of land, therefore, that should be reserved for road purposes, including the tree berms and side trenches, should not as a rule be under 66 feet, or 4 rods.   Where high embankments are necessary, a greater width may be required.   The following notes on road making will refer to a road of the width or dimensions just described.

THE ALIGNMENT.

Alignment of Road.—Having determined the points or places to be connected by a road, the country along the proposed line should be reconnoitered, and the general route of the road located by one or more reconnoissances.   Great discernment is requisite in this preliminary alignment, so as to combine shortness of route with the avoidance of swamps that would entail an unreasonable cost in securing a good road-bed, and hills that might entail either steep gradients or heavy cost in deep cuttings.   It should be remembered that in many cases it is of no greater length to go round a hill than

---

* *The Engineering and Building Record* Competition.—Second Prize Essay.

to go over it, thus securing both greater economy of construction and easier gradients.

Location of Road.—The preliminary route being thus determined, it must now be surveyed with transit and level, and the final alignment located along the selected line. It is of course expected that the person who undertakes the construction of a road has a thorough knowledge of the use of the transit and level, and the laying out of lines and curves and gradients.

Steepness of Gradients.—In laying out a road the gradients or longitudinal slope of the road should not be permitted, under ordinary circumstances, to exceed 3 feet vertical to 100 feet horizontal. Under exceptional circumstances 5 feet in 100 feet may be tolerated. A gradient steeper than 3 per cent. is an evil that should be avoided if it possibly can. In hilly country a gradient as steep as 5 per cent. is at times necessary, but endeavors should be made by judicious alignment to make the gradients as easy as possible. The powers of traction of a horse are much strained by gradients above 3 per cent. It would be far preferable, in many instances, to incur extra expenditure in keeping the gradient within this limit than to make the rate of ascent steeper, as the cost once incurred is incurred for all time; whereas, a steep gradient is for ever a tax on the power of the draught animals, on the harness, and on the vehicles, and, in a short time, the loss due to this tax will greatly overbalance the saving effected by the steeper gradient. A young engineer may pride himself on the straightness of his road, but if this straightness is obtained at the expense of easy gradients, that might have been attained by slight deflections and curves here and there, his straightness and pride will prove very expensive to the traveling community that uses his road. Let it then be remembered that directness of route and easy gradients must both control the alignment of the road, and that anything above 3 per cent. may be called steep; though, as I have said, in hilly country it may be necessary to go as high as 5 per cent. gradients.

On Road Gradients.—If the natural surface of the country through which the road passes has a natural rise or fall of less than 1½ feet per cent. the longitudinal run of the road surface may be kept parallel to this natural surface; though, at the same time, 6 inches above it. But if the natural surface of the country rises now and again above the grade of 1½ per cent. moderate cuttings and fillings up to 3 feet in depth or height may be resorted to if such will keep the gradients within the limit of 1½ per cent.; more especially if the cuttings will balance the fillings without an unreasonable length of haulage. At 3 per cent. grades the line of toleration should be drawn, and unless serious obstacles present themselves, such as would give an undue height of cutting or embankment, endeavors should be made, by judicious cuts and fills, not to exceed a 3 per cent. grade.

Gradients in Hilly Country.—In a hilly country, by going around a hill instead of over it, an easy gradient may at times be obtained without unduly increasing the length of the road, or it may be advisable to ascend the hill by a zigzag route, as represented by the capital letter S. But in whatever way it may be most advantageously done, the limit of gradient should should not exceed 5 per cent. Easy gradients permit the full utilization of the traction powers of a horse, steep gradients detract from it. A gradient as high as 3 per cent. may be tolerated, when, as I have said, undue depth of cut or height of embankment will result; but gradients of 5 per cent. should only be permitted where great extra cost or great extra length of road would result by adopting a lower gradient.

THE EARTHWORK OR FORMATION OF THE ROAD-BED.

Formation of Road-Bed.—The road being now aligned and its gradients fixed, the earthen surface or road-bed has now to be constructed. The width of this to be, as already stated, 34 feet.

Usefulness of Side Ditches.—Where the surface of the country has natural easy gradients, the road-bed should be raised six inches above it (as shown in the diagram), to keep the road dry. This raising of the road also effects the double object of forming a drainage ditch on each side to carry off the rain-fall to the nearest culvert or bridge. These ditches, from which the earth is excavated to form the raised road-bed, give an increased artificial height to the road to aid its subsoil drainage, and to prevent any tendency of its becoming sodden during a long continuance of wet weather.

Description of General Cross Section of Road in Embankment.—The cross section then of the earthen surface of the road, where it needs no action to improve on the natural easy gradient of the country, will be as follows:

Vertical scale of 2 feet, and horizontal scale of 12 feet to an inch.

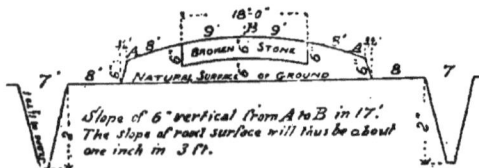

CROSS SECTION OF ROAD

To obtain the earth for raising the road-bed as above, the side ditches will have to be dug 7 feet broad at top, 1 foot at bottom, and 2 feet deep, with side slopes of 1½ feet horizontal to 1 foot vertical. This size trench will take up the balance of the 66 feet reserved for road purposes. The side trenches, which appear rather deep, will soon silt up somewhat. In no case is the earth for raising the road to be taken from the 8 feet berm reserved for trees, as this will make these berms lower than the surrounding country. All the earth required for the road must be taken outside the 8 feet berm.

Road in Marshy Land.—In low, marshy land, the height of the road-bed should not be under 1½ feet above the water line.

Side Slopes of Embankments, How Graded.—Where embankments are necessary, the side slopes of these embankments, if in ordinary soil, must be in the proportion of 1½ feet horizontal to 1 foot vertical, as shown in the diagram. Thus, an embankment 3 feet high, would have its toe 4½ feet beyond the vertical line of the edge of the 34 feet road surface. If 6 feet high, the toe would be 9 feet beyond; and so on; always adding half the height to the total height to find the spread of the toe of the embankment. In constructing embankments, all large clods of earth should be broken up, as otherwise, where these accumulate, depressions or hollows will form in the road surface when the embankment settles.

Extra Allowance for Settlement to be Made in Constructing High Embankments.—An extra allowance of from one to two inches (according to the nature of the soil) per foot of height of embankment must be made while under construction, so that when settlement occurs the embankment will have the full height originally intended, metal 6 inches thick.

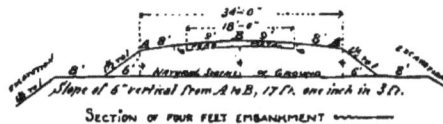

SECTION OF FOUR FEET EMBANKMENT

Road in Cutting.—Where the road, to obtain a proper gradient, is in cutting or excavated, the section of it will be as follows: The slope of the sides of the cutting to be according to the nature of the soil, from 1½ feet horizontal to 1 foot vertical in ordinary soil, to almost perpendicular when in rock. The side gutters to be 3 feet broad at top, and 1 to 1½ feet deep, with side slopes of 1½ horizontal to 1 vertical. If the cutting be of short length or in rock the gutters may be proportionately reduced in size.

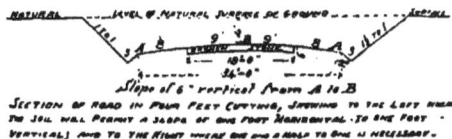

Road Along Face of Hill.—Where the road is in sloping or sidelong ground, as when it runs along the face or side of a hill, the

most economical form of construction is to have the cuts and fills of
the cross-section to balance one another. To effect this more of the
width of the road will be in cutting than in embankment, owing both
to the gutter and to the contents of the embankment portion being
of greater area than that of the cutting. The drainage slope of the
road surface to be made towards the hill side and the water thus not
permitted to run over the edge of the embanked portion. The
fall or slope of the road surface from the edge of the embankment
to the gutter on the inner or hill side will be 1 foot. The gutter to
be 3 feet wide at top and 1 foot deep, with side slopes of $1\frac{1}{2}$ to 1.

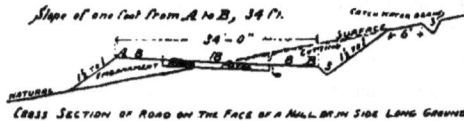

CROSS SECTION OF ROAD ON THE FACE OF A HILL OR IN SIDE LONG GROUND

Catch-water Drain.—In addition to the gutter on the inner side
of the road another gutter or catch-water drain should be made on
the upper side of the hill above the road and at least six feet from
the edge of the cutting, to receive and carry off the drainage of the
face of the hill above the road. These catch-water drains should be
continued to the culverts that should be constructed wherever there
is a necessity for cross-drainage, to relieve both the road gutter and
catch-water drain from an accumulation of water.

DRAINAGE.

Drainage of Road.—The manner of draining the 66 feet or
greater area, reserved for road purposes, should be under advise-
ment at the same time as the grading of the road is under consideration. In fact, the forming of the road embankment may be said
to be the same as forming the trenches for draining the area reserved for the road. The proper drainage of the area within the
road reservation is most important for securing a good road ; as all
water permitted to lodge within this area will tend to sodden the
road and render it incapable, unless by extra labor and expenditure,
of bearing a heavily laden wagon. Trenches of sufficient depth
and capacity should, therefore, be formed to carry off all water
from the road reservation to the nearest culvert or bridge. Those
latter should be provided at all points where the cross drainage of
the country impinges on the line of road. It is in the long run
cheaper to build a culvert to carry off this water than incur the
expenditure that will be necessary to keep the road in good order,
where water is allowed to lie and soak into the road by capillary
attraction. I have noticed the defective want of small culverts even
on otherwise well constructed railroads, causing the labor of fre-

quently lifting the rails to bring the iron-way to a proper level; this sinking being due to the softness of the earthen road-bed from the lodgment of water. The capacities of all culverts and bridges should be mathematically calculated to carry off all the water that impinges on the line of road. As instructions on the calculation of water-ways, and the designing of bridges and culverts, are outside the scope of this article, information on these should be obtained from treatises on these subjects. Any ordinary engineer by the use of a leveling instrument can give levels for the flow of water along a given line. The defective drainage of a road can consequently be only due to the want of knowledge of the value of efficient drainage to the well-being of a road.

THE METALING OR FORMATION OF A HARD ROAD SURFACE.

The Metaled Surface to be 18 Feet Wide.—As has been already stated an 18-foot width of metaling will suffice for the requirements of traffic on a country road. In the immediate vicinity of a large city, that is. for a mile or two out, accommodation may be necessary for the simultaneous passage of three vehicles. In such case the metaled surface should be made 25 feet broad and the earthen sides each 10 feet.

The Telford System of Providing a Hard Road-bed.—There are two systems in use in the formation of a hard surface to meet the requirements of heavy traffic on country roads—viz.: the Telford and the Macadam. The former consists of a rough pavement of more or less large flat rocks, varying in thickness from 6 to 12 inches, as a foundation to receive an uper coat of broken stone, 6 inches to 12 inches in thickness, which forms the road surface.

The Macadam System.—The Macadam system dispenses altogether with the foundation or rough pavement base, and places the broken stone, varying from 6 to 9 inches in thickness, immediately on the prepared earthen surface of the road. The Telford system may therefore be said to be the Macadam system with a foundation or hard base added. My own experience in India, in the Public Works Department of the Government, for whom I constructed over 400 miles of country road, and superintended the repair and maintenance at various times of over 1,500 miles of roads, is entirely in favor of the Macadam system, without the addition of a base. Where the method of constructing the earthen road-bed, as described by me, is properly carried out, the addition of a rock base under the macadamizing material is a useless waste of money. Where the road-bed is below the level of the country, and imperfectly drained, the Telford system may be necessary; but with the efficient drainage advocated by me, and the manner of laying the metal or macadamizing matter, carried out in the manner to be described hereafter, no base is necessary. It is more economical to expend a few hundred dollars per mile in securing efficient drainage in the original construction of a road than as many thousands in a rock base. I see that the Macadam system, in opposition to the double or Telford system, has also been introduced into America with good result by Mr. Pierce, of Bridgeport, Conn.

I am somewhat of the opinion of those who maintain that the broken stone or road surface material is more crushed and worn when between the hard, unyielding rock base below, and the heavy hammering weight of a horse and heavily laden wagon above, than where it rests on a more elastic earthen base, which is yet hard enough, by previous rolling with a 15-ton roller, to bear the weight of the traffic borne over it.

The material to be used to form the macadamized or metaled surface of the road will depend on what is locally obtainable suitable for such purpose. In some localities it is trap rock, in others limestone, etc. Again, on the prairie lands of Minnesota, Iowa, and parts of other States, there are no large masses of rock. In these gravel can be used as a substitute for stone. I have used broken stone, Kunker, a formation of limestone found in large quantities in India, gravel, and even coarse sand when nothing better was procurable at a reasonable cost. I will describe, seriatim, the best method, as it appears to me, for consolidating all these (Kunker excepted, which is not found in America), so as to form a hard surface suitable for traffic.

The Proper Thickness of Metaling.—It becomes now necessary to determine the thickness of metaling requisite to form the macadamized road surface. Various thicknesses have been used and have their advocates, of from 4 to 12 inches. The latter is, in my experience, excessive, and the former deficient. A 4-inch coat requiries constant supervision and expenditure in repairs to keep heavy traffic from breaking through so thin a crust when in any way worn. More frequent renewals are also necessary, causing both inconvenience to traffic and extra expenditure in supervision, labor, and rolling to lay a fresh coat of material. From my experience, I would say, a golden mean between excess and deficiency is arrived at by making the thickness 7 inches when spread, reduced by consolidation to a little over 6 inches.

Broken Stone—Its Quality and Size.—The hardest stone or rock locally procurable, or obtainable elsewhere at a reasonable cost, to be selected to form the macadamized or metaled surface of the road. This to be broken so that the pieces will pass through a ring $1\frac{3}{4}$ inches in diameter. This is the easiest and most practical way of testing the size of the pieces.

Slope of Cross Section of Metaled Surface, How Formed.—There are two methods in vogue in giving the requisite slope to the cross section of the metaled surface of the road to pass off the rainfall. In the one, the earthen surface is made level and the slope is given by a greater thickness of metaling at the center than at the sides. In the other the slope or camber is given to the earthen surface, and the metal put on of one uniform thickness.

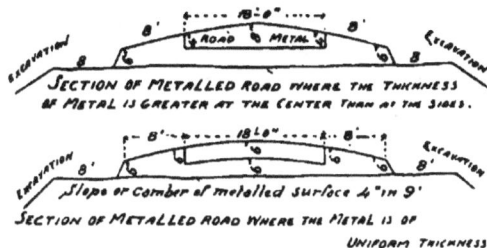

SECTION OF METALLED ROAD WHERE THE THICKNESS OF METAL IS GREATER AT THE CENTER THAN AT THE SIDES.

Slope or Camber of metalled surface 4" in 9'

SECTION OF METALLED ROAD WHERE THE METAL IS OF UNIFORM THICKNESS.

The Metal to be of Uniform Thickness.—The advocates of the first mentioned system say there is more wear at the center than at the sides, and consequently the metaling should be thicker at the former. Those in favor of the uniform thickness say that as an equal pressure should be sustained by the earthen base under all the wheels, so the resistance (indicated by the thickness of metal) should also be equal, and, that the extra wear at the center on a country road is inappreciable ; the weight thrown by the inclined plane of the road on the lower side wheels compensating, in great measure, for the slight extra traffic at the center. My experience is in favor of a uniform thickness of metal ; especially as it prolongs the time to a general renewal of the metaled surface, which has to be done sooner in the ununiform system, as the sides are soon worn too thin for traffic, and it is difficult to bring them to the proper thickness without a general renewal of the whole cross sectional surface. The cross slope, or camber of the road, should therefore be given to the earthen base, and the metal spread on it of a uniform thickness.

Preparing the Earthen Surface to Receive the Metal.—While the broken stone metaling for the road is being prepared, the earthen surface or base should also be prepared to receive it. If possible, the earthwork of the road, where in high embankment, should be exposed for a few months to the action of the rain-fall, which will tend to settle and consolidate it. Immediately before receiving the metaling, all hollows and depressions, etc., of the earthen road-bed, due to settlement or the action of the rain, should be filled up, and the earthen surface given the requisite camber or cross slope from the center to the sides, as shown in the diagram. The whole surface should then be rolled with a 15-ton steam road roller (all unevenness and derangement of shape due to the action of the roller being promptly made good), until the roller ceases to have any impression on it.   If the surface be of sand, or of such nature as to move or heap up before the roller, a thin layer of broken stone or gravel should be strewn on it.

Stone Bench Marks.—The earthen surface being now thoroughly rolled, and formed to the proper curved camber, stone bench marks 8′x8′x4′ should be sunk at 50-feet intervals along the edges and center of the proposed line of metaling, and set in position by a leveling instrument to indicate both the height of the earthen surface and the base of the metaling, and for giving the proper camber to the road.

Gauging Height of Metaling.—Temporary wooden gauges 7 inches high to be placed over these bench marks, to show the height to which the metaling must be raised as it is placed on the road.   The bench marks being placed exactly 50 feet apart can easily be found; and they will thereafter serve to indicate the depth of wear of the metaled surface.   The cost of these bench marks will be small, as their sides need not be cut to any regular shape; a flat upper surface to receive the bottom of the leveling rod being all that is needed.

The Earthen Sides to be Dressed off to the Level of the Spread Metal.—As soon as the metal is spread to the proper height, and before the consolidation of it is commenced, the earthen sides of the road are to be brought up to the proper level all along the edges of the metaling and given the prescribed slope to shed off the rain.'

Rolling.—The metal being now evenly spread (with the proper sectional camber) over a convenient length of road, and the earthen sides raised along its edges, the whole surface is to be rolled with a 15-ton steam road-roller, and the rolling continued until the roller produces an even, uniform, hard surface over the whole road.   Any depressions or unevenness occurring in the surface during the process of rolling to be promptly remedied by picking up with a pick the defective places and adding the requisite quantity of fresh stone to bring up the surface to the proper level.

The Metal Surface to be Watered while being Rolled.—After the road has been partially consolidated a water cart should precede the steam roller, as wetting the metaling causes the pieces to glide more readily together and to be more firmly bound together without crushing.   The cost of the water cart will be more than compensated by the greater expedition with which the consolidation can be completed.   This is my experience of the use of water.   When a hard surface has been effected by the roller, broken stone screenings passed through a ⅝-inch mesh screen, to be spread over the metaled surface to a depth of one inch, and the road again rolled, until its surface becomes such that water will run off it. The earthen sides being now finally dressed and rolled, the whole road is now completed.

Gravel.—Where broken stone is not procurable at a reasonable cost, or the exigencies of the local traffic does not oblige the use of so costly a material, while gravel is readily obtained, this material can be advantageously employed in making a good road. The difficulty in dealing with gravel is that it will not readily bind together under a roller, owing to the roundness of its particles, due to the attrition or manner in which nature has formed it.

Crushed Gravel.—If the gravel were crushed in a stone crusher, made to suit the smallness of its size, the difficulty due to its roundness would disappear; as the crushing machine would, by its action, produce enough of angularity or sharpness of sides in the particles to enable them to readily bind and not be easily displaced by the hoofs of horses. Thus crushed in a machine its consolidation would be effected in the same manner as described under the heading broken stone; with, however, this exception, that the consolidation of gravel must be effected in two layers of 3½ inches each, the first layer being in great part, though not thoroughly, consolidated before the second layer is added; the latter being then rolled until a hard compact surface results. A water cart should also precede the roller, as in the case of broken stone, after the surface has been partially rolled. For gravel a 10-ton steam road-roller will suffice, owing to the smallness of the particles. Gravel thus treated in a crusher would prove a fair substitute for broken stone where the traffic was not very heavy, and where it is readily procurable. It is far cheaper than stone, requiring in many cases to be but shoveled into a wagon, and it will cost less to crush and roll

If, however, the gravel is to be used in its natural state, as got from pits and the beds of rivers, a certain portion of dry, pulverized clay, in the proportion of one of clay to eight of gravel, must be mixed with it before it is spread on the prepared earthen surface or road-bed, on which it should be placed, as already directed, in two layers of three and a-half (3½) inches each, and consolidated in the manner described in the preceding paragraph. A layer of about one-third of an inch (not more) of sand should be strewn over the consolidated surface, and the roller finally passed over it until the sand is absorbed in the gravel surface.

Coarse Sand.—I have also used coarse sand as a road material, the largest particles of which were about the size of peas. This will not make a first-class road in very wet weather, but will make a good dry-weather road, and will render an otherwise impassable road, in wet weather, such that a team will pull a load through somehow, as the expression goes. It is a cheap method of paving a fairly good road for nine-tenths of the year. It may be the fore-runner of a good macadamized road, when the exigencies of traffic demand the latter. The sand should be screened so that a portion of the finer grains are eliminated from the mass. The road-bed should be prepared as previously described, and not more that two inches of sand spread at one time, and then rolled. The roller, a 10-ton steam roller, being preceded by a water cart. The road should then be open to traffic for a month or more, and then another layer of from one to two inches added, this again rolled, and so on, until some six or eight inches of coarse sand has been absorbed, when a fairly passable road in ordinary wet weather will result. Frequent rolling during the year will keep it a good road in fine weather and a passable one in wet. Where, for miles, a road runs along a river or sandy creek, the procedure above described will prove a cheap method of paving a fairly passable road even in ordinary wet weather.

Trees.—Suitable shade trees should be planted along the lower berms reserved for such purpose. They should be planted six feet from the toe of the earthen embankment of the road, and 30 feet apart.

Mile Posts.—A first-class road may be said to be incomplete without mile posts. The initial point of measurement for these should be some prominent building in the city from which the country roads radiate. The mileage numbers should be of a size so that he who rides may read.

THE REPAIR AND MAINTENANCE OF ROADS.

A road however well constructed will wear under traffic ; and, therefore, repairs, and in time more or less renewal, of its maca-damized or metaled surface is necessary.

The Effect of Traffic on a Metaled Road.—The first effect of traffic on a country road is the formation of ruts by the continuous passing of wheels over the same parts of the road. If these be not repaired or refilled they at length become so deep, and the crust of metal below proportionately so thin as to be unable to sustain the weight of a heavily laden wagon. The wheels then break through to the earthen base. The holes thus formed fill with water which softens the earth below, and every wagon wheel moving along the same line goes with a thud into these holes, increasing them in length, breadth and depth ; and the road thereafter becomes an impediment rather than an aid to traffic.   In a city there is not the same tendency to form ruts as on a country road, as in the former the frequent changes of direction to avoid other vehicles produces a more uniform wear over the whole surface of the road.

How to Make Repairs.—As soon, therefore, as a rut or hole, or depression becomes one and a half (1½) inches deep on a broken stone road, or one inch deep on a graveled road, the rut or depression should be repaired or refilled in the following manner :   The part affected should be cut out square (to use a well understood expression), as shown by the dotted lines, to the depth worn out, and

the space thus excavated refilled with the same description of fresh material to a height a little above the level of the sides of the excavation, so that it will sink down to that level when consolidated.   If the rut or hole thus repaired be of small extent the section hand in charge of the repairs of that portion of the road should go over it with a stone or iron rammer, about 7 inches diameter, and weighing about fourteen (14) pounds, and ram it to a hard surface, using the material dug out of the hole, instead of screenings, for the top coat. If the rut repaired be of any appreciable length it should be gone over by a small road iron roller drawn by a single or double team according to the width of the rut, and the size or weight of the roller necessary to consolidate it.   For the purpose of these repairs there should be a two-ton and four-ton roller on the road establishment, to draw which teams should either be kept or hired when necessary. The section man in charge of repairs should always have the means at hand of repairing the road.

Repair Material.—To enable him to conveniently do so a certain quantity of broken stone should be stacked at every 100 feet along the lower earthen berm reserved for trees.   The quantity thus stacked at intervals of 100 feet should not be less than a cubic yard of road metal, or 52 cubic yards to the mile.

Repair Implements.—The section hand should also be provided with a wheelbarrow, a pick, shovel, spade, four-pound hammer, and a road rammer, as implements for the execution of his work. When the repair material thus conveniently placed at his disposal has been reduced by use to about one-third of a cubic yard the stack should be renewed to the full cubic yard.

Repairs of Gravel Roads.—By the means just mentioned the section man will always have at hand the means of repairing the road the moment he considers such necessary. When a graveled road shows signs of roughness or disintegration of surface, the best remedy is to fill up all unevenness with the stated mixture of clay and gravel, and to pass a steam roller over it, preceded by a water cart or after a shower of rain.

Repair of Earthen Sides.—The earthen sides to be kept in repair by excavating the earth required from the side trenches. By this means the double object will be effected of keeping the trenches clean and free from drainage, and the earthen sides of the road in proper order.

Removal of Dust.—In France it has been found by experiment that a road wears better by removing the dust off it; this dust acting as emery powder does in the hands of a jeweler. Where labor is cheap, the removal of dust might be effected by human labor, but I doubt whether it would be advantageous to do so in America. I have never seen a sweeping machine, such as is used on stone and asphalt pavement, tried on a macadamized or broken stone road; but if it took off the dust without disintegrating the road surface, it might be advantageously used on that portion of a road within four or five miles of a city.

Renewal of Metaling.—When the general surface of any length of road is so worn as to be reduced to about 4¼ inches in thickness, a renewal of its surface is necessary by the addition of a coat of as much new metal as will bring it to the original thickness. The reduction of thickness from wear can be determined by exposing the stone bench marks placed, as has been directed, every 50 feet along the center and edges of the metaling. The old surface should be roughened up by means of a pick, and the new metal added and consolidated in the manner already described for the original metaling of a road.

County Engineer and Assistant Engineer.—For each county there should be an engineer and assistant, whose duties should not be solely confined to the earthwork and metaled surface of the road, but who should also have the designing and construction of all culverts and bridges required for the cross drainage that impinges along the road alignment. Both these officials to be appointed by the County Commissioners from qualified engineers, of not less than eight years' professional standing in the case of the engineer, and five years of his assistant. The salary of the former should not be under $1,800 a year, and of the latter not under $1,200, with ten cents per mile added in each case for every mile traveled on duty; with the exception that no mileage allowance be granted for distances under five miles from headquarters or place of residence, permanent or temporary.

State Engineer.—There should also be a State or chief engineer, whose jurisdiction should extend over all the county engineers, who should submit to him all their plans for both road and bridge construction. The chief engineer should be competent to advise the county engineers on all matters pertaining to their duties. There would thus be a guarantee for efficiency and uniformity in all works undertaken throughout the State.

Minor Establishment.—For road repairs there should be a section hand in charge of the repairs of two to four miles of road according to the amount of traffic; and a foreman, with extra salary for a horse, for every 30 miles of road. When special repairs were needed, the petty establishment should of course be increased. There should be at least one steam road roller for every county, and a 2-ton and 4-ton horse iron roller for every 20 miles of road.

The present system of repairing and maintaining country roads, by exacting so many days' labor in the year from the farmers, should cease. It is obvious from what has been stated that roads cannot be maintained in good order under such a system.

Cost of Road Earthwork.—The cost of constructing roads in the manner directed in this treatise will vary, in the case of the earthen road-bed, according to the undulating or hilly nature of the country passed through. In the prairie lands of Illinois, Minnesota, Iowa, Texas, etc., the cost of the earthen road-bed will be from $500 to $800 per mile; in more undulating land, from $700 to $1,500, and in hilly country from $1,500 to $2,500 per mile. Special miles will in each cost above these figures.

Cost of Broken Stone Metal.—The cost of the macadamized or metaled surface will vary according to the material and the distance it is transported. It will be from 50 to 75 cents per consolidated square yard, for a 6-inch consolidated thickness, if of broken stone; or from $5,000 to $7,500 per mile.

Cost of a Graveled Road.—If the material be gravel, the cost will vary from 30 to 50 cents per consolidated square yard of 6-inch thickness, or $3,000 to $5,000 per mile.

APPENDIX.

The preceding notes and remarks on the construction of macadamized roads is the result of many years' experience in their construction in British India, in the Public Works Department of the Government of India. The country roads there are all macadamized or metaled, to use a more familiar term in use there. They are also thoroughly maintained in good order; and, taking the vastness of the country into consideration, and the efficient state in which the roads are kept, they may be said to form the most magnificent system of internal communication of that nature in the world. The Government of British India holds and acts up to the principle that it is as necessary to keep the country roads in good order as it is to maintain railroads. They each have their value, and in their way should be equally efficient. The first journey taken by all commodities is from the farm or field along a wagon road; and it is a just and right principle that the men (farmers) who convey these commodities to market and to the railroad depot should have every facility and convenience of doing so, which can be provided by good roads. The time and labor lost by the farmers in hauling their goods along bad roads adds to the price of their commodities—that is, to the food of the whole population, and to all the raw products of commerce; for the farmers must compensate themselves in some measure for the diminished power of their draught animals, and the extra wear and tear of their harness and vehicles, caused by bad roads. In some transactions the loss suffered by one man is the gain of another, but in the case of bad roads there is a general yearly loss to the whole community of many millions of dollars. The Legislature of each State should therefore take up the matter, and establish such laws and regulations, and make monetary provision for the construction of so necessary a benefit to the general community as good roads. For further reference regarding roads, see *Harper's Weekly* of August 10, 1889.

An article by Jeremiah W. Jenkins, published by the American Economic Association.

An article in *Scribner's Magazine*, by N. S. Shaler.

Governor Hill's message to the New York Legislature on the necessity for constructing good roads.

The Engineering and Building Record, of New York and London, for the latter half of the year 1889 and beginning of 1890.

CONSTRUCTION OF ROADS.*

BY JOHN P. PRITCHARD ("A SIMPLE SCRATCH"), QUINCY, MASS.

The first thing to do is to have the road properly laid out by a competent engineer, making lines and grades which include all earthworks, cutting and filling, culverts, drains, bridges, etc.

The earth road-bed to be free from loam ; excavate to the required depth, then grade to the proper shape ; to be well watered with a one-horse watering cart, and then rolled with the steam-roller; fill all depressions which shall appear with the same material as the road-bed, roll until it is compact and solid, then spread two inches of sand over all for a bed for the paving. On this road-bed set the paving stone, set a line of paving through the center of the road 9 inches in depth, then gradually diminishing to 5 inches to line of curb, the stone always to be laid with the best bed down, laid close and in parallel lines, as near together as possible, across the road, breaking joints. The stone for each section wants to be as near of a size as possible ; no stone should exceed 15 inches, except through the middle of the road, then a single line can be laid parallel with the road ; after having paved 100 feet the stone must be wedged up tight with spalls, chips, etc. It is not necessary that the wedges should be driven to the bottom of the paving ; after wedging, all the projections sticking up above the grade line of the paving should be broken off ; no wedging to be done within 15 feet of the face of the paving ; almost any kind of stone will do for the paved bottom if they are carefully and well set ; after the bottom is thus prepared the steam-roller can be run over it, commencing at the curb ; work towards the center ; the amount of rolling for this bottom will have to depend on the superintendent of the work ; never roll within 20 feet of the face of the paving ; never use screenings from gravel to wedge or surface up the paved bottom.

On top of this foundation lay 4 inches of broken stone (when rolled), not to exceed 3 inches in size ; the tailings and spalls from the breaker can be carted on to the road and men with hammers can break them to the required size, or, they can be thrown on top of the paving and used for wedging. Ballast and stone chips from stone sheds can be used for this course if care is taken to have them broken up to the proper size ; in spreading this course never spread within 18 inches of the curb line ; the roller will take care of this space by crowding it down to the curb line. There should not be any stone less than 2½ inches in this course ; great care should be taken to pick out all the round stone that may appear,

---

* *The Engineering and Building Record* Competition.—Third Prize Essay.

if not they will work up through the surface. This course is now ready for the roller, and a great deal depends on properly rolling this course. The flat stones must be broken or picked out as fast as they appear; the water cart can now be put on to the road; it will do more good than to put on binding, which should be avoided if possible, and should never be resorted to unless the stone round up and will not bind ; such places will occur when very hard and angular stone are used, but a few shovels full of binding and a little water will make them bind all right.

On top of this course lay 3 inches of broken stone that shall have passed through a 2-inch circular hole, and spread with a shovel; never use a ʃrake ; be careful and pick out all flat and round stones. Use the water cart freely on this course if dry weather.

On top of this course lay enough broken stone that shall have passed through a 1½-inch circular hole, and spread with a shovel thick enough to fill all interstices ; to make it almost smooth this course wants more rolling than either of the others.

On top of this course comes the binding. Use the screenings from the breaker that have passed through a 1-inch circular hole, 2 inches thick. This course contains stone and dust, and when it is well watered and rolled, the dust and loam washed out of it, you will have a good surface, and it will be ready for public travel.

Sand can be used for binding instead of screenings, but in my experience screenings are more satisfactory for our roads.

The stone used in the second, third and fourth layers should be of one texture, near as possible ; the wear will be more even.

A great saving can be made in this class of road by paving a space through the middle wide enough for teams to pass both ways. The sides can be macadam or gravel.

GRAVEL ROADS.

In building gravel roads commence and screen through a 2½-inch screen ; then screen this through a 1½-inch screen ; then this through a ½-inch screen; when the bed is properly shaped, wet and rolled, put on the 2½-inch, wet and roll; then put on the 1½-inch, wet and roll ; then put on the ½-inch, wet and roll. The required thickness of each course will depend on the kind of road you are building. For main or side roads the custom in most towns is to take the gravel from the bank, dump it on the road, throw the loose stones into the bottom, and cover them up with the small ones ; in a few years the small ones are on the bottom and the large ones are kicking about the top of the road.

MACADAM ROADS.

Have the bed properly shaped, wet and rolled ; coarse stone ballast tailings from the breaker will do for the foundation course, but they must be well broken so they will all be of even size.   Next a course of 2½-inch from the breaker, four inches thick ; next course 2-inch from the breaker, three inches thick ; next course surface up with 1½-inch, well rolled ; cover with screenings from the breaker that have passed through the 1-inch circular hole. Great care should be taken to have the flat and round stones, as well as all rotten stone excluded ; all courses to be well watered and rolled.   Have the two last courses of one texture, as near as possible.

MAINTENANCE OF ROADS.

After a road has been properly constructed it should always be maintained, no matter what the amount of travel there is on it. Ruts and hollows must be filled up as soon as they appear.   No water should be allowed to stand on top of the road ; keep the road as free from dust and mud as possible.   When filling ruts and holes good hard metal that has passed through a circular hole, one and a half or two inches, should be used, loosening the bottom with a pick to facilitate the binding.   To repair a macadam road put the spikes into the roller ; go over it three or four times, pick out all the large stones or break them up, then run the roller over just enough to smooth it up a little, then put on the stone, never over 2½-inches in size, and for side streets 2 inches will do ; then surface with 1½-inch size, wet and roll.   You will not have to put any binding on this, for enough binding will work up through it if it is properly wet and rolled.   The great trouble in regard to the maintenance of roads in Massachusetts is that in the fall, winter and spring, when the roads need looking after the most of any time in the year, the appropriations are all spent, the men discharged, and nothing can be done until after the annual elections ; then they have no time to look after repairs, but get to work on new work, and before they get half done on the work laid out to do the money is all gone.   We shall never have good roads in Massachusetts until the business of road-building is taken out of politics and taken in charge of by the State.

Good roads can be constructed by towns, if they are not rich enough to buy a steam roller, but no town can afford to be without a stone-breaker ; and because a town does not own a steam roller or stone-breaker it is no excuse that they cannot build good roads. With forty or fifty stone-breaking hammers, an iron sectional roller weighing two tons, and a competent man as superintendent, good roads can be built.   I built a mile and a half of Telford road before I even knew how to use a steam roller or a stone-breaker. They will cost more though and take more time to build.

A good road plant consists of a 10 or 15 ton steam road roller, a 9x15 stone breaker, and a revolving screen with 1-inch, 1½-inch, and 2-inch circular holes.   A great saving can be made in breaking stone by having the breaker set up 15 or 20 feet above grade, or an elevator can be used to convey the stone, and have the top of the mouth to the breaker even with the floor.   This saves a great deal of handling the stone.   One man can do as much as two where they have to be lifted from the platform.   Another advantage of setting up the breaker is that hoppers can be constructed and the stone drop direct into the mouth of the screen from the breaker, then three different sizes into the hoppers.   Each hopper contains 40 tons.   These hoppers can rest on a heavy frame of hard pine timber, high enough from the ground so that a cart can be backed under any of them, and by the simple motion of drawing a lever allows the contents of the hoppers to fall directly into the carts, and a reverse motion of the lever closes the slide tight.   A cart can be filled in one minute.   Three men can break 100 tons a day.   Cities and towns will find it to their advantage to look into this mode of arranging their stone breakers. There is a stone breaker at work in a city near Boston that takes nine men and two horses to keep it a going.   The reason of this is because it is set on the ground, and the stones have to be handled over twice before they are put on to the road.   They are shoveled into the carts by hand—but don't it make work for the poor laboring man? and don't we want his vote?   That is the whole secret of bad roads in Massachusetts.   Potter says bad roads have a tendency to make the country disagreeable as a dwelling place, and a town which is noted for its bad roads is shunned by people in search of rural homes.   There is no kind of work done in the New England States that there is so much money wasted on as there is in building and taking care of roads.   Shaler says, in no phase of public duties does the American citizen appear to such disadvantage as in road building.   Learned says towns complain of the first cost of macadam roads, while annually spending millions of dollars and moving countless tons of earth without having a good road.   Whether I get the prize or not for this essay, if it can be called so, if my instructions are followed in regard to Telford road building by any competent superintendent, I shall have no fears as to the result, for I have been building just these kinds of roads since 1877, and some of them are having the heaviest traffic passing over them of any place in this country; talk about your thin roads, we would cut them up in one

week with the rain-fall we have had this winter. I can't say anything about asphalt ; I wish I could ; I have never had anything to
do with it, but I have great faith in it, and it will be the coming
paving ; it is only a question of time.

The roads that I have built are open for inspection. I have
not given the cost of any of the roads that I have built, because the
prices must vary according to the length of haul and price of stone.
I built a mile of Telford road, 40 foot roadway, for \$7,000. I have
built a half a mile that cost \$8,000. The first one, some of the
material was hauled over half a mile ; the other, some of the material
was hauled three miles. Seventy-five cents per square yard is the
price quoted by the different road builders as a fair price for Telford
roads.

Good sidewalks this season of the year is what we want. The
best material for muddy walks is coal ashes ; mix a little loam and
loose fine salt ; have it mixed up in large piles, when you have soft
weather and the frost is coming out of the ground, you will find it a
good deal better than gravel ; it will bind and become compact with
a good smooth surface and will shed the water. Every town should
have a large shed where they could have the good clean ashes that
is collected and deposited ready for emergencies.

NOTE.

In preparing the following abstracts of the essays given Honorable Mention, the purpose of the Committee of Award was to
confine them to such propositions as were not substantially covered
in the three essays published—and thus avoid unnecessary repetition. The omission of parts will, therefore, be understood as no
reflection on the essays.

BY PROF. JOHN V. HAZEN ("GRANITE STATE"), HANOVER, N. H.

(*Abstract.*)

Services of an Engineer Should be Secured.—The skill of a competent engineer is of great value in laying out roads. Figure 1
shows an old and a new road laid out in writer's native town. The
new road is only 234 feet longer, and an elevation of 60 feet is saved
and maximum grades are reduced from 1 in 4 to 1 in 10, and might
have been reduced more.

Curving Roads not so Much Longer as They Seem.—Curving
roads on a flat country are not so much longer as they seem.   A
road between two places, five miles apart, curving so that the eye
can nowhere see further than a quarter of a mile, will only be 225

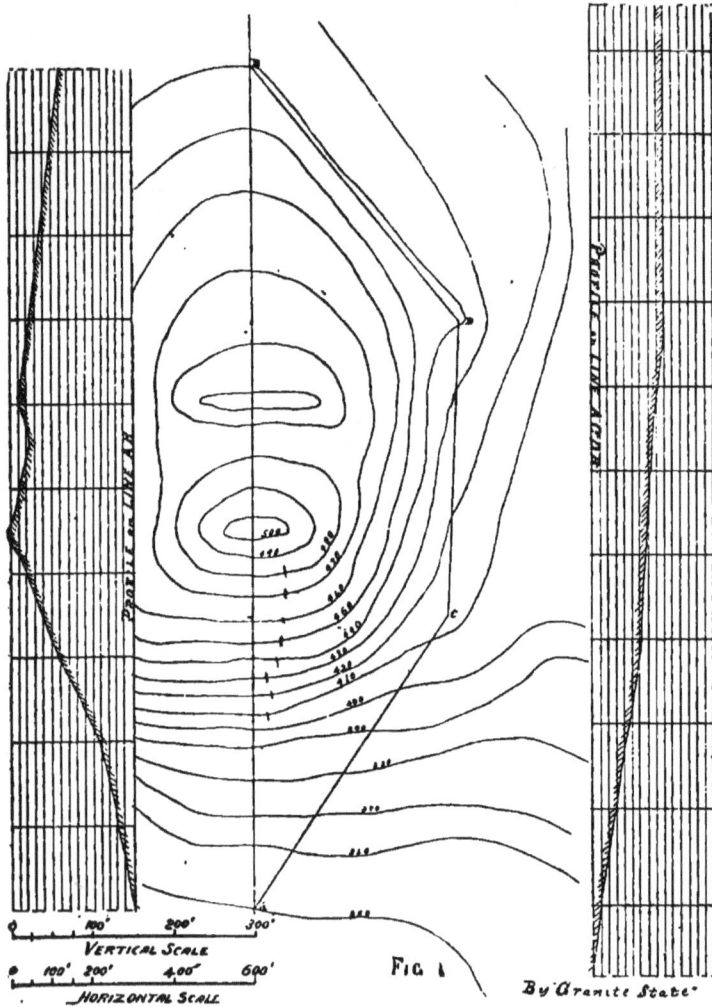

Fig. 1

feet longer than a straight one joining the same termini.   Within
proper limits a road may wisely be increased in length by 15 times
the vertical height avoided by the detour.

A Narrow Road Frequently the Best.—A well kept narrow road,
when width is sufficient for traffic, is much better than a broad one
with no greater amount expended for repairs.

Earthwork Paid for in Excavation.—Earthwork and rockwork should be paid for in excavation on account of shrinkage in the former and increase in volume of the latter when placed in embankment.

The Best Embankment Built in Layers.—Embankments are usually made by dumping the earth from their ends, keeping them up to their ultimate height, but wider at the top and narrower at base than when finished. The better system is to make them in layers of a foot thick, each successive layer being compacted by depositing those which follow.

Method of Underdraining.—For underdrainage in heavy soil ditches are dug transversely of the streets at intervals of 20 to 30 feet, and with a fall of from 1 in 130 to 1 in 20. The drain may be of stone or brick, built like a small culvert, or of agricultural tile. The latter are the most effective and generally cost no more. Two-inch tiles cost from $11 to $20 per 1,000 feet. They should be laid with above fall, with the closest possible joint, end to end, in a narrow trench of uniform slope carefully aligned. Such a drain placed 2 feet deep will cost, with tiles at $15 per 1,000, about 45 cents per rod when completed.

The English Method of Laying.—J. Bailey Denton, an English authority, advocates two lines of 2-inch tiles beneath the two edges of the traveled road longitudinally. Others place a line of tiles directly beneath the side ditches relieving them of much surplus water. This has the disadvantage of not much removing the moisture from center of the roadway. In another system one line of 3-inch tile is placed directly beneath the center of traveled road. This has the advantage of being cheaper than the others; of conveying away the moisture from center of the roadway, and, on the whole, is the most satisfactory. Cross drains connect at intervals the second and fourth systems with the side ditches, the latter being sufficiently deep to receive the outflow. Well built gutters, 18 to 24 inches deep below bed of road, with finished surface of side ditch above, kept clean, free from grass, weeds and stones, will frequently render underdrainage unnecessary even in quite moist soils.

Culverts.—Culverts for transferring the water from the higher to the lower gutters should be put in at every depression. These may be of vitrified stone 20 to 24 inches in diameter, oval or egg-shaped cement pipe of stone, brick, or of wood if constantly under water. Culverts should have a fall of at least 1 in 100. Care should be taken to make them large enough to carry away the water without overflow. A rule sometimes used is $A = C\sqrt{M}$ where A = area of opening; M = drainage area in acres, and C = a coefficient depending on country; as 1 for a flat country, $1\frac{6}{10}$ for hilly, and 4 for mountainous.

Earth Roads.—While earth roads are not ideal roads, they are frequently the best a town's finances will allow. Their ordinary condition may be much improved by careful attention to grades, surface drainage, underdraining, and keeping the surface as nearly homogeneous and hard as possible. An occasional rolling will have a marked effect.

Amount of Rolling.—The total amount of rolling necessary will vary with materials and soil. From 30 to 60 hours per 1,000 yards is desirable. Silicious sand, clay or clean sharp gravel are sometimes used as a binding material. The amount of crowning in use varies from $\frac{1}{18}$ to $\frac{1}{80}$ of the chord, or a slope of 1 in 24 to 1 in 40. Some engineers increase the slope with the grade. In Providence grades of

0.5 to 4 feet per 100 have a transverse slope of .04 per foot.
4 to 6   "   "   "   "   .05   "
6 to 9   "   "   "   "   .08   "

This provides for a rapid conveyance of surface water to side ditches. The total thickness of macadam varies with the locality. Six to ten inches is the usual thickness. The latter being used in or near cities. Good roads have been made only 4 inches thick, and as it is often a question of thin roads or no improved roads, they are worthy of trial. A 10-inch road costs from $1 to $1.75 per square yard, while good 4-inch roads have been built for 28 to 38 cents. Where the traffic increases, or sub-soil is soft, the thickness should be increased. Some engineers vary the thickness with the grades, decreasing it as the grades increase; 4 to 6-inch roads recently built near Plainfield, N. J., are said to have cost from $3,000 to $8,000 per mile; some 4-inch roads in Bridgeport, Conn., about $3,000 per mile.

Quality of Stones of Sub-pavement.—In Telford paving the spaces between the blocks should not be filled with anything smaller than stone chips, otherwise an essential characteristic, capacity to drain off rapidly whatever surface water works through, would be destroyed.

Total Thickness of Pavement.—The total thickness of Telford pavement will vary much with the locality and expected traffic. It should never be less than 6 and seldom more than 16 inches, and it may decrease as the grade increases. In cost they vary from 90 cents to $1.75 per square yard. Some roads recently built in New Jersey cost from $8,000 to $10,000 per mile.

Advantages of Telford Roads.—The advantages claimed for Telford roads are that the sub-pavement needs no renewal, and provides for thorough drainage. That in clayey, compressible soils the clay does not work up into the voids; that the broken stone does not work down into the soil; that extreme winter weather has no effect upon it nor does the ensuing thawing out. The advocates of macadam roads claim that the earth does work up into the voids of a Telford pavement; that the broken stone works down into the same voids, and that the broken stone wears out much more rapidly on a Telford sub-pavement than on compressed earth.

Good Foundation Important.—Foundations may be made of sand or blast furnace slag compacted by rolling.

Repair of Earth Roads.—Earth roads should be kept smooth, hard, up to grade and cross-section by the addition of suitable materials at frequent intervals, and in small quantities at a time, on all places out of grade, securing a surface such as shall quickly convey the water to side ditches. The latter should be kept open, of uniform and sufficient slope, free from rocks, ridges, depressions, and continuous to some natural or artificial outlet. Sprinkling and rolling are valuable adjuncts of repair, especially in dry weather, and a thorough rolling in spring, after ground has settled, is of marked benefit.

Sprinkling and Rolling Advantages.—In dry weather, sprinkling and running the road-roller backwards and forwards over a macadam road will preserve its surface, or if water is not at hand a coating of sand or other binding material does the same.

Repair of Large Areas.—When large areas are repaired, the work, if possible, should be done in damp weather.

Thickness Worn Out per Annum.—French experience shows that, measured by thickness, the annual wear on ordinary country roads is seldom over $\frac{1}{4}$ inch, and on the most frequented roads 1 inch. Systematically maintained English roads confirm these results.

Cost of Breaking Stone and Repairing Roads.—It costs 70 to 80 cents per cubic yard to break trap rock with crushers; the cost of delivery varies with the distance, about \$1.10 for two miles, or a total of \$1.90. For $\frac{1}{4}$-inch wear, road 20 feet wide, it will take about 164 yards, costing \$300 per mile for material.

Plainfield, N. J., roads are said to have cost last year, with careful watching and immediate repair of weak spots, \$1,000 for 40 miles. Time will show annual cost for complete maintenance.

Cost of Transportation on Poor Roads.—On a macadam road, recently built in Connecticut, it is said that 5,000 pounds can be hauled where 1,200 to 2,000 was a good load before rebuilding for the same team. A farmer living ten miles from market, and having 100 tons annually to sell, on the unimproved road would have to make at least 100 trips, on improved road 40, a saving of 60 days' time, which, at \$2.50 per day, would be \$150; probably several times his tax for the improvement.

Low Rates Follow Good Roads.—The introduction of macadam roads in parts of New Jersey is said to have doubled, in some cases, the value of real estate.

COST OF SOME ROADS RECENTLY BUILT, ETC.

Locality.	Total thickness of covering.	Method.	Cost per mile.	Remarks.
Kingston, R. I.	8'	3 layers: 1st, 4" of 1" to 2½" stone; 2d, 3" of 1" stone; 3d, screenings.	$5,500	1¼ mile, 20 feet wide. ½ " 16 " "
Linden Township, N. J...	12'	Telford, 8'; 2 layers broken stone; 1st, 2½"; 2d, 1½" thick.	11,600	16 feet wide.
City of Rahway, N. J....	12'	"	9,349	" "
Union Township, N. J...	12'	"	11,900	" "
Westfield, N. J...	12'	"	9,640	" "
Fanwood, N. J...	12'	"	9,530	" "
Bridgeport, Conn	4'	Macadam.	3,000	18 feet to 20 feet.
Plainfield, N. J..........	4" to 6'	"	3,000	16 feet wide.
Fairfield, Conn	4'	"	5,000	20 " "
Franklin Township, N. J.	4'	"	4,700	15 " "
Hoboken to Paterson. ..	.....	.....	5,000	....

Kansas City has just renewed five miles at prices varying from $1.17 to $1.85 per yard.

SAMUEL L. COOPER ("PALISADES"), NEW YORK CITY.

(*Abstract.*)

Grades on First Class Roads.—It costs more to maintain a grade of 1 in 20, than a grade of 1 in 40.

Width of Roads.—The metal should be 16 feet wide for ordinary country roads, and as much wider up to 25 feet as the money available will permit; but for village roads and suburban pleasure drives, the metal should extend across the road to the gutter pavement on each side. Expense of making the metal 16 feet wide will sometimes compel a narrower width, but it should not be made less than 13 feet, as that is the minimum width on which wagons can pass each other with safety. The following are suggestions for dimensions of roads of different classes :

	Total Width.	Width Roadway.	Metaled Width.
For a First-Class Road	75 to 100	40 to 60	25 to 35
For a Second-Class Road	50 to 66	25 to 35	16 to 25
For a Third-Class Road......	40	.25	13

Crown.—When the surface of a road is poor the crown should be greater.

Consolidation vs. Mass.—Perfection in roads cannot be obtained without proper consolidation of the materials that make the road. This is more important than mere mass without consolidation.

Steam Roller.—It can best be obtained by a steam roller of about ten tons weight, and any community that seeks the best roads for the least outlay can get the most for its money by using the best materials, buying a steam roller, and employing an intelligent road builder to make the roads and maintain them.

For Macadam.—Each layer of stone must have a certain amount of binder to make it a compact mass. The best material for the purpose is screening from the crusher, but when that is not available, clean gravel and sharp sand should be used.

The rolling should be continued on the finish until water will flush over the entire surface.

Sides of Road to be Rolled.—The sides of the road between the gutters and the metal should be carefully graded and rolled to ultimate resistance and a true surface with a 5-ton roller

The cost of macadam roads in the vicinity of New York, by contract, to prepare the surface, provide good trap rock, spread and roll the same ready for use, is about :

50 cents per square yard for 6 inches of metal.
65 "   "   "   "   " 8   "   "
80 "   "   "   "   " 10   "   "

Foundation for Telford Road.—The foundation and broken stone will seldom be required to be thicker than 12 inches, and unless the width to be metaled is more than 16 feet, it should be of uniform thickness across the section. When the width is over 16 feet, the thickness at the sides may be reduced to 10 inches.

Foundation Stone, Size.—The foundation stones for a 12-inch pavement should be 7 inches deep, not over 4 inches wide on top, and from 8 to 12 inches long, and fairly uniform and regular. They should be of the hardest and toughest stone available, preferably trap rock, and laid with the joints widest on top, in parallel courses, breaking joints by at least 1 inch across the road. After being set, they should be wedged firmly by inserting and driving down with a bar, spalls of the same stone, in every possible place, until the whole foundation is solid and firm. All irregularities and projections above the 7-inch line should then be broken off carefully with a hammer, so done as not to loosen the foundations. All the rest of the joints not then filled with spalls, should be filled with spalls and chips pounded in with a hammer, so that the top is a regular but not too smooth surface.

An 8-inch pavement should have a 5-inch foundation. The bottom course of stone should be broken to pass a 2-inch ring, and the top a 1-inch ring. The binding and rolling should be the same as above described. In the vicinity of New York the cost of a Telford macadam road 12 inches thick is about $1.10 per square yard; 10 inches thick is about $1 per square yard; 8 inches thick is about 90 cents per square yard.

Six-inch Roads.—We are prepared to say that success is entirely possible with 6-inch macadam roads, properly made and maintained.

Such roads can be built in the vicinity of New York for from $5,000 to $6,000 a mile, with a metaled roadway 16 feet wide, and when skillfully made, and thoroughly maintained, they answer all the purposes of a more expensive road; if neglected, however, they will soon go to pieces.

Essentials for Success.—The following are essential to success with 6-inch roads:

(1) Good stone, preferably trap rock, and screenings or gravel as a binder.

(2) A steam roller of about ten tons weight.

(3) The services of a man who knows how to make such a road.

(4) Constant and intelligent treatment in maintenance, particularly during critical periods.

Proper maintenance includes:

(1) The removal of the wear of the road, and its prompt replacement by new material.

(2) The immediate repair of ruts, holes and washouts.

(3) The keeping clear of the gutters and culverts.

(4) To these should be added additional care during critical periods, when the frost is leaving the ground, and during prolonged seasons of wetness or drought.

General Repairs.—In the fall of the year, the road should be put in first-class shape, to better resist the damaging effect of frost during the coming winter.

More general repairs, such as an entire resurfacing, will be necessary periodically, according to the thoroughness of the construction and the amount of traffic.

FRANK B. SANBORN ("ROY"), BROOKLINE, MASS.

(*Abstract.*)

Records.—The expenses incurred by the details of carefully located side-lines will not usually equal the cost of one of the many law-suits that the town will be obliged to withstand on account of indefiniteness of street lines.

Drains.—In an address before Maine Board of Agriculture, C. B. Stetson said: "It is clear that neither our farmers nor our road makers half appreciate the wonderful results which can be secured by thorough drainage of fields and highways—by keeping them, so far as water is concerned, in fit condition for perpetual service."

For the drainage of ordinary highways the writer recommends tile drains laid at least three feet deep in trenches 20 inches wide. After the tiles have been laid and the joints well protected with tarred paper or cheese-cloth the trench should be filled with small stones not more than three inches in diameter—the stones laying near the pipe should be carefully placed so as to protect it from the superimposed weight. Large stones which nearly fill the trench exert simply a downward thrust and should not be used. In ordinary soil a trench can be dug, the tiles laid and covered with stone, as described above, for 25 to 50 cents per linear foot. The writer has had charge of the construction of two roads the past year which were drained by 4, 6 and 8-inch tiles for 35 cents per linear foot. Henry F. French says in his book on Farm Drainage: ". . .

drainage with tiles will generally cost less than one-half the expense of drainage with stone drains and be far more satisfactory in the end."

Road-bed.—It is preferable to make the width of drive-way proper (which includes the gutters) some multiple of 8 feet, which is about the width necessary for each team.

Requisites.—It is better to bring good material, if necessary, from a distance rather than make use of poor material for road making.

Conclusion.—Finally, engineers of vast experience have concluded, that: A perfectly good road should have a firm and unyielding foundation, good drainage, a hard and compact surface, free from all ruts, hollows or depressions. The surface neither too flat to allow water to stand, nor too convex to be inconvenient to the traffic; and free from loose stones.

A. T. BYRNE, C. E. ("ZAMORA"), BROOKLYN, N. Y.

(Abstract.)

It is advisable for road constructors to abandon precedents and build roads which will best suit the requirements of the traffic.

The subject of the proper construction and maintenance of roads has never been considered by the people in more than a limited and superficial way. A lack of public appreciation of their true value, a mistaken idea of economy, and, in most States, the existence of laws that make and permit bad work, has made their condition a conspicuous blot upon the page of our general progress as a nation.

The system of requiring personal service upon our country roads by our rural population is unsound in principle, unjust in its operation, wasteful in its practice, and unsatisfactory in its results.

Road Laws.—New legislation is needed, but no legislation can produce good roads until the people are willing to pay for them.

New Laws.—First, the burden of the leading lines of communication should be borne by the whole community. Second, the employment of skilled engineers to superintend the road making and repairs. Third, the abolishment of personal labor, and the levying, instead, of a money tax. The money tax will be found to be not only more equitable than the labor system, but even less burdensome. None of it will be wasted, and those who have the skill and strength for road-work will receive back in wages more than their share of it.

Roads and Morality.—Bad roads are the cause of more profanity and ill-nature than any other trial to which human nature is subjected.

Effect of Bad Roads.—Nothing has led farmers' children to a dislike of the country, and a desire for city life, more than the monotony of the rural districts in the winter months, produced by the lack of social intercourse, for the want of good roads.

Advantages of Good Roads.—By the improvement of our roads every branch of our agricultural, commercial and manufacturing industries would be materially benefited. Every article brought to market would be diminished in price; and the number of horses would be so much reduced that by these and other retrenchments many millions of dollars would be annually saved to the public.

All the produce and industry which by these improvements finds for the first time a market, is, as it were, a new creation.

Cost of Roads.—The road which is truly cheapest is not the one which has cost the least money, but the one which makes the most profitable returns in proportion to the amount which has been expended upon it.

Maintenance of Natural Soil Roads.—In the maintenance of *sand* roads the aim should be to have the roadway as narrow as possible, so as to have all the vehicles run in the same track ; to have an abundant growth of vegetation on each side of the rut, for by these means the sheering of the sands is in a great measure avoided. Ditching beyond a slight depth to carry away the water is not desirable, for it tends to hasten the drying of the sands, which is to be avoided.

Where possible, the roads should be overhung with trees, the leaves and twigs of which, catching in the roadways, will still further serve to diminish the effect of the wheels in moving the sands about. If clay can be obtained within a moderate distance, a coating six inches thick will be found a most effective and economical improvement ; four inches of loose straw will, in a few days' travel, grind into the sand and become as hard as a dry clay road.

Clay roads can only be made into satisfactory ways by means of effective drainage, so contrived that the least possible amount of water will remain in the material. Deep side-ditches are absolutely necessary. Cross and sub-drains may be employed to great advantage. The narrower the roadway the more effective will be the drainage. If sand can be obtained within a moderate distance, a coating three inches thick will form a very beneficial improvement. Trees should be removed from the borders of the road, so as to expose its surface to the drying effects of the sun and wind. Neither sods nor turf should be used to fill holes or ruts, for though at first deceptively tough, they soon decay and form the softest mud; neither should the other extreme be reached, by filling up the ruts with stones. They will not wear uniformly with the rest of the road, but will produce hard ridges.

Coal-slack, ashes and cinders, judiciously applied, make fair roadways; yet in many parts of the coal regions the people are pulling through the mud in plain sight of heaps of that material sufficient to cover all the roads in their neighborhood, and are heard to lament the bad condition of the roads.

Stone.—The qualities required in a good road stone are hardness, or that disposition of a solid which renders it difficult to displace its parts among themselves, toughness, or that quality by which it will endure light but rapid blows without breaking.

Chemical Qualities.—The porosity or water-absorbing capacity is of considerable importance. Of two rocks which are to be exposed to frost, the one most absorbent of water will be the least durable.

For light traffic the carboniferous and transition lime-stones are sufficiently durable, make the smoothest and most pleasant roads, and possess the quality of forming a mortar-like detritus which binds the stones together, and enables it to wear better than a harder material that does not bind.

For heavy traffic the lime-stones are too weak, the sand-stones too soft, the green stones too variable, the gneiss, quartz, and silicious rocks, though hard, are too brittle and deficient in toughness. The slates are inadmissible, the quartzose, feldspathic and micaceous granites are bad, the quartzose is too brittle, the feldspathic too easily decomposed, and the micaceous too easily laminated. The sienitic granites, which contain hornblendes in place of feldspar, are good, and better in proportion to their darkness of color, the traps and basalts are the best, though most difficult to break up. The softer rock may be used for the lower course in a pavement.

Rolling.—The amount of rolling, with a heavy roller, should be about ten hours for each 1,000 square yards of surface for each layer of stone, but it must be continued until all motion of the stones has ceased. After several passages of the roller, any hollows which appear must be filled up with small material and the rolling continued. Each course is to be spread and treated in the same manner.

Watering is desirable where it can be done from the commencement of the rolling. It is done best by sprinkling. Excessive watering is to be avoided, especially in the earlier stages, as it tends to soften the foundation.

Binding.—Binding should be spread dry and uniformly in small quantities over the surface, and rolled into the interstices with the aid of watering and sweeping. By using binding materials in large quantities the amount of rolling is lessened, but at the expense of durability.

Wear and Maintenance.—When the wear on a road is confined to the crushing and grinding at the surface, it is the least possible, but when a road is weak from insufficient thickness or solidity on a yielding foundation, bending and cross-breaking of the covering take place under passing loads, in addition to surface wear, and the effects are aggravated by the softening action of water finding its way into the road-bed through cracks formed in the surface, and by the disintegrating action of frost. Wear is measured by the loss of thickness in the covering. It is seldom found to exceed one inch per year on the most frequented roads.

Effect of Wheels.—Legislation is needed for regulating the width of wheel tires. With the same burden, a two-wheeled cart does far more damage than one of four wheels. Wheels with $2\frac{1}{2}$-inch tires cause double the wear on a road than those do which have $4\frac{1}{2}$-inch tires. No greater width is useful, as a wider tire does not bear evenly. The following proportions have been advised:

Load on each wheel.	Vehicle without Springs.	With Springs.
$\frac{1}{2}$ to $\frac{3}{4}$ of a ton...	6 inches.	1 inch.
$\frac{3}{4}$ to 1 ton..........	4 "	3 "
1 to $1\frac{1}{2}$ tons.....................	6 "	4 "

Wheels of large diameter do less damage than small ones, and cause less draught for the horses.

AN ESSAY ON ROAD MAKING AND MAINTENANCE IN LESS THAN SIXTY WORDS.

A. L. PHILLIPS ("TO THE POINT"), PENCOYD, PA.

First.—No unnecessary roads.

Second.—A competent engineer in charge of survey, location, construction and maintenance.

Third.—Proper macadamized roadway from eight to four inches depth as needed, if cost of stone not prohibitory.

Fourth.—Use of roller in construction.

Fifth.—Complete drainage, sub and surface.

Sixth.—Constant, intelligent watching and repairing by permanent, responsible employees.

A PLEA FOR ÆSTHETIC CONSIDERATIONS IN ROAD BUILDING.

BOSTON, April 7, 1890.

To the Editor of THE ENGINEERING AND BUILDING RECORD:

SIR: In your issue of March 29, you publish the first part of the First Prize Essay on Roadmaking in which I was greatly struck by two rules that are laid down, which, from an artistic standpoint, would be more honored in the breach than in the observance. I call attention to them because they are an epitome of the standpoint of modern engineers with regard to æsthetic considerations. The two rules are: (In making a road) "make the line as nearly straight as practicable, and when changes of line occur connect them by regular curves of proper radius. When the line is intended to be straight make it so absolutely." These rules are, I am aware, considered obligatory by most engineers, for which reason an engineer may, in most cases, be depended upon pretty certainly to do a good deal toward spoiling the beauty of any landscape through which he may be called upon to make a road, with his "absolutely straight lines" and "regular curves of proper radius." It seems a pity that a few elementary ideas as to what constitutes beauty might not be instilled as part of the training of a profession that has in its hands the making or marring of so much natural beauty. But probably to most engineers the idea of considering the possible effect of the line of a road on the beauty of the landscape, and planning the road with regard to this consideration among others, seems ridiculous and quixotic. I fear that the training of most engineers tends to blunt and deaden such ideas of beauty as they may have by nature, and leads them to regard beauty as a thing unworthy of serious consideration. And so it comes about that fine trees of a century's growth are ruthlessly cut down and hillsides marred by deep and ugly cuttings, when a slight bend in the road would not only preserve the trees but add to the beauty of the road; or when by following the contour of the hill, with its natural and *irregular* curve, the road would be given some beauty and expense could often be saved. Among the rules laid down by the prize essayist I do not find any which speaks of the necessity of considering the natural conditions which *ought* to be among the determining elements of a line of road, when yet the capability to seize upon and make the most of these natural conditions ought to be one point of distinction between a good and an inferior engineer.

The æsthetic elements in the problem of making a road are not necessarily at variance with practical considerations. On the contrary, it will often be found that a consideration of these æsthetic elements will lead to practical benefits and economics. To consider these æsthetic questions, which are involved in a large proportion of engineering undertakings, would doubtless add to the engineer's difficulties. He would be less dependent on mere rule. It would give in each case opportunity for the exercise of judgment and skill in the harmonizing of sometimes apparently contradictory requirements; but it would add greatly to the interest and individuality and value of his work.

I appeal to the engineers themselves to consider this question for a moment, not as engineers, but as individuals who may be affected by engineering operations, to free themselves for the moment from the strait-jacket of rules of their engineer's training and to consider how much beauty might be preserved, nay, might be added to our landscape, especially in suburban communities, by even the slightest consideration or forethought for beauty, for its own sake, and this often without any added expense or the sacrifice of any reasonable utilitarian requirement.

It is the wanton disregard of beauty without any corresponding gain against which I protest.

When our civilization reaches a point at which it is willing to sacrifice some utilitarian and material considerations for the sake of beauty it will mark a distinct advance. At present this is hardly to be expected.   H. LANGFORD WARREN.

COMMENTS BY THE COMMITTEE OF AWARD.

The following comments are made by the Committee of Award, upon various propositions made in the several essays :

Traction.—Refinements as to traction are of little practical value. In every case it is desirable to fix upon a maximum grade. This being done, no exceptions should be allowed other than for very short distances. Authorities differ as to the power of horses, and as to the traction resistance of various road surfaces.

Alternations in Grade.—It will not do to say that these must not be allowed. The "lay of the land" and other circumstances will often compel their adoption; the only alternatives often being either a great increase in expense, or great detriment to adjacent property. In connection with the subject of grades and alignment, we commend the communication of H. Langford Warren, printed above.

The "Angle of Repose of Vehicles."—Much stress has been laid on the requirement that descending grades shall not exceed this. We believe the refinement an impossible one. What will be true for one vehicle with a certain width of tire, size of wheels and axles, state of lubrication and amount of load carried will be different for another vehicle, varying essentially in these particulars.

Again, what will be true for a given vehicle with the road dry and in perfect condition, may not be true when wet, frosty, etc.

These latter variations will be less the harder and more perfect the road surface.

Width of Road-Bed.—We think the width of road metal or other covering should not be less than 16 feet, except where all traffic is very slow, where 12 feet may answer. This is a matter for the judgment of the engineer.

Cost of Handling Earth.—No fixed rules can be given as to this, as it depends so largely upon location, cost and kind of labor and teams, and thorough management. The remarks on the cost of handling earth, it should be remembered, are based on wages paid "before the war," when 75 cents a day was nearer an average than one dollar. It is a matter of regret that in this connection no mention was made of the cost of scraper work, and that no essayist described the working of a "road machine," an extremely valuable implement in forming and maintaining earth roads.

Weight of Roller.—It is obvious that no road roller can be made heavy enough to give the same pressure per inch as the heaviest loaded wheel, but we think the use of a 10 to 15 ton roller advisable where it can be had.

Binding Material.—The question of the amount of binding to be used is largely dependent on the amount of rolling done, and the subsequent care of the roads. It is admitted that a considerable saving can be made by using more binding and less rolling, and where money must be made to go as far as possible this expedient is justifiable.

We think, however, it cannot be questioned that heavy rolling with a moderate amount of binding, preferably screenings, gives a more durable and cleaner surface under heavy traffic.

Thickness of Pavement.—This is to be decided by a due consideration of all the circumstances in each particular case, and should be left to the judgment of the engineer. It is rarely the case that every part even of the same road needs the same treatment. Retentiveness of the soil, gradients, thoroughness of drainage possible, climate, character and extent of traffic, and amount of money available, all have to be considered, and we deem it unwise to try to lay down hard-and-fast rules on this all-important matter.

Asphaltic Pavements.—The remarks of " Cesa " on asphalt pavements are unfortunate, as there is an entire lack of distinction between the various kinds of pavements going under this name, and the essayist, by implication, condemns pavements that are satisfactory, on purely theoretical grounds.

There is one direct error in statement, also, by this writer, that attention should be called to. He states that in England roads are under national supervision. The English highways are under supervision that varies in extent from the bounds of a single parish to that of a county, and while the great mass of American roads are bad enough, it is thought that no roads have been anywhere built that were better than those built by the New York Central Park Engineers in and about the Park, and we doubt if any European country can show better surfaced country roads than many in New Jersey, Massachusetts and Connecticut.

Berms and Ditches.—The arrangement of berms and ditches in the second prize essay is questionable, except on sandy soils.

Foundation of Telford Roads.—As there seems to be some discrepancy of statement in the several papers as to the foundation blocks in Telford pavement, we desire to state that in no case must such stone lie on the flat or broad side of the stone. The broadest edge should be down, the stone set lengthwise across the street, breaking joints in the several rows. The stones in any one row should be approximately of the same thickness. Very thick stones are bad for durability. Every needless refinement in road-making costs money, and while engineers should exact close attention to details upon which durability depends, it is not wise to expend the minute care upon them that might be bestowed upon a city street.

Repairs.—Where money is scarce, repairs may be made more cheaply and so as to answer a good purpose by the use of road metal, with enough clay or loam with it to make it compact quickly. One of the greatest enemies of Macadam roads is the wind, and to prevent the effects of winds the rolling and repairing of the surface as soon as is practicable in the spring, is one of the most effective means.

Water, sand, clay, or loam should be promptly used if a road begins to break up in the spring winds.

In making repairs, we doubt whether it is advisable (in the case of trap macadam at least) to break up the surface with spikes or with picks, where water for softening the road is available.

The circumstances under which it is advisable to use material from ditches, a practice usually condemned, are not fully brought out. Washed coarse sand and gravel from side ditches may make a valuable addition to the road-bed, but the mud and fine sand washed from this will not add to the wear of the road-bed.

Stone Breakers.—Where suitable stone is available for road-making, a town owning a stone-breaker is more apt to have a sufficiency of road metal than where a vote has to be passed for the expenditure of a definite sum of money at intermittent times ; but if the demagogue could be eliminated from the town meeting, it would sometimes be the case that the broken stone could be purchased more cheaply from a large establishment making the breaking of stone a business.

County and Assistant Engineers.—We consider the remarks on this subject in the second prize essay ill-advised. The question of time has less to do with the capabilities of a man for such work than natural ability and habits o reasoning observation.

No engineer should be required to waste his employer's time by walking five miles. The better arrangement is to furnish at all times such horses and conveyance as may be required.

F. COLLINGWOOD.
EDWARD P. NORTH.
JAMES OWEN.

IF you are charged with the construction of a large Building of any kind, a Public Institution, a Bridge, a Water-Works, Sewerage, Gas or Electric-lighting system, Pavements, or require any work done, we suggest that you consider the desirability of advertising for proposals for the work and material in THE ENGINEERING AND BUILDING RECORD, which makes a speciality of publishing information that is valuable to contractors, a class of business men among whom it circulates in every part of the United States and Canada, and who look to its columns regularly for news that suggests to them possible business.

If any cause induces the restriction of competition to home parties entirely, you will no doubt follow the practice of some municipalities which advertise such matters in local papers only. If, on the other hand, it is your desire to obtain the best possible results by reaching parties in every part of the United States and Canada whose special business it is to do any form of contracting work, and who are familiar with most modern methods, you will follow the course first suggested above. While the original outlay to do so is a mere trifle, the experience of others shows that the return in suggestions, advice, first-class workmanship, etc., as a result of large facilities and special adaptability—not to speak of possible advantages in prices obtained—will prove a handsome investment. The various departments of the United States Government, Municipal Authorities, Water-Works, Building Committees, and others intrusted with Public Work, thus use our columns each week.

The rate is 20 cents per line (about 8 words to the line).

— · —

THE ENGINEERING AND BUILDING RECORD,

Prior to 1887, THE SANITARY ENGINEER.

DEVOTED TO

Engineering, Architecture, Construction and Sanitation.

MUNICIPAL ENGINEERING A PROMINENT FEATURE.

Published Saturdays at 277 Pearl Street, New York.

SOLD BY ALL NEWSDEALERS.

$4 Per Year.   **10c. Per Copy.**

A COLLECTION OF DIAGRAMS

Representing the General Plan of

Twenty-Six Different Water-Works,

Contributed by Members of the

NEW ENGLAND WATER-WORKS ASSOCIATION,

And Compiled by a Committee.

1887.

INTRODUCTION.

OFFICE OF SECRETARY,
NEW BEDFORD, MASS., November 1, 1887.

THIS collection of diagrams is the result of the persistent efforts of Messrs. William B. Sherman, of Providence, R. I , and Walter H. Richards, of New London, Conn., who, as a Committee on Exchange of Sketches, have secured these drawings from members of the Association. The following extract from a report presented by these gentlemen at the Manchester, N. H., meeting in June, 1887, will explain in part the origin of the collection :

" In answer to circular letters sent out to members, there were received rough sketches of general plans of twenty-three water works represented in the Association. Having this data on hand, though crude in many particulars, it was decided to put the same into available shape for the benefit of the members. This has been accomplished by the Committee without cost to the Association. From these rough sketches—revised, reduced to uniform size of 10 by 15 inches – a set of tracings has been made, and a sample folio of blue prints prepared. This folio and set of tracings are herewith presented as forming the main part of this report."

Since the Manchester meeting three more subjects have been received and subscriptions for sets of reproductions from the tracings have been called for. The ready response to the call is evidence of the value of the Committee's work, and arrangements were made with *The Engineering & Building Record* for publication in this present form.

R. C. P. COGGESHALL,
Secretary, New England Water-Works Association.

Published by THE ENGINEERING & BUILDING RECORD.

INDEX.

Address, BOOK DEPARTMENT,

THE ENGINEERING & BUILDING RECORD,

P. O. Box 3017.   No. 277 Pearl Street, New York.

GEO. F. BLAKE MANUFACTURING COMPANY,

Builders of Steam and Power

For Water Works Service.

THE BLAKE IMPROVED
Compound Condensing Duplex Pumping Engines,

Geo. F. Blake Manufacturing Company,

95 and 97 Liberty St., New York.   111 Federal Street, Boston.   535 Arch Street, Philadelphia.

Some Details of Water-Works Construction.

By W. R. BILLINGS, Superintendent of Water-Works at Taunton, Mass.
WITH ILLUSTRATIONS FROM SKETCHES BY THE AUTHOR.

INTRODUCTORY NOTE.

Some questions addressed to the Editor of *The Engineering and Building Record and The Sanitary Engineer* by persons in the employ of new water-works indicated that a short series of practical articles on the Details of Constructing a Water-Works Plant would be of value; and, at the suggestion of the Editor, the preparation of these papers was undertaken for the columns of that journal. The task has been an easy and agreeable one, and now, in a more convenient form than is afforded by the columns of the paper, these notes of actual experience are offered to the water-works fraternity, with the belief that they may be of assistance to beginners and of some interest to all.

TABLE OF CONTENTS.

HANDSOMELY BOUND IN CLOTH.

Sent (post-paid) on receipt of $2.00.   Address,

BOOK DEPARTMENT,   THE ENGINEERING AND BUILDING RECORD.

THE HARRISBURG PATENT

IMPROVED

DOUBLE ENGINE STEEL STEAM

ROAD ROLLER.

Unequaled in Economy, Efficiency and Workmanship.

MANUFACTURED BY

FOUNDRY & MACHINE DEPARTMENT,

HARRISBURG, PA.

BUILDERS OF THE

IDE AND IDEAL AUTOMATIC ENGINES,

BOTH PLAIN AND COMPOUND.

Also Boilers and Tank Work of all Descriptions.

CIRCULARS ON APPLICATION.

WATER-WASTE PREVENTION:

Its Importance and the Evils Due to its Neglect.

With an account of the Methods adopted in various Cities in Great Britain and the United States.

By HENRY C. MEYER, Editor of THE ENGINEERING & BUILDING RECORD.

With an Appendix.

EXTRACT FROM PREFACE.

During the summer of 1882 the Editor of THE SANITARY ENGINEER carefully investigated the methods employed in various cities in Great Britain for curtailing the waste of water without subjecting the respective communities to either inconvenience or a limited allowance. The results of this investigation appeared in a series of articles entitled " New York's Water-Supply," the purpose being to present to the readers of THE SANITARY ENGINEER such facts as would stimulate public sentiment in support of the enforcement of measures tending to prevent the excessive waste of water so prevalent in American cities, and especially the city of New York, which was then suffering from a short supply Numerous requests for information, together with the recent popular agitation in connection with a proposition to increase the powers of the Water Department of New York City with a view to enabling it to restrict the waste of water, have suggested the desirability of reprinting these articles in a more convenient and accessible form, with data giving the results of efforts in this direction in American cities since the articles first appeared, so far as they have come to the author's notice

Large 8vo. Bound in Cloth, $1.00.

*** Sent post paid on receipt of price. Address,

BOOK DEPARTMENT,

THE ENGINEERING AND BUILDING RECORD,

No. 277 Pearl Street, New York.

WATER-WASTE PREVENTION.

By HENRY C. MEYER, Editor of THE ENGINEERING & BUILDING RECORD.

PRESS COMMENTS.

8vo., bound in cloth, $1.00. *Sent post-paid on receipt of price.*

Address,   BOOK DEPARTMENT,
THE ENGINEERING AND BUILDING RECORD,

P. O. Box, 3037.   No. 277 Pearl Street, New York.

Obtainable at London Office, 92 and 93 Fleet Street, for 5s.

PORTER'S
West Virginia Vitrified Paving Blocks,
FOR STREETS AND ROADWAYS.

Manufactured by

THE JOHN PORTER COMPANY.

MAIN OFFICE AND WORKS,

NEW CUMBERLAND, W. VA.

Pittsburg Office, 708 Penn Avenue.
Philadelphia Office, 1345 Arch St.

WHEELING, WEST VA., May 26th, 1885.

JOHN PORTER, New Cumberland, W. Va.

DEAR SIR:—I do most cheerfully testify to you and those whom it may concern, that your Vitrified Paving Block, now in use in this city, is the best paving for general use that we have, both on account of it being smooth and of its noiselessness; and I will say further, that for the two years we have used it on the Street Railway it has not cost the Company one cent for repairs, nor, judging by present appearances, will it do so for some years to come, not one block being removed. And it has not worn in the least where the horses travel, nor do the horses slip on it, or snow remain on it any length of time. I consider it the cheapest material a railroad company can use.

C. A. WINGERTER,
President Citizens' R. R. Co.

AN INDEX TO MATTER PERTAINING TO

Sewerage and Sewage Disposal
IN VOLUMES V—XVII.
(December 1881—June 1888.)

OF

THE ENGINEERING AND BUILDING RECORD,
(PRIOR TO 1887, THE SANITARY ENGINEER.)

Compiled by DANA C. BARBER, C. E.

Large 8vo., Cloth, $2.00.

Address, Book Department, THE ENGINEERING AND BUILDING RECORD,
Sent post-paid on receipt of price.   277 PEARL STREET, N. Y.

PUBLIC WATER SUPPLIES;
THEIR
Collection ; Storage ; Distribution ; Engineering ; Plant ; Purity and Analysis.

A DIGEST AND INDEX TO VOLUMES V-XVIII., OF

The Engineering and Building Record,
(PRIOR TO 1887, THE SANITARY ENGINEER.)

Compiled by D. WALTER BROWN, Ph. D.

8vo.   242 *Pages.*   *Price,* $2.00.

Address, BOOK DEPARTMENT,

THE ENGINEERING AND BUILDING RECORD,
277 PEARL STREET, NEW YORK.

BUILDERS' AND CONTRACTORS' ENGINEERING AND PLANT.

Now appearing in THE ENGINEERING AND BUILDING RECORD.

ADDRESS, BOOK DEPARTMENT,

THE ENGINEERING AND BUILDING RECORD,

277 PEARL STREET, NEW YORK.

MAY 15, 1890.

www.ingramcontent.com/pod-product-compliance
Lightning Source LLC
Chambersburg PA
CBHW021346210326
41599CB00011B/764